T0400509

Detection of Biological Agents
for the Prevention of Bioterrorism

NATO Science for Peace and Security Series

This Series presents the results of scientific meetings supported under the NATO Programme: Science for Peace and Security (SPS).

The NATO SPS Programme supports meetings in the following Key Priority areas: (1) Defence Against Terrorism; (2) Countering other Threats to Security and (3) NATO, Partner and Mediterranean Dialogue Country Priorities. The types of meeting supported are generally "Advanced Study Institutes" and "Advanced Research Workshops". The NATO SPS Series collects together the results of these meetings. The meetings are co-organized by scientists from NATO countries and scientists from NATO's "Partner" or "Mediterranean Dialogue" countries. The observations and recommendations made at the meetings, as well as the contents of the volumes in the Series, reflect those of participants and contributors only; they should not necessarily be regarded as reflecting NATO views or policy.

Advanced Study Institutes (ASI) are high-level tutorial courses intended to convey the latest developments in a subject to an advanced-level audience

Advanced Research Workshops (ARW) are expert meetings where an intense but informal exchange of views at the frontiers of a subject aims at identifying directions for future action

Following a transformation of the programme in 2006 the Series has been re-named and re-organised. Recent volumes on topics not related to security, which result from meetings supported under the programme earlier, may be found in the NATO Science Series.

The Series is published by IOS Press, Amsterdam, and Springer, Dordrecht, in conjunction with the NATO Public Diplomacy Division.

Sub-Series

A.	Chemistry and Biology	Springer
B.	Physics and Biophysics	Springer
C.	Environmental Security	Springer
D.	Information and Communication Security	IOS Press
E.	Human and Societal Dynamics	IOS Press

http://www.nato.int/science
http://www.springer.com
http://www.iospress.nl

Series A: Chemistry and Biology

Detection of Biological Agents for the Prevention of Bioterrorism

edited by

Joseph Banoub

Fisheries and Oceans Canada, Science Branch, Special Projects
and
Memorial University of Newfoundland, Chemistry Department, St. John's,
Newfoundland, Canada

 Springer

Published in cooperation with NATO Public Diplomacy Division

Proceedings of the NATO Advanced Research Workshop on
Detection of Biological Agents for the Prevention of Bioterrorism:
Terme di Spezzano, Italy
June 26 – July 2, 2009

ISBN 978-90-481-9879-5 (PB)
ISBN 978-90-481-9814-6 (HB)
ISBN 978-90-481-9815-3 (e-book)

Published by Springer,
P.O. Box 17, 3300 AA Dordrecht, The Netherlands.

www.springer.com

Printed on acid-free paper

CONTENTS

Preface

Detection of Biological Agents for the Prevention of Bioterrorism

As concerns about biological and chemical security threats increase worldwide, so does the need for early warning systems capable of providing timely and accurate intelligence. Biological and chemical threats are significant challenges which are very difficult to predict or prevent, they come in many forms, and they can spread quickly without warning. Little information is available regarding the incidence and distribution of diseases caused by critical chemical/biological agents, toxins, explosives and other hazardous materials used in selected acts of terrorism. The terrorism threat has driven the demand for timely techniques that can quickly detect the agent(s) used in an attack. Biological agents typically require culturing prior to identification and at present no technology can rapidly distinguish one or more characteristic microbial components released during a terrorist incident from those of other organisms already present in the environment.

The proceedings of this book entitled: "NATO-Science for Peace and Security-Chemistry and Biology Series: Detection of Biological Agents for the Prevention of Bioterrorism" result from the NATO-ARW meeting held from June 26 to 2 July, 2009 in the American Hotel, at the Albanian Terme di Spezzano, under the auspices of NATO Science for Peace Program (SPS).

Mass spectrometry has emerged as an important tool for the characterization of chemical/biological agents, toxins, explosives and other hazardous materials that may be used in a terrorist event. The terrorism incidents of October 2001 have focused and elevated our level of concern, in particular our vulnerability to biological agents. Biological agents and toxins are easy to manufacture, conceal, and release, making bioterrorism difficult to prevent. As a result, the practical front line of biodefence is to determine the presence of an attack as quickly as possible, enabling effective decontamination and treatment.

Although 50 participants had agreed to come to this NATO-ARW, the actual number of attendees shrunk to 35 participants, they well represented the state-of-the-art in the world community including both the US, Canada, Europe, Eastern Europe and North Africa. The effects of the financial crisis were also evidenced by the complete lack of sponsoring funds that had been expected from industrial scientific vendors.

The meeting was a major success and communication was open and discussion detailed. As noted, the state-of-the art in biodetection was presented by its practitioners. Clearly the meeting was of great utility in mutual education.

Summary of the NATO-ARW: The first lecture of this NATO-ARW program, on June 27, 2009, was initiated by Professor Catherine Fenselau of the University of Maryland (USA), which presented the complexity and the problems associated with the determination of bacteria directly for mass spectrometry. Professor Fenselau is also the chief editor of the prestigious American review Analytical Chemistry.

The first day of this NATO-ARW continued with the participation of the president, Professor Marine Gennaro, of the Faculty of Biotechnologies of the University of Naples, and of Prof. the Pierluigi Reschiglian of the University of Bologna, which contributed deeply to the methodology for the survey of pathogenic bacteria. Dr. Suzane Kalb of the "Centers for Disease Control" of Atlanta Georgia (USA) concluded the intense activities of the morning by discussing the problematics of the proteases "the botulinum neurotoxins" (BoNTs), whose toxic action produces the destruction of some essential cellular proteins and the release of neurotransmitters. In the afternoon session Professor Bogumila Szponar, Immunologist from the Academy of Polish Sciences, discussed the survey of endotoxins in the atmosphere by mass spectrometry. The last the two contributions of the day were dedicated to two experts on soldiers' field exposure to biological hazards. Dr. John. Smith, US Army Medical Research Institute of Chemical Defence Medical Diagnostic and Chemical Br, Aberdeen Proving Ground, MD, USA and Dr. R. Read, of the Defence Science and Technology Laboratory, Porton Down, UK discussed the uses of the mass spectrometry of useful markers for the identification of the human crew exposure.

On June 28, 2009, Professor Alvin Fox Department of Pathology, Microbiology, and Immunology, USC School of Medicine, Columbia, USA, which is also Editor-in-Chief of the Journal of Microbiological Methods, joint-Editor-in-Chief of Molecular and Cellular Probes and is on the Editorial board of the Journal of Clinical Microbiology, gave an excellent presentation on the Protein Markers for Biodetection using Tandem Mass Spectrometry; a Twenty-First Century Challenge. This was followed By Professor Joseph Banoub, Special Projects, Fisheries and Oceans Canada and Chemistry Department, Memorial University, St John's, Newfoundland, Canada, which introduced the tandem mass spectrometry quantification of a variety of fish species Vitellogenin serum proteins, as a biomarker for xenobiotic chemical endocrine disruptors and discussed the potential hazard of bioterrorism in the marine environment. Dr. Erica M. Hartmann, Civil and Environmental Engineering, Biodesign Institute Arizona State University, Tempe AZ, USA presented the tedious challenges of Detecting Bioterrorism Agents in Complex Matrices. The morning session finished with the excellent presentation of Professor Guenter Allmaier, Institute of Chemical Technologies and Analysis, Vienna University of Technology on the GEMMA (gas-phase electrophoretic mobility molecular analyzer) and PDMA (parallel differential mobility analyzer as analysis and collection device) of viruses and bionanoparticles. The afternoon session continued with the participation of Dr Eric Ezan, CEA, Direction Des Sciences du Vivants, Saclay, France on the Detection of Functional Ricin by Liquid Chromatography/Tandem Mass Spectrometry. This was followed by Professor Bruce McCord form the Department of Chemistry, Florida International University, Miami, FL, USA on the Applications of Electrospray Time of Flight Mass Spectrometry in the analysis of low explosives. The evening session ended with dr. Paul D'Agostino , DRDC Suffield, Canada on the LC-ESI-MS/MS and DESI-MS/MS Analysis of Chemical Warfare Agents and Related Compounds.

The third day of this workshop, June 29, 2009 was initiated by Professor Robert J. Cotter, Professor of Pharmacology and Molecular Sciences at the Johns Hopkins

University School of Medicine which presented the novel uses of Low powered ion trap system that we are designing for Mars Rover and as a field instrument for biodetection. This was followed by Professor Lennart Larsson, Department of Laboratory Medicine, Division of Medical Microbiology, Lund University, Sweden on the Identification of Mycotoxins in the Indoor Environment. Professor Guenter Allmaier discussed the intact cell mass spectrometry of fungal spores for fast isolate and species differentiation. Colonel Levent Kenar MD, PhD from the Department of Medical CBRN Defence, Gulhane Military Medical Academy, Ankara presented the Detection of Bacillus anthracis spores using monoclonal antibody immobilized QCM biosensor.

The afternoon session was started by Dr. Adrian R. Woolfitt, CDC/USA which gave an excellent presentation on MALDI-MS Analysis of Bacillus anthracis: From Fingerprint Analysis of the Whole Organism to Quantification of its Toxins in Clinical Samples. This was followed by Dr. Ben Van Barr which presented and described the novel state-of-the art instrumentation concerning the Direct Aerosol MADLI-TOF-MS of Bacteria. The afternnon session ended with the lecture of Professor Enzo Mollace, CETA-ARPACAL, Catanzaro, Italy on the Risk Assessment in Environmental Toxicology.

The fourth day of this of this workshop, was initiated by Professor Richard Caprioli, Stanley Cohen Professor of Biochemistry, Director of the Mass Spectrometry Research Center, VICC Member, and Chief Editor of the Journal of Mass Spectrometry Researcher, which gave an elegant lecture on Tissue Imaging Using Matrix-Assisted Laser Desorption/Ionization Mass Spectrometry. This was followed by the lecture of Professor Karen Fox, Department of Pathology, Microbiology, and Immunology, USC School of Medicine, Columbia, USA which discussed on the Proteomics from the perspective of molecular microbiologist. This was followed by Professor Anna Napoli University of Calabria, Italy on the Proteomic Approach to Cattle Diseases. Professor Eugenio Parente, Department of Agri-food Biology, Defence and Biotechnologies – Università della Basilicata, Italy presented the Chemometrics Methods for the Identification of Microorganism. The morning session ended with the excellent presentation of Professor Mark Duncan of the School of Medicine, Universiy of Colorado USA on the Mass Spectrometry in the Qualitative & Quantitative Analysis of Proteins.

The afternoon session was initiated with Professor Chris Cox, Chemistry and Geochemistry Department, Colorado School of Mines, CO, USA on the Phage amplification with MALDI-MS for the Identification of Bacteria. This was followed by the presentation of Professor John S. Fletcher, Manchester Interdisciplinary Biocentre, University of Manchester, UK. on the 2 and 3D TOF-SIMS Imaging for Biological Analysis. The afternoon session ended with Workshop on Mass Spectrometry vs. Genetics vs. Microbiology vs. Animal studies chaired by Professor Karoly Vekey (Hungary), Alvin Fox and Adrian Woolfitt.

Finally, last day of this workshop, started with a lecture of Professor Luigi Mondello, School of Pharmacy, University of Messina on the Identification of Cellular Lipid Fraction of Bacteria by GCxGC/MS using an Innovative MS Library with the use of Linear Retention Indices. This was followed by Dr. Thomas Elssner,

CBRNE Application, Bruker Daltonik GmbH, Leipzig, Germany on the Microorganisms Identification Based on MALDI-TOF-MS Fingerprints.

The NATO-ARW closed with a Workshop on Food security and Safety, Giovanni Sindona (chair), University of Calabria, Karoly Vekey, Hungarian Academy of Science, Luigi Mondello of the University of Messina.

Perceived Problems Requiring Solution

Bacteria: It was agreed that different strategies may be necessary for biodetection (environmental, e.g. airborne dust), clinical diagnosis (human body fluids) or protecting the food and water supply. Biodetection involves detecting one prokaryotic species (e.g. anthrax spores) among a plethora of other species (sometimes their close relatives [e.g. *B. cereus* and *B. mycoides*]) in close to real-time in a samples with a great deal of variation in the biological matrix (e.g. airborne dust). For a clinical sample (e.g. blood) there will often be only the species of interest and the matrix is defined (e.g. larger eukaryotic particles [red and white blood cells and soluble proteins). Often for a clinical sample culture will have been performed. In this instance the sample will have been received days or possibly weeks earlier and one is merely carrying out verification. For the food supply, large quantities of material may contain only small pockets of contamination, so sampling the needle may not be characteristic of the haystack.

Viruses: The problem for viruses is similar to bacteria except virus particles are much smaller than bacteria which changes the type of biodetection technology that can be used. Furthermore viral diseases are much harder to diagnose since they only grow within cells. Thus diagnosis generally relies on molecular biology (e.g. the polymerase chain reaction (PCR) or microscopy.

Bacterial Exotoxins: These are generally proteins which are released by bacterial cells (e.g. botulinum toxin). Exotoxins fall between the realms of bio- and chem-detection.

New Detection and Identification Tools

Bacteria Particles: It is a simple problem to identify bacteria once they have been biologically amplified by culture. There were a number of presentations where scientists presented MALDI-TOF-MS profiling of extracted bacterial proteins from cultured bacteria. For example the commercial Bruker instrumentation and software packet was presented. Unfortunately in a complex matrix (e.g. airborne dust) or a body fluid this is totally inadequate. The state of the art for was presented for analysis of a single bacterial cell (van Baar). In this instance, while in flight, the bacterial particle is coated with a MALDI (matrix assisted laser desorption/time of flight mass spectrometry) matrix and a mass profile is obtained one particle at a time. No data was presented yet for analyses from real environmental samples but this technique shows great promise. As regards biological fluids, pilot work for separation of bacteria by manual means was presented (Elssner). Independently others presented one possible solution (field flow fractionation) for isolating biological fluids prior to analysis (Perluigi Rischiglian). Unfortunately current MALDI-tandem mass spectrometer instrumentation (QqTOF) can only analyze small proteins (up to 2000–3000 MW) thus more specific tandem mass spectrometry (MS/MS) for biomarker analysis are needed. Other instrumental configurations (involving electrospray (ESI) e.g. coupled with an Orbitrap) may be essential and can be also used.

Tissue Imaging: Tissue imaging is in its infancy as regards viewing bacteria in mammalian tissue. However the state-of-the art in 2D tissue imaging (Caprioli, MALDI-TOF-MS) was presented particularly with application to cancer. More basic research (secondary ion mass spectrometry (SIMS) with potential for 3D imaging was also presented (Fletcher).

Proteins: An alternative strategy was presented by several scientists for isolating proteins from a complex matrix for detection of a specific protein marker to xenobiotic chemicals. This primarily involves liquid-chromatography-electrospray ionization tandem mass spectrometry (e.g. Banoub).

Xenobiotic Chemicals: Similar strategies were also presented for explosives (Mccord) mycotoxins (Read, Larsson) and chemical agents (Ezan, Read, Smith). Currently biomarkers consist of peptides released by tryptic digest. However, with further increase in instrumental sensitivity it is anticipated a simpler approach may involve MS/MS analysis of intact proteins without the need for tryptic digestion. Current technology (e.g. electron transfer dissociation) appears adequate for the purpose.

Cotter presented his work on development of a "tiny" MALDI ion trap for use in future MARS missions. This instrument may have utility in helping replace previous generations of instrumentation (e.g. the CBMS1 [pyrolysis] and CBMS2 [fatty acid profiles]). However, as noted by numerous participants to have the required level of specificity in a complex matrix some type of separation (either at the particle or protein level) and possibly MS/MS may prove essential.

Viruses: For a number of years the Voorhees group has explored the use of bacterial phages (viruses) to infect bacterial cells. It is the amplified phage proteins, not the bacterial proteins that are detected. Phage amplification of proteins acts like in PCR (DNA) in providing more protein for analysis and thus providing greater instrumental sensitivity. However, phages for all organisms of interest are not well defined. An update was provided by a member of the Voorhees team.

Allmeir presented a device designed to detect airborne particles of the size of viruses and accurately determine size. This was yet another example of the high level of sophisticated technology presented at this meeting.

Exotoxins: The standard approach for detecting exotoxins is to see if they kill a mouse. Woolfitt and Kalb representing the Barr team presented a clever approach they have implemented in which instead of detecting the toxin protein directly one detects the enzymatic product (resulting from digestion of a defined peptide substrate). This amplifies the range of detection logarithmically allowing practical detection of botulinum, anthrax and other toxins in body fluids in a few hours. This approach is in routine use at the CDC.

Biology: The meeting was dominated by chemists which represents the biodetection field accurately. Fox (A) presented his teams work on combining taxonomy, molecular biology and mass spectrometry in defining protein biomarkers (e.g. spore glycoproteins and small acid soluble proteins [SASPs] that fit with the instrumental developments. The SASP methodology was originally developed by Fenselau who also gave a presentation. Fox (K) presented the perspective of a molecular biologist that current DNA technology gives 100% DNA sequence with 100% certainty. Protein markers must be as reliable and have distinct advantages (e.g. a. speed and b. simplicity by avoiding the use of fragile enzymes [such as DNA polymerase used in PCR] in field samples). As noted above

Conclusions: The meeting was a major success and communication was open and discussion detailed. As noted the state-of-the art in biodetection was presented by its practitioners. Clearly the meeting was of great utility in mutual education. It is anticipated this book and future training programs in the form of a prospective NATO-ASI will be of similar quality and will benefit a much wider audience within NATO and the defence community.

USA   Alvin Fox
Morocco   El Mokhtar Essassi
Canada   Joseph H. Banoub

Broadband Analysis of Bioagents by Mass Spectrometry

Catherine Fenselau, Colin Wynne, and Nathan Edwards

Abstract Mass spectrometry was first reported to provide analysis of intact metabolite biomarkers from whole cells in 1975.[1] Since then advances in ionization techniques have extended our capabilities to polar lipids and, eventually, to proteins.[2, 3] Mass spectrometry provides a broadband detection system, which, however, has great specificity. Bioinformatics plays an important role in providing flexible and rapid characterization of species, based on protein and peptide mass spectra collected in the field.

Keywords Intact microorganisms • Broadband detection • MALDI • Time-of-flight • Ion traps • Bioinformatics • Background clutter • Genetic engineering

1 Introduction

Mass spectrometry is of considerable interest as a technique for detection and identification of biological agents, including both microorganisms and toxins, because of its speed, sensitivity and capacity for automation. Mass spectrometry brings to these tasks the yin and yang of both broad band detection and high specificity. The former reflects the fact that every atom and molecule has a mass and thus a mass spectrum. Other chemical and biological detectors ask: *is it there?* The mass spectrometer asks: *what is there?* Specificity is achieved because both molecular and fragment masses provide detailed *fingerprints*, which can also be *interpreted*. An additional attraction of mass spectrometry is that both bioagents and chemical agents can be analyzed on the same instrument. In summary, mass spectrometry is a physiochemical method that is orthogonal and complementary to biochemical and morphological methods used to characterize microorganisms and toxins.

C. Fenselau (✉), C. Wynne, and N. Edwards
Departments of Chemistry and Biochemistry, College of Chemical and Life Sciences,
University of Maryland, College Park, MD 20742
e-mail: fenselau@umd.edu

J. Banoub (ed.), *Detection of Biological Agents for the Prevention of Bioterrorism*,
NATO Science for Peace and Security Series A: Chemistry and Biology,
DOI 10.1007/978-90-481-9815-3_1, © Springer Science+Business Media B.V. 2011

The primary interest of this laboratory has been the development of a fieldable system that uses mass spectrometry to characterize proteins and their host organisms in less than 5 min. We and others have chosen matrix assisted laser desorption ionization (MALDI) as a robust ionization method that requires little sample preparation and tolerates contaminants relatively well. MALDI has been field-tested with both time-of-flight (TOF)[4] and ion trap analyzers.[5] We are also interested in contributing new methods for rapid mass spectrometry-based analysis of microorganisms and toxins in well equipped reference laboratories.

Design of a fieldable system must include consideration of sample collection and preparation as well as measurement and computer-supported interpretation. Selective collection from air has been achieved with inertial impactors, constructed to discard particles larger than 1 μm and smaller than 0.1 μm. Enrichment from water, food, urine, etc. is commonly facilitated with antibodies. However the use of antibodies, phage or bioassays discards the broadband approach and limits the analysis to pre-determined targets. The use of antibodies limits the query to: *is it there?* Broadband affinity agents are needed to extend detection to more complex media than air. Sample preparation includes the addition of the photon-absorbing matrix required for MALDI, and, in the case of spores and viruses, lysis of the sample. No fractionation is envisioned in the field portable system, in the interest of time and ruggedness.

MALDI spectra of intact or lysed microorganisms contain strong signals that comprise ionized proteins, usually in the range of 4,000–20,000 Da. These are judged to be excellent biomarkers, in part because they are directly related to the genome. Proteins in prokaryotes undergo minimal post-translational modifications. Various groups have observed that ribosomal proteins are readily detected in vegetative bacteria.[6,7] These are highly abundant and strongly basic, thus readily ionizable. The abundant small acid soluble spore proteins have been shown to be readily released by acid.[8,9] Limited work with viruses suggests that abundant capsid and some membrane proteins may be suitable biomarkers.[10,11] Ions with masses below 4,000 Da are not expected to be translated proteins, but rather cyclic peptides and other secondary metabolites, or polar lipids, whose presence may reflect health and environmental status more than genomic identity.

2   Identification of Microorganisms

Several approaches have been taken to provide identification of microorganisms based on mass spectra of their proteins or peptides.

2.1   *Library Searching/Pattern Recognition/Spectral Matching*

In this approach a mass spectral library is constructed with spectra for every targeted bioagent. Usually both library and target spectra must be measured using reproducible conditions. A large number of software programs are available to compare the mass spectrum of a suspected agent with this library. This approach is now offered commercially for medical diagnosis using MALDI-TOF mass spectrometry.

2.2 Machine Learning

In this strategy Bayesian belief networks based on probabilistic analysis of reference mass spectra offer a more sophisticated approach to establishing library spectra for matching.

2.3 Proteomics

This is an interpretive approach that provides probability assessments for matching suites of measured protein masses to protein masses predicted from sequenced genomes. Computer programs reference the protonated molecular ions in the target spectrum to a database derived from sequenced prokaryote genomes.[12, 13] In an alternative strategy, in situ proteolysis is accomplished, chemically or enzymatically, and tandem mass spectrometry measurements of peptide products provide microsequences to support identifications of proteins and thus of microorganisms.[14, 15] Advantages of the proteomic method include:

- This approach is not dependent upon growth conditions, sample preparation methods or MALDI matrices.
- The prokaryote genome database is under development around the world.
- One database supports all ionization techniques.
- Proteomic analysis enables the identification of specific biomarkers and determines the uniqueness of biomarkers.
- This strategy has been proven for spores, viruses, bacteria and toxins.

Both public and private databases can be used in proteomic strategies. RMIDb[16] is one publically accessible database that contains prokaryote protein sequences. An algorithm has been reported to search for masses of proteins with and without N-terminal methionine[17] and this simple strategy can also be applied to N-acetylation and phosphorylation.

However, several challenges remain to be met before any mass spectrometry-based system can be confidently deployed for automated broad band surveillance, detection and identification of bioagents. These include (1) recognition and analysis of cluttered samples, or of mixtures, (2) recognition and analysis of genetically engineered organisms, and (3) characterization of proteins and organisms in the absence of a sequenced genome. An approach to each of these issues will be summarized here.

3   Proteomic Analysis of Mixtures

The natural background of microorganisms in the atmosphere in many regions of the planet is expected to contaminate bioagents that might be collected for analysis. *Bacillus thuringiensis* is widely deployed in the pesticide BT, and surveys of background usually identify *Bacillus* species as well as a high percentage of unknown microorganisms. Historically, components of mixtures were recognized by analysis of single colonies of culture plates, or by the targeted use of antibodies or phage. We proposed several years ago that mixtures of bacteria and/or spores can be analyzed rapidly in the field by automated proteomic strategies. Microorganisms are lysed and proteolysed in situ to provide peptides for identification by MS/MS methods.[9, 14, 15, 18] If many peptides are automatically selected for analysis, it is likely that proteins will be identified from all the species present. When species selective proteins are identified, the species are as well.[19] A schematic of this approach is shown in Fig. 1. An example of analysis of the minor component in a 10:1 mixture of *B. thuringiensis* Kurstaki and *B. subtilis* is illustrated in Figs. 2 and 3. Figure 2 shows protonated peptide ions formed by tryptic digestion in the MALDI sample holder. Most of the peptides were analyzed by tandem mass spectrometry and their bacterial sources are indicated in the figure. Figure 3 shows an MS/MS spectrum obtained on a Kratos TOF^2 instrument of a protonated peptide of mass 1,880.8 Da. When the MS/MS spectrum was searched against entries in the NCBInr database taxonomically restricted to eubacteria, the peptide was identified, proteins that contain this peptide were identified, and the bacterial source of these proteins was identified as *B. subtilis*. The analysis of several peptides from the spectrum of the mixture is summarized in Fig. 3.

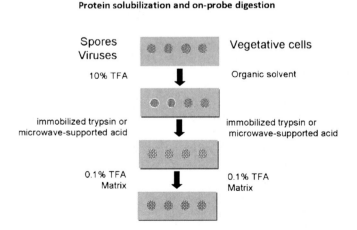

Fig. 1 MS-MS analysis of bacterial digests and identification of proteins through bioinformatics

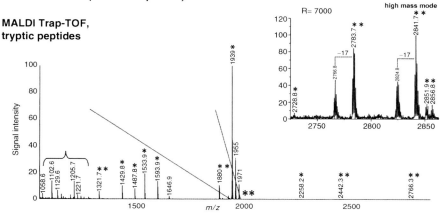

Fig. 2 Analysis of peptides from a 10:1 *Bacillus* spore mixture [14]

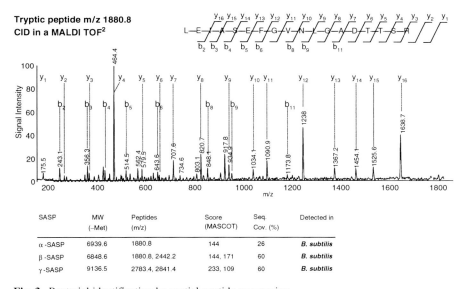

SASP	MW (−Met)	Peptides (m/z)	Score (MASCOT)	Seq. Cov. (%)	Detected in
α-SASP	6939.6	1880.8	144	26	*B. subtilis*
β-SASP	6848.6	1880.8, 2442.2	144, 171	60	*B. subtilis*
γ-SASP	9136.5	2783.4, 2841.4	233, 109	60	*B. subtilis*

Fig. 3 Bacterial identification by partial peptide sequencing

4   Genetically Engineered Bacteria

Genetic engineering might be employed to enhance activity in a known pathogen, or to introduce a toxic protein in an otherwise harmless species. The major interest may be in identifying the payload protein. however, its expression will likely be delayed for some time after pathogen release or infection. To answer the more fundamental question – has the bacterium been genetically engineered? – it may be advantageous to target a systematic plasmid protein. Among the available plasmids, a very limited number of resistance genes is used to allow purification of transformed organisms. Among these, the *bla* gene that encodes β-lactamase is most commonly used, and we recently reported a targeted microsequencing strategy for rapid and automatic detection of plasma-borne β-lactamase in *E. coli* cells[20]. In that work a commercially available *E.coli* was studied, which contained the plasmid illustrated in Fig. 4. The same bacterium missing the plasmid was used as an ampicillin susceptible control. Figure 5 illustrates a traditional

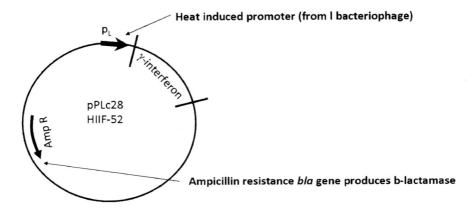

Fig. 4 The pPLc28 plasmid carried in our *E. coli* model system (ATCC 39278)

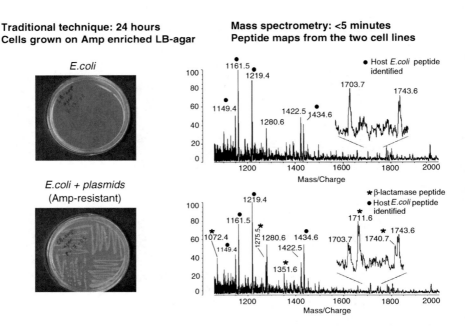

Fig. 5 Two methods to characterize successfully engineered *E. coli*[20]

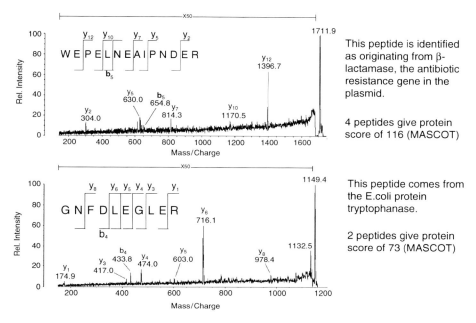

Fig. 6 Identification of plasmid insertion and of the host organism in TOF[2] spectra acquired after on-probe tryptic digestion of genetically engineered *E.coli*[20]

method to test for resistance to antibiotics, cell growth in the presence of ampicillin. The figure also shows mass spectra of peptides produced by tryptic digestion in situ from ampicillin susceptible (top) and resistant (bottom) *E. coli*. Peptides from β-lactamase are starred in the bottom spectrum. In both spectra peptides are detected from proteins characteristic of the host cell *E. coli*. Figure 6 presents MS/MS spectra obtained on a MALDI-TOF[2] instrument of a peptide identified as originating from β-lactamase and a peptide identified as originating from a tryptophanase unique to *E. coli*. Various methods can be envisioned for automating the interrogation of peptide spectra like those shown in Fig. 5. In a hypothesis driven approach, conserved peptide digestion products from the resistance protein(s) can be targeted for interrogation. In a threshold approach all peaks (in the m/z range 1,000–3,000) with relative intensities above, for example, 20% can be interrogated by TOF[2] analysis. In this approach analysis is biased to the most intense precursor ions, which tend to be host-related. Finally, a combination of these two approaches can be used to identify both the host cell and the genetic insert.

5 Identification of Proteins from Organisms without Sequenced Genomes

One limitation that exists for proteomic strategies for analysis of microorganisms is that not every species has a sequenced genome. We have evaluated the possibility of identifying proteins from bacteria with unsequenced genomes, initially with the objective of identifying the biomarkers in MALDI signatures of such organisms, and, subsequently, with the objective of partially characterizing the organism itself. A signature is shown in Fig. 7 for *Yersinia rohdei*, for which no annotated genome sequence was available at the time of our investigation.[21] This approach is not (yet) envisioned as a rapid fieldable method, but rather as suitable for a high end reference laboratory. The work was carried out on an LTQ Orbitrap, an instrumental system with a high resolution analyzer, using electrospray ionization. A sample of *Yersinia rohdei* was lysed and the lysate was fractionated by C-8 HPLC. Proteins with masses below about 15,000 Da were collisionally activated in the LTQ module. Masses of both precursor and product ions were measured with 30,000 resolution in the Orbitrap, and spectra were decharged and searched with ProsightPC 2.0[22] against custom and public databases. Finally, BLASTp and ClustalW similarity searches were used to confirm that the identified proteins are strongly conserved in related species. An example of the kind of fragmentation that can be obtained in such top-down analyses[23] is shown in Fig. 8. The top panel shows the MS/MS spectrum of the 9+ precursor ion at m/z 807.80. The middle panel shows the decharged MS/MS spectrum, which was searched using ProSightPC 2.0 software against a custom database. The third panel shows the sequence of the protein identified as S50 ribosomal protein L29, and summarizes the peptide bond fragmentation assigned. This protein is present in the related bacterium *Yersinia enterocolitica*, and the identification is assigned an expect value of 6.79×10^{-24} by the software.

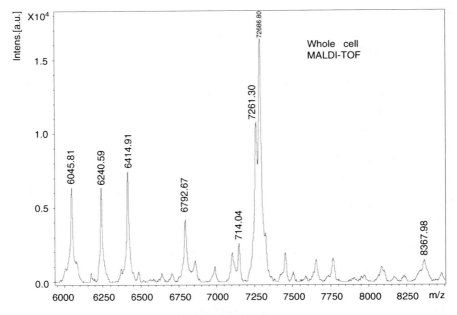

Fig. 7 MALDI Signature spectrum for *Yersinia rohdei*[21]

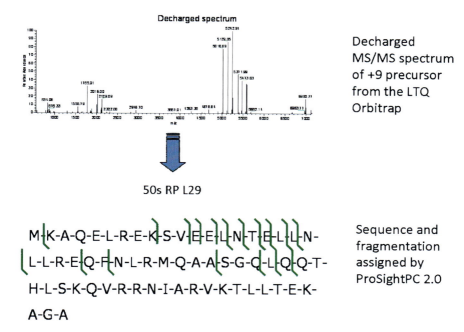

Fig. 8 Identification of ribosomal protein L29 as a biomarker for *Y. rohdei*[21]

(An Expect value smaller than 10^{-4} is considered reliable.) A total of ten proteins from *Y. rohdei* were identified on the basis of homology with proteins from one or more enterobacteria.[21] Nine of these are basic and abundant proteins from the ribosome. This approach is expected to be more rapid than the more rigorous approaches of sequencing the entire genome, or purifying sufficient amounts of these proteins for complete de novo sequencing.

In silico informatics using the RMIDb indicates that the ribosomal proteins, in addition to being readily observable by mass-spectrometry, exhibit very high conservation between species. Figure 9 shows the extent of cross-species sharing of ribosomal proteins with appropriate molecular weights, with nodes representing species and edges representing shared ribosomal protein sequences with molecular weight between 4 and 16 kDa. Edge thickness is proportional to the number of sequences shared between species. To simplify the figure, edges representing less than five proteins are hidden and resulting components of three or more species retained. Cross-species sharing shown in this figure range from *Shewanella sp. W3-18-1* and *Shewanella putrefaciens* in the *Shewanella* genus component, which share all 38 of their potential ribosomal biomarkers, to the three neighbors of *Haemophilus influenzae*, all from different genera, which share five of their potential ribosomal biomarker sequences.

This and other experiments support the conclusion that unknown protein biomarkers can be identified by homology to sequences of known proteins in related species, and support the proposal that phyloproteomic classifications can be

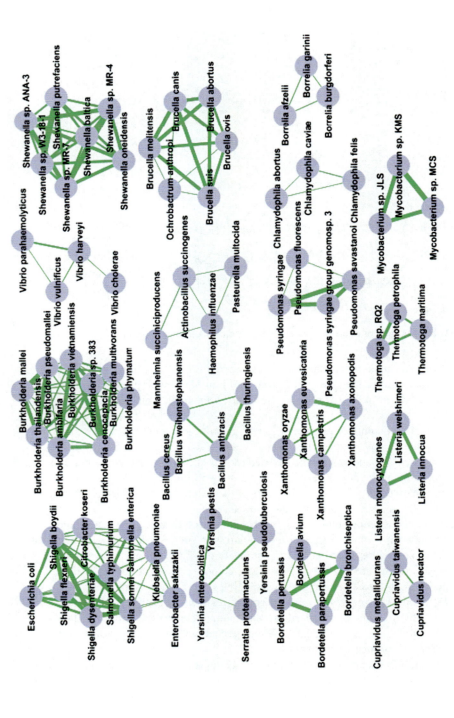

Fig. 9 Network of cross-species shared ribosomal protein sequences with molecular weight between 4kDa and 16kDa. Edges represent at least five potential protein biomarkers, components of less than three species removed. Edge thickness varies with number of shared sequences

deduced for many unknown or unsequenced microorganisms by identifying the related organisms that contain matching, homologous, protein sequences.[21]

6   Conclusions

We have undertaken to make the argument that an interpretive, proteomic analysis of mass spectra of whole organisms will provide the most reliable, yet flexible, method for rapid identification of bioagents. We have presented several case studies as illustrations of how to meet the challenges of analyzing mixtures, recombinant bacteria and species without sequenced genomes.

References

1. Anhalt JP, Fenselau C (1975) Identification of bacteria using mass spectrometry. Anal Chem 47:219–225
2. Demirev P, Fenselau C (2008) Mass spectrometry for rapid characterization of microorganisms. Ann Rev Anal Chem 1:1–93
3. Demirev P, Fenselau C (2008) Mass spectrometry in biodefense. J Mass Spectrom 43:1441–1457
4. Ecelberger SA, Cornish TJ, Collins BF, Lewis DL, Bryden WA (2004) Suitcase TOF: a manportable time-of-flight mass spectrometer. Johns Hopkins APL Tech Digest 25:14–19
5. Sundaram AK, Gudlavalleti SK, Oktem B, Razumovskaya J, Gamage CM, Serino RM, Doroshenko VM (2008) Atmospheric pressure MALDI-MS-MS based high throughput automated multiplexed array system for rapid detection and identification of bioagents. In: Proceedings of the 56th Conference of the American Society for Mass Spectrometry, Denver, CO, 1–5 June 2008
6. Ryzhov V, Fenselau C (2001) Characterization of the protein subset desorbed by MALDI from whole bacterial cells. Anal Chem 73:746–750
7. Pineda FJ, Antoine MD, Demirev PA, Feldman AB, Jackman J, Longenecker M, Lin JS (2003) Microorganism identification by matrix-assisted laser/desorption ionization mass spectrometry and model-derived ribosomal protein biomarkers. Anal Chem 75:3817–3822
8. Ryzhov V, Hathout Y, Fenselau C (2000) Rapid characterization of spores of *Bacillus cereus* group bacteria by MALDI-TOF mass spectrometry. Appl Environ Microbiol 66:3828–3834
9. Warscheid B, Fenselau C (2003) Characterization of *Bacillus* spore species and their mixtures using post-source decay with a curved-field reflectron. Anal Chem 75:5618–5627
10. Thomas JJ, Falk B, Fenselau C, Jackman J, Ezzell J (1998) Viral characterization by direct detection of capsid proteins. Anal Chem 70:3863–3867
11. Kim YJ, Freas A, Fenselau C (2001) Analysis of viral glycoproteins by MALDI-TOF mass spectrometry. Anal Chem 73:1544–1548
12. Demirev P, Ho YP, Ryzhov V, Fenselau C (1999) Microorganism identification by mass spectrometry and protein database searches. Anal Chem 71:2732–2738
13. Pineda F, Lin JS, Fenselau C, Demirev P (2000) Testing the significance of microorganism identification by mass spectrometry and proteome database search. Anal Chem 72:3739–3744
14. Warscheid B, Jackson K, Sutton C, Fenselau C (2003) MALDI analysis of *Bacilli* in spore mixtures by applying a quadrupole ion trap time-of-flight tandem mass spectrometer. Anal Chem 75:5608–5617

15. Warscheid B, Fenselau C (2004) A targeted proteomics approach to the rapid identification of bacterial cell mixtures by MALDI mass spectrometry. Proteomics 4:2877–2892
16. RMIDb: Rapid microorganism identification database. URL: http://rmidb.org. July 15, 2009
17. Demirev PA (2004) Enhanced specificity of bacterial spore identification by oxidation and mass spectrometry. Rapid Commun Mass Spectrom 18:2719–2722
18. Pribil P, Patton E, Black G, Doroshenko V, Fenselau C (2005) Rapid characterization of *Bacillus* spores targeting Species-unique peptides produced with an atmospheric-pressure MALDI source. J Mass Spectrom 40:464–474
19. Fenselau C, Russell S, Swatkoski S, Edwards N (2007) Proteomic strategies for rapid characterization of micro-organisms. Eur J Mass Spectrom 13:35–39
20. Russell S, Edwards N, Fenselau C (2007) Detection of plasmid insertion in *E.coli* by MALDI-TOF mass spectrometry. Anal Chem 79:5399–5409
21. Wynne C, Fenselau C, Demirev PA, Edwards N (2010) Top-down identification of protein biomarkers in bacteria with unsequenced genomes. Anal Chem, in press
22. Boyne MT, Garcia BA, Li MX, Zamdborg L, Wenger CD, Babai S, Kelleher NL (2009) Tandem mass spectrometry with ultrahigh mass accuracy clarifies peptide identification by database retrieval. J Proteome Res 8:374–379
23. Kelleher NL (2004) Top-down proteomics. Anal Chem 76:A197–A203

Hollow-Fiber Flow Field-Flow Fractionation for Mass Spectrometry: From Proteins to Whole Bacteria

Pierluigi Reschiglian, Andrea Zattoni, Diana Cristina Rambaldi, Aldo Roda, and Myeong Hee Moon

Abstract Mass spectrometry (MS) provides analyte identification over a wide molar-mass range. However, particularly in the case of complex matrices, this ability is often enhanced by the use of pre-MS separation steps. A separation, prototype technique for the "gentle" fractionation of large/ultralarge analytes, from proteins to whole cells, is here described to reduce complexity and maintain native characteristics of the sample before MS analysis. It is based on flow field-flow fractionation, and it employs a micro-volume fractionation channel made of a ca. 20 cm hollow-fiber membrane of sub-millimeter section. The key advantages of this technique lie in the low volume and low-cost of the channel, which makes it suitable to a disposable usage. Fractionation performance and instrumental simplicity make it an interesting methodology for in-batch or on-line pre-MS treatment of such samples.

Keywords Field-flow fractionation • Hollow-fiber flow field-flow fractionation • Mass spectrometry • Proteomics Bacteria identification • Forensic analysis

---

P. Reschiglian (✉), A. Zattoni, and D.C. Rambaldi
Department of Chemistry, "G. Ciamician", Via Selmi 2, 40126 Bologna, Italy
and
I.N.B.B. Consortium, Rome, Italy
e-mail: pierluigi.reschiglian@unibo.it

A. Roda
Department of Pharmaceutical Science, Via Belmeloro 6 40126, Bologna, Italy
and
I.N.B.B Consortium, Rome, Italy

M.H. Moon
Department of Chemistry, Yonsei University, Seoul, South Korea

J. Banoub (ed.), *Detection of Biological Agents for the Prevention of Bioterrorism*,
NATO Science for Peace and Security Series A: Chemistry and Biology,
DOI 10.1007/978-90-481-9815-3_2, © Springer Science+Business Media B.V. 2011

1   Introduction

It is widely acknowledged that when sample complexity exceeds the resolution capabilities of most sophisticated mass spectrometry (MS) techniques, the availability of so-called "pre-MS" methods is required. When MS methods are applied to complex protein samples, for instance, many efforts are devoted to mining the low-abundance proteins (LAP) among the huge wealth of high abundant proteins (HAP) possibly present in the sample. This can be also the case of proteomics or bacterial fingerprinting in forensic applications.

The increased threat of biological warfare and the strategies to counteract bioterrorism are widespread reminders of the urgent need of robust methods to identify and characterize protein toxins or pathogenic microorganisms in complex samples. MS has proved to be powerful tool for the rapid identification and characterization of microorganisms.[1] Fenselau and co-workers have pioneered development and application of matrix-assisted laser desorption/ionization time-of-flight MS (MALDI-TOF-MS) for the characterization of intact microorganisms.[2] This is because species desorbed from bacterial cells are intact proteins in the molar-mass (M_r) range 4,000–15,000, and the proteins coded by bacterial genomes fall within this M_r range.[3] Biomarkers can then be found in this range, and bacteria can be identified through proteomic database searching algorithms.[4,5] Proteomic approaches for MS-based bacteria identification do not suffer from limitations due to spectra reproducibility issues. Nonetheless, protein databases are as yet available only for a limited number of bacterial species. For this reason, the most common approaches so far employed to identify unknowns by MALDI-TOF MS analysis of intact bacteria are based on the similarity between the spectra of the unknown bacteria and those in MALDI-TOF-MS libraries of reference bacterial species.[6] Nonetheless, for these methods to be valid, a high degree of reproducibility is required. This is a particularly critical aspect in the identification of bacteria mixtures with high differences in the relative percentage of the different strains, because the resulting spectra are highly complicated. MALDI-TOF-MS of bacteria mixtures is not only complicated by the high number of ion signals in the spectra, but also by the fact that MALDI is a competitive ionization process, and the spectra of bacteria mixtures can be quite different from the linear combination of characteristic signals obtained for each individual bacterial species. As a consequence, comparing the characteristic signals obtained from bacteria mixtures with the ion signal databases obtained with individual bacterial species could give inaccurate results. Sample preparation methods able to enrich the sample in one bacterial species can potentially reduce the analytical complexity and difficulties in interpreting spectra obtained for bacteria mixtures.

1.1   pre-MS Separation Methods

Few methods are available for the separation of whole bacterial cells, while separation techniques have been most natural complements of MS to increase the information obtained from MS analysis of complex protein samples. They however have "pros & cons".

Two-dimensional polyacrylamide gel electrophoresis (2D PAGE) is still the most applied pre-MS separation method in proteomics.[7–9] The two orthogonal separation mechanisms on which 2D PAGE is based greatly enhance the final resolution capacity. 2D PAGE is also a relatively inexpensive method, and it has the unparallel advantage to give immediate visualization of the differences in protein composition/expression. However, when separation is completed the proteins are left in the gel matrix. This makes it difficult to retrieve protein spots in their intact conditions, and with total recovery. The latter constitutes a serious limitation for the isolation and characterization of LAP, even in case of highly-sensitive MS-based proteomic methods.[10,11] Moreover, the process of separation, spot isolation, and sample preparation for further MS analysis is time-consuming, and difficult to automate.

Free-flow electrophoresis (FFE) is becoming popular for semi-preparative scale separation of proteins because, in principle, it provides unlimited throughput.[12,13] However, separation is based only on pI differences, the carrier is expensive, and before MS it must be removed from protein fractions.

Capillary zone electrophoresis (CZE) is far more efficient than other separation methods. When coupled with MS, CZE shows very promising for "micro-scale" proteomics, like in the case of single-cell proteomics.[14] Sensitivity issues related to the limited amount of separated samples can be faced using nanospray interfaces. However, some technical aspects still limit routine application of CZE-MS systems. For instance, without a proper on-line desalting device the non-volatile salts possibly present in the buffer can cause ionization problems in MS analysis. Moreover, possible issues due to electrical interferences caused by differences in the voltages applied to CZE and to the ion source are not, as yet, completely solved.

Reversed-phase (RP) HPLC competes with PAGE as mostly applied separation method for proteomics, though it does not provide comparable resolution. Narrow-bore, long columns under high or ultra-high pressure conditions (RP UPLC) provide higher efficiency. However, upon high pressure conditions, and/or using organic modifiers in the mobile phases, protein degradation may occur. Multidimensional HPLC increases separation performance but undesired interaction between proteins and stationary phases have more chances to occur.

Multidimensional HPLC and CZE are high-resolution methods but they are not particularly selective with respect to changes in the protein diffusion coefficient due to changes in protein structure. SEC separation depends on the protein structure. SEC is therefore used for functional proteomics, and it is applied also in preparative scale. However, interaction between protein and packing material might occur, which can cause protein entanglement and, consequently, affect the native conformation.

Separation methods are often combined with other pre-analytical steps to remove highly abundant components. Immunoaffinity separations still are the leading methods, which however suffer from intrinsic limitations such as samples dilution, and unwanted co-depletion of analytes associated to the removed components. Immunosorption can also reduce depletion specificity. In immunoaffinity chromatography, the risk of run-to-run sample carry-over is also present, and depleted sample desalting is also necessary. Immunoaffinity methods can be also employed to separate and selectively capture bacteria. However, the presence of antibodies bound to bacterial membrane antigens can affect accuracy of MS-based bacteria identification.

A single "does-it-all" method is still far to be available. It therefore appears interesting the availability of novel separation methods able to fill limitations of current methodologies. When combined with most-established techniques, comprehensive pre-MS methodologies can be eventually developed to enhance MS application to the identification of pathogen proteins and/or bacteria in complex samples.

1.2   Flow Field-Flow Fractionation

Among separation techniques, field-flow fractionation (FFF) has shown broadest M_r application range. In the bio-analytical field, applications spanning from proteins to whole cells have been reported.[15,16]

In common with LC, FFF uses similar experimental setup. The separation run consists of the injection of a narrow sample band into a mobile phase stream that sweeps sample components down the separation channel to finally reach a detector and/or collection device. Very differently from LC, the FFF mechanism is not based on interaction of the analyte with a stationary phase, but with an external field that is applied perpendicularly to the mobile phase flow. Because of different M_r, size, and/or other physical properties, the different analytes are driven by the orthogonal field into different velocity regions within the parabolic flow profile of the mobile phase across the channel.

Different field types have originated different FFF variants. In flow FFF (F4), the field is a second stream of mobile phase that is applied across the channel (crossflow). The force driving separation then is the viscous force exerted on the analyte by the crossflow stream. Because of the most universal field, F4 is capable of separating almost all macromolecules and particles (e.g. from proteins to whole cells) from 1 nm to more than 50 μm in size. The lower size limit is related to the M_r cut-off of the accumulation wall, which is usually constituted of an ultrafiltration membrane able to retain the macromolecular analytes inside the channel. F4 is flexible in channel design, and we can have flat channels, with the cross flow applied either in a symmetrical or asymmetrical configuration, or tubular channels with a radial cross-flow configuration. In the latter case, a hollow-fiber (HF) membrane for micro-dialysis makes the accumulation wall of the channel (HF5).[17]

2 Hollow-Fiber Flow Field-Flow Fractionation

The idea of using a HF membrane for FFF operations dates back 1974.[18] More than 10 years later, a significant effort on the development of HF5 fundamentals was reported.[19]

In a typical HF5 arrangement, the HF is connected to a pump that generates along the HF a longitudinal flow of mobile phase. A pressure drop between the inner and outer wall of the HF generates the radial crossflow. The radial flow velocity at the HF wall (u_r) is expressed as[19]

$$u_r(R,z) = u_r(R,0)\exp(-\alpha z) \tag{2.1}$$

where R is the HF inner radius, z the axial coordinate, and α is a constant including HF membrane permeability and mobile phase viscosity. Assuming laminar flow conditions, a parabolic flow profile is established in the HF, with the axial velocity u_z given by[19]

$$u_z(r,z) = u_z(0,z)(1 - x^2) \tag{2.2}$$

where r is the radial coordinate, and $x = r/R$. The average u_z decreases along the HF due to the loss of mobile phase through the HF wall.

By integration of Eq. 2, the void time t_0 can be calculated as[20]

$$t_0 = \frac{V_0}{F_{rad}} \ln\left(\frac{F_{in}}{F_{out}}\right) \tag{2.3}$$

where V_0 is the channel void volume ($\pi R^2 L$, with L the fiber length), and the average flow rates (F_{in}, F_{out}, F_{rad}) are expressed in terms of volumetric flowrates.

2.1 Normal Retention Mode

The normal retention mode (Fig. 1) takes place when displacement of the sample components across the HF channel due to their Brownian diffusion is comparable to the displacement due to the field generated by the radial flow. This is generally true in the case of macromolecules (e.g. proteins) and relatively small (e.g. nano-sized) particles.

When such sample components are introduced in the HF channel, they undergo to a process of relaxation/focusing to achieve, before separation, a steady-state concentration profile across the HF. During such a relaxation/focusing process, the migration of sample components towards the channel wall due to the radial flow generates a concentration gradient in the radial direction. Otherwise, sample particles diffuse in a direction opposite to the radial flow until they reach an equilibrium

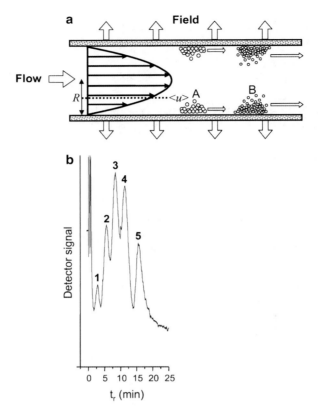

Fig. 1 (**a**) The normal elution mode. Sample particles are driven by two opposite transport processes to a dynamic equilibrium position of characteristic average elevation from the HF inner wall: the cross-flow (radial flow) field drives sample particles towards the HF wall while Brownian diffusion drives the particles toward the HF center. Retention time of smaller particles (A) is then shorter than that of bigger (B) particles because A particles are eluted by faster streamlines. (**b**) Normal-mode HF5 of polystyrene, standard nanobeads: (1) 50 nm, (2) 102 nm, (3) 155 nm, (4) 200 nm, (5) 300 nm. HF channel: $R = 0.41$ mm, $L = 240$ mm, 30,000 M_r pore cut-off. Experimental conditions: $F_{in} = 1.45$ mL/min, $F_{rad} = 0.075$ mL/min; mobile phase: 0.1% FL-70, 3.2 mM NaN$_3$, 2.5 mM Tris. Sample load: 0.1 mg (Reprinted with permission from ref. [17], © 2007, Bentham Science Publishers)

condition. In this case, the concentration profile in the radial direction can be expressed as

$$C = \beta \exp\left(Pe\left(x^2 - \frac{x^4}{4} \right) \right) \tag{2.4}$$

where β is an integration constant, and Pe is the Peclet number defined as

$$Pe = u_r\left(R, z \right) R / D \tag{2.5}$$

where D is the analyte diffusion coefficient. Combining Eqs. 2 and 4, and assuming an ideal Gaussian concentration profile for the sample band in the longitudinal direction, the exact expression for t_r can be obtained[19].

Simplified treatments for the expression of retention time (t_r) were proposed.[20,21] Equation 1 in fact states that the value of u_r decreases along the fiber due to the pressure drop existing between the inlet and outlet extremities of the fiber. However, under typical pressure conditions in HF5 the decrease of u_r between the HF inlet and outlet is lower than 2%. As a consequence, a uniform radial flow velocity can be assumed along the HF channel. Under these conditions (i.e. with $Pe > 50$), the t_r of highly-retained species takes the form

$$t_r = \frac{R^2}{8D} \ln\left(\frac{F_{in}}{F_{out}}\right) \qquad (2.6)$$

The relaxation/focusing process is usually realized around a certain position L_0 of the HF channel (i.e. at $z = L_0$), said the focusing point. When the relaxation/focusing process is completed, the elution process actually starts from the focusing point. Equation 6 was then modified as

$$t_r = \frac{R^2}{8D} \ln\left(\frac{F_{in} - \left(\frac{L_0}{L}\right) F_{rad}}{F_{out}}\right) \qquad (2.7)$$

From Eq. 7, and from the well-known Stokes-Einstein expression for D, one finally gets the relationship between t_r and the hydrodynamic diameter (d) of the analyte

$$d = \frac{8kT}{3\pi\eta R^2} \ln\left(\frac{F_{in} - \left(\frac{L_0}{L}\right) F_{rad}}{F_{out}}\right)^{-1} t_r \qquad (2.8)$$

where k is the Boltzman constant, T the absolute temperature, and η is the mobile phase viscosity.

2.2 Hyperlayer Retention Mode

The hyperlayer retention mode (Fig. 2) is known to govern retention of analyte particles the size of which is sufficiently high (e.g. micronsized particles like whole cells) to make negligible the effect of Brownian diffusion.[22]

In hyperlayer HF5, the t_r of such sample components can be expressed as[23]

$$t_r = \frac{t_0}{2\gamma} \frac{R}{d} \qquad (2.9)$$

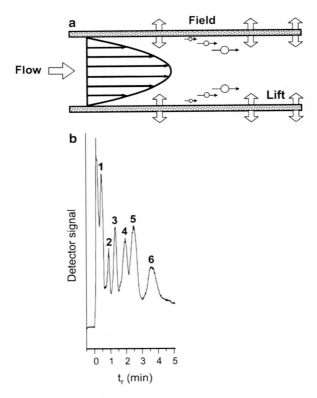

Fig. 2 (**a**) The hyperlayer retention mode. Micron-sized particles, which have negligible diffusion transport, are driven by the cross-flow (radial flow) field towards the HF inner wall and form a thin layer hugging the wall. During elution, the field force is opposed by flow-induced lift forces, the intensity of which also depends on physical features of the sample particles. Since larger particles are driven by the lift forces away from the HF wall into faster streamlines, the retention order is reversed with respect to the normal mode: larger particles elute first. (**b**) Hyperlayer HF5 of PS standard micronsized beads: (1) 15 μm, (2) 10 μm, (3) 7 μm, (4) 5 μm, (5) 4 μm, (6) 3 μm. HF channel: $R=0.44$ mm, $L=240$ mm, 50,000 M_r pore cut-off. Experimental conditions: $F_{in}=2.0$ mL/min, $F_{rad}=0.235$ mL/min; mobile phase: 0.1% FL-70, 3.2 mM NaN_3, 2.5 mM Tris; Sample load: 22.5 mg (Reprinted with permission from ref. [17], © 2007, Bentham Science Publishers)

where γ is a correction factor that depends on physical features of the sample particles such as shape, flexibility, and surface features, and on the flow conditions. The relationship between d and t_r can be determined by calibration with standard particles of known physical features[23]

$$\log t_r = \log t_{r1} - S_d \log d \qquad (2.10)$$

where S_d is the diameter-based selectivity, and t_{r1} is the extrapolated retention time for a particle of a unit diameter. Using polystyrene (PS), spherical bead mixtures, S_d values were found in the range 1.2–1.7, and to depend on the membrane type and on F_{rad}. The experimental plate height values were found to be higher than 700, and to increase with increasing F_{out}/F_{rad}[24].

2.3 Methodology

A HF5 run typically involves two steps: sample injection/focusing/relaxation, and sample elution.

The first step is realized by delivering inside the HF channel the sample from the injection port using a longitudinal flow of mobile phase towards the HF channel outlet, and a second flow of mobile phase in opposite direction, from the outlet to the inlet of the HF channel. At the focusing point, the resulting longitudinal flow rate is zero, while a given cross-flow goes through the pores from the inner to the outer wall of the HF to make the injected sample focus and achieve its steady-state condition, said relaxation. The focusing point position then depends on the ratio between the inlet and outlet flowrate values (F_{in}, F_{out}).

When the injection/focusing/relaxation process is completed, the flow pattern is changed to set on the sample elution step. For the elution, a single mobile phase flow stream is longitudinally applied inside the HF towards the channel outlet.

2.3.1 The HF5 System

The first HF5 system was described in 1989.[19] The HF module was built by inserting a piece of HF membrane into an empty glass tube equipped with a radial flow outlet, and inserted into an LC-like apparatus consisting of a HPLC-type injection port, two pumps, a UV/vis detector, and a control unit for flow rate management. The first pump delivered the mobile phase fluid inside the HF5 channel. The second pump was connected to the radial flow outlet of the HF5 module, and it worked in "unpump" mode. This pump in fact drew the mobile phase fluid from the inner wall of the HF channel through the HF pores.

Ten years later, a second scheme using a single pump was proposed.[21] The injection/focusing/relaxation step was carried out by splitting the pump flow into two streams that were applied in opposite direction to the inlet and outlet extremities of the HF5 module. Using a single pump gave advantages in terms of simplicity in the system operations, and lower costs.

It was later proposed a third scheme using two pumps: a HPLC pump to generate the required flow rate during sample elution, and a syringe pump to generate, in combination with the first pump, the opposite flows required for the sample injection/focusing/relaxation step.[25] This scheme allowed for more efficient operations when HF5 was coupled with ESI/TOF MS or MALDI-TOF-MS.[26,27] The general schematic of this HF5 system is reported in Fig. 3.

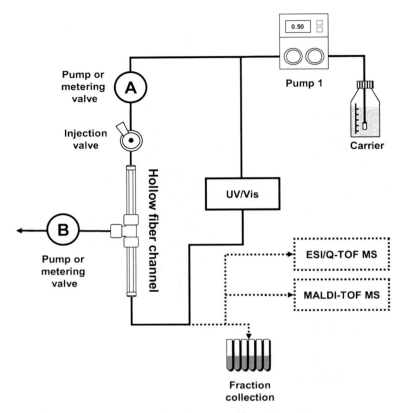

Fig. 3 General, schematic diagram of the HF5 system. Module A is either a pump or a metering valve to regulate flowrates in the sample injection/focusing/relaxation step. Module B is either a metering valve or a pump working in "unpump" mode, which is used to regulate the radial flow rate during the elution step. (Adapted with permission from ref. [17], © 2007, Bentham Science Publishers)

Most recently, HF5 was implemented into a commercial system for asymmetrical F4 (AF4) with multi-angle laser scattering (MALS) detection by replacing the standard, flat-type channel with the HF5 module.[28] This HF5-MALS system has been applied to fractionate and size/shape characterize lipoproteins in whole blood serum. The promising results allow to foreseeing use of HF5 as an alternative to flat channels in commercial systems for AF4.

2.3.2 HF Membrane Type

HF membranes made of polymeric materials such as polysulphone (PSf), poly-acrylonitrile (PAN) and chlorinated polyvinylchloride (cPVC) have been employed. A pressurized, relatively flexible polymeric HF membrane should naturally tend to a perfectly cylindrical geometry. As a consequence, with respect to the commercial, flat channels, a polymeric HF channel is, in principle, more suitable to the establishment of an ideal F4 mechanism. On the other hand, to ensure good channel-to-channel fractionation reproducibility the HF wall must be perfectly smooth, and the pore-size distribution must be homogeneous to have a known inner diameter (I.D.) and a well-defined cross-section profile.[29,30] Size and chemical properties of the analytes also influence the choice of the best HF surface composition and pore cut-off value (generally in the 6–100 KDa range) because best fractionation performance is achieved if interactions between analytes and the HF inner wall are minimized.

PSf HF5 channels have been used for the analysis of diverse samples, including water-soluble synthetic polymers and biological samples such as proteins, viruses, bacteria and yeast cells.[18,25–27,31,32] PSf HF membranes are however relatively soft, and they may swell during usage causing an increase in retention time.[21,33] HF swelling during usage increases with increasing backpressure generated by the system connected downstream to the HF module. Using low back-pressure detector cells like light-pipe cells, an increase in retention time reproducibility can be observed.[25]

Because of their rigidity, PAN HF membranes can be used at relatively high flow-rate conditions for hyperlayer HF5.[24] PAN HF membranes also show chemical resistance to different organic solvents. For this reason, they have been employed also for the analysis of organic-soluble polymers.[34]

Possible adsorption of the analyte on the HF inner surfaces can severely reduce sample recovery. The injection/focusing/relaxation step is recognized as one of prime cause of sample losses during HF5 operations. This is commonly observed by visual inspection of the inner side of used HF modules. However, X-ray photoelectron spectroscopy (XPS) analysis of the HF inner wall after several protein sample runs showed homogeneous protein layers present all along the HF.[35] The HF composition and, consequently, surface polarity is therefore to be carefully considered. Relatively polar HFs made of cPVC have been successfully used for HF5 of human red blood cells.[31]

Ceramic, rigid fibers were also proposed to improve robustness and chemical inertness of the channel.[36] Ceramic HF channels for the analysis of standard proteins and protein aggregates showed promising results in terms of recovery, run-to-run repeatability and long-term stability.[37]

2.3.3 The HF5 Module

Over more than 25 years, many efforts have been focused on the optimization of the HF5 module design and construction. Special emphasis has been given to define best design for easiest implementation of the HF5 module into a chromatographic-like system, for easy replacement of the HF channel, and for an optimized flow-rate management. Only in relatively recent times some significant advances on these aspects have been reported.

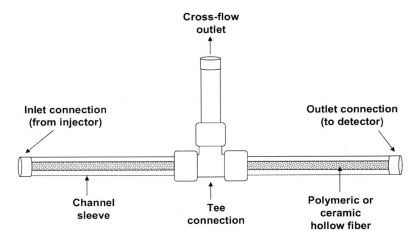

Fig. 4 Typical HF5 module design (Reprinted with permission from ref. [17], © 2007, Bentham Science Publishers)

The most general design currently used for a polymeric HF5 module is schematized in Fig. 4.[17] A piece of HF membrane is inserted into two 1/8"-O.D. (outer diameter) Teflon sleeves of equal length. The sleeves are connected by a tee union to make the radial flow exit. Standard, plastic fittings are placed to make inlet and outlet connections. This simple design shows the advantage to employ commercially-available and relatively inexpensive components, and to standardize connections to an LC-like system. Typical HF channel dimensions are about 20 cm in length and 0.80–1.00 mm I.D.

Miniaturization in analytical separation techniques generally provides several advantages such as decrease in sample injection amount, and the enhancement of separation speed, efficiency, and detection limit. Using microbore, polymeric HF channels (450 μm I.D. × 25 cm length) with geometrical volume of only ~40 μL, HF5 was scaled down to a microflow rate regime.[38]

Scale-up of the HF5 process for semi-preparative purposes has been recently described.[39] A multiplexed (Mx) HF5 device has been made with six parallel HF5 modules connected to seven-port manifolds. This assembly shows the possibility of handling up to 50 μg of proteins without incurring overloading. Mx HF5 offers a unique advantage of scaling up HF5 separation without using wide-bore HF membranes, which would increase separation time.

3   HF5 and MS

Off-line and on-line coupling to MS has made F4 enter the field of proteomics.[40,41] HF5 was the first micro F4 variant used for MS-based protein characterization. HF5 shows unique, intrinsic features for either off-line or on-line coupling to MS:

(a) low channel volume (in the order of 100 μL), which reduces sample dilution; (b) low flow rate conditions (as low as 200 μL/min) which, in case of on-line coupling to MS, does not require high split ratios between the channel outlet and the ionization source; (c) possible disposable usage, which eliminates the risk of run-to-run sample carry-over and, then, spectra contamination; (d) on-the-fly sample desalting.

3.1   Proteins

For the analysis of intact proteins and protein complexes under native conditions HF5 shows additional features. Without a stationary phase, mechanical or shear stress on the analyte caused by packing material, which can cause entanglement or alter the native protein conformation, is very little (if any). As mobile phase, HF5 can utilize any solution. Other separation techniques need to use mobile phases that may cause proteins to lose their three-dimensional conformations, or they may unexpectedly induce dissociation of protein complexes during separation, or they can be incompatible with further MS characterization. Finally, D-based selectivity of HF5 is particularly high, and from retention-based measurements of D, interaction/aggregation between fractionated proteins can be deduced.

The analysis of protein complexes is important, for instance, in the study of functional drugs because self-association or dissociation phenomena are strongly related to the protein drug activity. Coupling HF5 with MALDI-TOF-MS and with a chemiluminescence (CL) enzyme activity assay has allowed to relating the supramolecular structure of an enzyme drug (uricase) with its enzymatic activity.[42]

HF5 shows interesting also as a "smart" method to reduce sample complexity. For instance, when applied to untreated, whole human blood serum, HF5 is able to effectively fractionate serum proteins, as shown in Fig. 5 (from ref. [43]). Effective fractionation of albumin and other serum HAPs under native conditions allows, in perspectives, to using HF5 for proteomic studies on peptides/proteins associated to HAPs.

HF5 can be online coupled to ESI/TOF MS by simply connecting the detector outlet to the ion source via a splitting valve. Preliminary results have shown spectra able to demonstrate that during fractionation the proteins can maintain their native structure, and be effectively desalted.[27] Possible correlation between the M_r values independently measured by ESI/TOF MS spectra and from HF5 retention time measurements can then produce significant information on the quaternary structure of the fractionated proteins. In Fig. 6 (adapted from ref. [27]) it is reported an example of HF5-ESI/TOF MS of human hemoglobin (Hb). The representative spectrum shows the presence of three species with M_r values corresponding to the M_r values of the α and β subunits ($M_r = 15,126.5 \pm 0.3$, $M_r = 15,867.3 \pm 0.5$, respectively), and of the α-heme complex ($M_r = 15,741.5 \pm 0.7$). This corresponds to the spectrum of native Hb, and the M_r value obtained from HF5 retention of Hb at pH 7.0 (data point Hb1 of the regression plot inset in Fig. 6) corresponds to the M_r

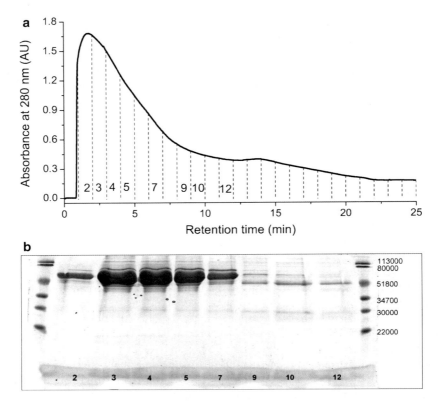

Fig. 5 HF5 of human blood whole serum 1:5 v/v diluted in the mobile phase (5 mM NH$_4$Ac). (**a**) HF5 fractogram and fractions collected for SDS PAGE. (**b**) SDS PAGE of the collected fractions. PSf HF membrane: 30,000 M_r nominal cut-off, $R = 0.040$ cm (referred to dried conditions), $L = 24$ cm. Experimental conditions: $F_{in} = 0.7$ mL/min, $F_{rad} = 0.4$ mL/min (Reprinted with permission from ref. [43], © 2008, Elsevier Publishers)

value of the tetramer. Since similar mass spectrum was obtained also at pH 8.2, the difference in retention observed by increasing pH (data point Hb2 of the regression plot) could be ascribed exclusively to the conformational changes that are known to occur in Hb with increasing pH.

It is known that in ESI/MS an increase in detection sensitivity is possible by reducing the source inlet flow rate. Under micro/nanoflow regimes, miniaturized separation methods are generally characterized by high efficiency. Miniaturized (microbore, μ) HF5 can improve protein identification using shotgun proteomics. Using μHF5, the *Corynebacterium glutamicum* proteome was fractionated, and fractions of different M_r values were run through nanoLC-ESI/MS.[44] Ionization suppression and MS-exclusion effects from spectral congestion were also observed.

A 2D, rapid, gel-free separation method for nanoLC-ESI/MS-based proteomics was developed by the hyphenation of μHF5 with capillary isoelectric focusing

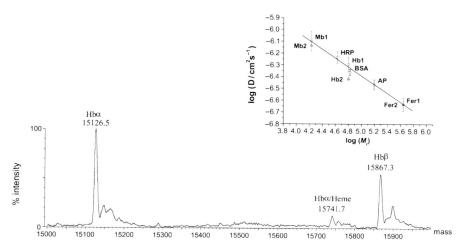

Fig. 6 HF5-ESI/TOFMS of human hemoglobin (Hb): molar mass spectrum. F_{in}=0.70 mL/min; F_{rad}=0.38 mL/min. In the inset: regression plot log D vs. log M_r: ($\circ$) 30,000 M_r HF cut-off; (Δ) 6,000 M_r HF pore cut-off; Mb=horse heart myoglobin, BSA=bovine serum albumin, AP=calf intestine alkaline phosphatase, HRP=horseradish peroxidase, Fer=horse spleen ferritin (Adapted with permission from ref. [27], © 2005, American Chemical Society)

(CIEF).[45] CIEF-µHF5 maintains the advantage of µHF5 to carry on separation in empty ducts, and it also provides, during second-dimension µHF5, removal through the HF wall of the ampholyte solution used for first-dimension CIEF.

3.2 *Whole Bacteria*

F4 has been successful for the separation of whole cells.[15,16] First example of F4 as pre-MALDI-TOF-MS of whole bacterial cells employed a commercial, macro-column F4 channel.[46] The work threw light on three issues that could limit the effective use of F4 for MALDI-TOF-MS of whole bacteria. Firstly, possible run-to-run sample carry-over due to incomplete sample recovery from F4 could affect spectra reproducibility and, thus, reduce fingerprinting capabilities of MALDI-TOF-MS. Secondly, the relatively high sample dilution reached after the F4 step could result in cell concentrations that are below the detection limits for MALDI-TOF-MS. Thirdly, the time required by the F4 step could affect the intrinsic rapidity of MALDI-TOF-MS analysis.

When compared to commercial, flat-type F4 technology, potentially disposable usage of HF5 shows not only the advantage to eliminate risks of run-to-run sample contamination, but also the advantage to reduce possible sterilization issues. This is a key point when living, toxic microorganisms need to be analyzed. In common with applications to protein analysis, the reduced channel volume (typically less than 100–150 µL) can reduce sample fractionation volume, which in turn can reduce sample dilution. Such low sample dilution means that fewer concentration steps have to be performed on the fractionated samples to maintain MALDI-TOF-MS detectability. This contributes to decrease total analysis time, and to simplify possible system automation. HF5 actually shows to be an effective method to fractionate whole prokaryotic and eukaryotic cells,[25,31] and it has been applied to MALDI-TOF-MS of whole bacteria.[26]

3.2.1 HF5 Sorting

Whole cell retention in HF5 is governed by the steric/hyperlayer mechanism,[47,48] so that cells are fractionated according to differences in their physical features.[49] Figure 7 reports two examples of HF5 of different bacterial species.

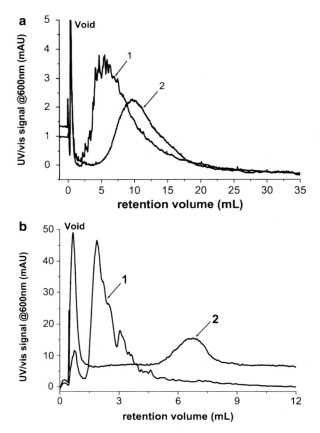

Fig. 7 HF5 of different bacteria. (**a**) Lyophilized *E. coli* (fractogram 1) cells, *B. clausii* spores (fractogram 2); F_{in}=3.0 mL/min, F_{rad}=0.3 mL/min. (**b**) Lyophilized *B. subtilis* (fractogram 1) cells, cultured *B. subtilis* (fractogram 2) cells; F_{in}=4.0 mL/min, F_{rad}=0.7 mL/min. Mobile phase: 1 mM ammonium cholate, pH=9.2 (Adapted with permission from ref. [26], © 2004, American Chemical Society)

Figure 7a shows that spores of *Bacillus clausii* (*B. clausii*) (fractogram 2) are more retained than lyophilized *E. coli* cells (fractogram 1). The *B. clausii* spores have, in fact, spherical shape, with size comparable to that of rod-shaped *E. coli* cells. The latter cells are ca. 2 μm in length and ca. 0.7 μm in width, which correspond to an aspect ratio of about 2.7.[50] The shorter retention time observed for the lyophilized *E. coli* cells is, therefore, a consequence of the higher aspect ratio. Figure 7b shows that the retention of lyophilized *Bacillus subtilis* (*B. subtilis*) cells is completely different from retention of the cultivated *B. subtilis* cells. This is because the cell physical features depend on the cell growth stage.

HF5 is not only able to distinguish different bacteria under different living conditions, but it can also sort bacteria of same species. *E. coli* cells can be fimbriated, i.e. some cells can show "pili" on their membrane.[31] Figure 8 shows an example of HF5 of a fimbriated (CS5 0398) and of a non fimbriated (XC113 A2) *E. coli* strain. A very large difference in retention between the strains can be observed, with separation between fimbriated and non fimbriated *E. coli* cells due to differences in surface features rather than to differences in the aspect ratio.

3.2.2 Bacteria Mixture Analysis

Table 1 reports an example of two databases of most characteristic ion signals obtained for lyophilized *E. coli* and *B. subtilis* cells, which were created by run-to-run and day-to-day replicates of MALDI-TOF-MS spectra.[26] For each bacterial species were used the *m/z* values obtained with a deviation of less than ca. 5 *m/z* units between repeated MALDI-TOF-MS spectra. Comparison of the signals

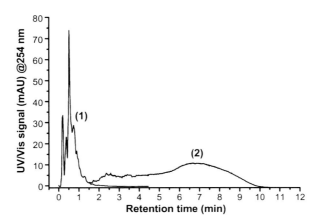

Fig. 8 HF5 of *E. coli*. (1) Fimbriated strain C55 0398. (2) Non-fimbriated strain XC113 A2. Mobile phase: 0.1% v/v FL-70, 0.002% w/v NaN_3, 0.125% w/v TRIS. Experimental conditions: $F_{in}=3.00$ mL/min, $F_{rad}=0.44$ mL/min. PSf HF channel. (Adapted with permission from ref. [31], © 2003, Elsevier Publishing)

Table 1 Database of most characteristic MALDI-TOF-MS signals found with run-to-run and day-to-day variations of less than ca. 5 m/z units

E. coli	B. subtilis
Mean m/z values ($N=30$)	Mean m/z values ($N=22$)
6,109.0	4,260.3
7,273.8	4,445.1
9,063.5	5,638.2
9,225.0	7,328.0
9,535.4	7,369.9
9,737.0	7,538.1
10,741.7	7,549.8
12,199.7	9,078.6
12,211.1	11,255.2
15,406.1	11,262.9

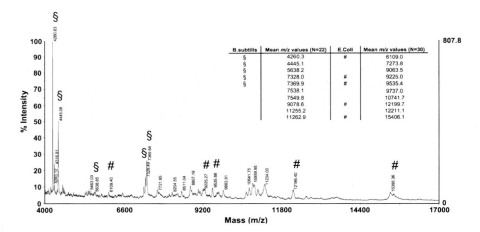

Fig. 9 MALDI/TOF m/z spectrum of a 1:1 mixture of lyophilized *E. coli* and *B. subtilis* cells. #: Most characteristic *E. coli* peaks recovered; §: Most characteristic *B. subtilis* peaks recovered. Adapted with permission from ref. [26], © 2004, American Chemical Society

reported in Table 1 with *E. coli* identification data reported in the literature,[51] and with data obtained from a model-derived protein biomarker search[5] makes it possible to assign at least one of these signals to one *E. coli* protein: the 50 S ribosomal unit L29, SwissProt #P02429 ($m/z=7,273.8$).

Figure 9 reports a representative MALDI/TOF spectrum obtained by mixing an approximately equal amount of lyophilized *E. coli* and *B. subtilis* cells. Both for *E. coli* and *B. subtilis*, only five of the most characteristic ion signals in Table 1 are

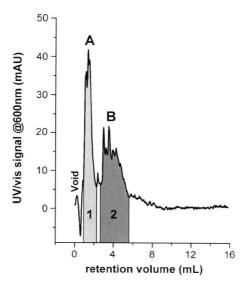

Fig. 10 Fractogram of a 1:1 mixture of lyophilized *E. coli* and *B. subtilis* cells. F_{in} = 4.0 mL/min; F_{rad} = 0.8 mL/min. Mobile phase: 1 mM ammonium cholate, pH = 9.2. Band **A**: *B. subtilis*; band **B**: *E. coli*. Fraction collection times: fraction **1** (*B. subtilis*) from 20 to 40 s; fraction **2** (*E. coli*) from 50 to 100 s. Adapted with permission from ref. [26], © 2004, American Chemical Society

recovered, and the *E. coli* signal likely assigned to the ribosomal protein SwissProt#P02429 is no longer recognized.

Representative results obtained from applying HF5 to MALDI-TOF-MS are shown in Figs. 10 and 11. In Fig. 10 it is reported an example of a fractogram obtained for the same 1:1 mixture of lyophilized *E. coli* and *B. subtilis* mixture the MALDI/TOF spectrum of which is reported in Fig. 9. A complete separation between the two species is achieved. As representatively shown in Fig. 11a and b, two completely different spectra can be obtained from the fractions collected in correspondence to band **A** and **B** (shaded fraction **1** and **2**, respectively), with the spectral features found for each individual bacterial species that is found in each spectrum. None of the most characteristic *E. coli* ion signals in Table 1 was found in spectra obtained from bands of type **A**, and none of the most characteristic *B. subtilis* ion signals was found in spectra from bands of type **B**. It is also evident that fractionation can significantly increase the analytical information obtained from the spectra. The number of characteristic signals of each species found in the spectra from each band always increased with respect to the number of characteristic signals found in the spectra of the unfractionated mixture. This is evident comparing the representative spectra in Fig. 9 and in Fig. 11a and b. It is also worth noting that in the representative spectrum in Fig. 11b, the *E. coli* ion signal likely assigned to the protein #P02429 is now recovered, while it was lost in the spectra of the unfractionated mixture (Fig. 9).

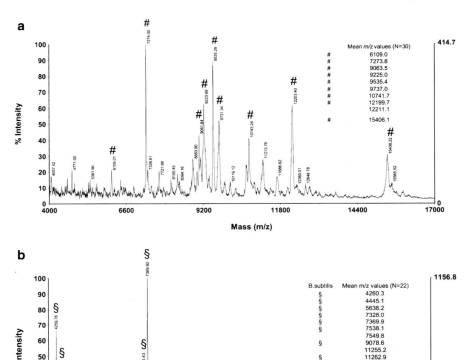

Fig. 11 (a) MALDI/TOF *m/z* spectrum of the collected fraction **1** in Fig. 10; §: Most character-istic *B. subtilis* peaks recovered. (b) MALDI/TOF *m/z* spectrum of the collected fraction **2** in Fig. 10; #: Most characteristic *E. coli* peaks recovered. Adapted with permission from ref. [26], © 2004, American Chemical Society

4 Perspectives

HF5 shows interesting features as a pre-MS, separation method for the analysis of proteomes or whole cells. Its separation performance cannot be however compared to that of electrophoretic techniques or HPLC. Nevertheless, HF5 should not be seen as a competing but rather as a complementary, separation technique within a comprehensive pre-MS analytical platform. The unique advantage of HF5 in fact lies in the "gentle" mechanism able to keep the analytes in their native conditions. The low channel volume and the potential disposable usage are also key features able to maintain low limits of detection, and to avoid run-to-run sample carry-over. These are very critical points when accurate and meaningful MS identifications want to be obtained.

To increase HF5 performance, some HF5 evolutions are on progress. Multiplexed HF5[39] can be applied to large proteome samples to obtain subproteome fractions to be submitted to MS-based analysis. Tandem HF5 is currently under study to allow for in-line re-injection of a selected band slice of the eluate, which is "trapped" at the HF5 channel outlet. Because of in-channel sample focusing after re-injection, tandem HF5 is not expected to significantly increase the final sample dilution and, then, to affect the MS limit of detection.

HF5 still is, however, a prototype technology whose performance shall be improved also by development of some important technical aspects. Rugged channel design, the use of HF membranes specifically made for HF5 usage, channel cartridge engineering, and system operation optimization and automation are necessary to make HF5 a ready-to-market technology for routine-based applications. We wish that such a technological gap could be filled up in a relatively short time.

Automated HF5 procedures shall become even more necessary than high fractionation efficiency if HF5 wants to be applied to fast MALDI-TOF-MS methods for bacteria mixture identification in complex samples. To this end, in fact, rapidity is essential while total HF5 separation of different bacteria species, as obtained in the cases of model-mixtures here described, should not be considered as strictly necessary. Instead of a "chromatography-like", baseline separation between two populations of cells, HF5 shall most frequently involve enrichment of cells with some given characteristics with respect to the entire population of different types of cells. This is because HF5-based cell sorting in fact is the result of a continuous distribution of multi-polydispersity in the different physical indexes of the cells. The high selectivity of HF5 makes cells elute as relatively broad bands, because the multi-polydispersity in the different physical features of the cells is translated into a continuous difference in cell retention. As a consequence, HF5 band slicing before MALDI-TOF-MS should be sufficient to decrease the complexity of cell mixtures and to adequately enrich the HF5 fractions in a specific bacterial species. Thanks to such enrichment, the fractionated bacteria species could subsequently be better identified by MALDI-TOF-MS.

The results here presented show HF5 able to discriminate between cultured bacteria and spores of the same species. The technique shall be likely able to discriminate between living and dead bacteria of same species because living and dead microorganisms are known to originate differences in FFF retention due to differences in the hydrodynamic properties.[52] HF5 can then assist MS methods to ascertain if an obtained profile can be assigned to spores, living or dead bacteria of same species.

Acknowledgments Work supported by the University of Bologna, the Italian Ministry of Education, University and Research, the Italian Ministry of Foreign Affair, and by the Korea Foundation for the International Cooperation of Science & Technology (KICOS). MHM thanks for the grant support from Korea Research Foundation (KRF-2008-313-C00567).

References

1. Anhalt JP, Fenselau C (1975) Identification of bacteria using mass spectrometry. Anal Chem 47:219–225
2. Fenselau C, Demirev PA (2001) Characterization of intact microorganisms by MALDI mass spectrometry. Mass Spectrom Rev 20:157–171
3. Demirev PA, Ho YP, Ryzhov V, Fenselau C (1999) Microorganism identification by mass spectrometry and protein database searches. Anal Chem 71:2732–2738
4. Demirev PA, Lin JS, Pineda FJ, Fenselau C (2001) Bioinformatics and mass spectrometry for microorganism identification: proteome-wide post-translational modifications and database search algorithms for characterization of intact H. pylori. Anal Chem 73:4566–4573
5. Pineda FJ, Antoine MD, Demirev PA, Feldman AB, Jackman J, Longenecker M, Lin JS (2003) Microorganism identification by matrix-assisted laser/desorption ionization mass spectrometry and model-derived ribosomal protein biomarkers. Anal Chem 75:3817–3822
6. Jarman KH, Cebula ST, Saenz AJ, Petersen CE, Valentine NB, Kingsley MT, Wahl KL (2000) An algorithm for automated bacterial identification using matrix-assisted laser desorption/ionization mass spectrometry. Anal Chem 72:1217–1223
7. Anderson L, Anderson NG (1977) High resolution two-dimensional electrophoresis of human plasma proteins. Proc Natl Acad Sci USA 74:5421–5425
8. Klose J, Kobalz U (1995) Two-dimensional electrophoresis of proteins: an updated protocol and implications for a functional analysis of the genome. Electrophoresis 16:1034–1059
9. Righetti PG, Castagna A, Herbert B (2001) Prefractionation techniques in proteome analysis. Anal Chem 73:320A–326A
10. Gygi SP, Corthals GL, Zhang Y, Rochon Y, Aebersold R (2000) Evaluation of two-dimensional gel electrophoresis-based proteome analysis technology. Proc Natl Acad Sci USA 97:9390–9395
11. Smith RD (2000) Probing proteomes – seeing the whole picture? Nat Biotechnol 18:1041–1042
12. Hannig K (1982) New aspects in preparative and analytical continuous free-flow cell electrophoresis. Electrophoresis 3:235–243
13. Krivankova L, Bocek P (1998) Continuous free-flow electrophoresis. Electrophoresis 19:1064–1074
14. Hu S, Zhang L, Newitt R, Aebersold R, Kraly JR, Jones M, Dovichi NJ (2003) Identification of proteins in single-cell capillary electrophoresis fingerprints based on comigration with standard proteins. Anal Chem 75:3502–3505
15. Reschiglian P, Zattoni A, Roda B, Michelini E, Roda A (2005) Field-flow fractionation and biotechnology. Trends Biotechnol 23:475–483
16. Roda B, Zattoni A, Reschiglian P, Moon MH, Mirasoli M, Michelini E, Roda A (2009) Field-flow fractionation in bioanalysis: A review of recent trends. Anal Chim Acta 635:132–143
17. Zattoni A, Casolari S, Rambaldi DC, Reschiglian P (2007) Hollow-fiber flow field-flow fractionation. Curr Anal Chem 3:310–323
18. Lee HL, Reis JFG, Dohner J, Lightfoot EN (1974) Single-phase chromatography: solute retardation by ultrafiltration and electrophoresis. AIChE J 20:776–784
19. Jönsson JA, Carlshaf A (1989) Flow field flow fractionation in hollow cylindrical fibers. Anal Chem 61:11–18
20. Wijnhoven JEGJ, Koorn JP, Poppe H, Kok WTh (1995) Influence of injected mass and ionic strength on retention of water-soluble polymers and proteins in hollow-fibre flow field-flow fractionation. J Chromatogr A 699:119–129
21. Lee WJ, Min BR, Moon MH (1999) Improvement in particle separation by hollow fiber flow field-flow fractionation and the potential use in obtaining particle size distribution. Anal Chem 71:3446–3452
22. Giddings JC (1983) Hyperlayer field-flow fractionation. Sep Sci Technol 18:765–773

23. Moon MH, Lee KH, Min BR (1999) Effect of temperature on particle separation in hollow fiber field-flow fractionation. J Microcolumn Sep 11:676–681
24. Min BR, Kim SJ, Ahn KH, Moon MH (2002) Hyperlayer separation in hollow fiber flow field-flow fractionation: effect of membrane materials on resolution and selectivity. J Chromatogr A 950:175–182
25. Reschiglian P, Roda B, Zattoni A, Min BR, Moon MH (2002) High performance, disposable hollow fiber flow field-flow fractionation for bacteria and cells. First application to deactivated Vibrio Cholerae. J Sep Sci 25:490–498
26. Reschiglian P, Zattoni A, Cinque L, Roda B, Melucci D, Dal Piaz F, Roda A, Moon MH, Min BR (2004) Hollow-fiber flow field-flow-fractionation for whole bacteria analysis by matrix-assisted laser desorption/ionization time-of-flight mass spectrometry. Anal Chem 76:2103–2111
27. Reschiglian P, Zattoni A, Roda B, Cinque L, Parisi D, Roda A, Moon MH, Min BR, Dal Piaz F (2005) On-line hollow-fiber flow field-flow fractionation-electrospray ionization/time-of-flight mass spectrometry for the analysis and characterization of intact proteins. Anal Chem 77:47–56
28. Rambaldi DC, Zattoni A, Casolari S, Reschiglian P, Roessner D, Johann C (2007) Analytical method for size and shape characterization of blood lipoproteins. Clin Chem 53:2026–2029
29. Carlshaf A, Jönsson JA (1991) Effects of ionic strength of eluent on retention behavior and on the peak broadening process in hollow fiber flow field-flow fractionation. J Microcolumn Sep 3:411–416
30. Carlshaf A, Jönsson JA (1993) Properties of hollow fibers used for flow field-flow fractionation. Sep Sci Technol 28:1031–1042
31. Reschiglian P, Zattoni A, Roda B, Cinque L, Melucci D, Min BR, Moon MH (2003) Hollow-fiber flow field-flow fractionation of cells. J Chromatogr A 985:519–529
32. Wijnhoven JEGJ, Koorn JP, Poppe H, Kok WTh (1996) Influence of injected mass and ionic strength on retention of water-soluble polymers and proteins in hollow-fibre flow field-flow fractionation. J Chromatogr A 732:307–315
33. Shin SJ, Chung HJ, Min BR, Park JW, An IS, Lee K (2003) Separation of proteins mixture in hollow fiber flow field-flow fractionation. Bull Korean Chem Soc 24:1339–1344
34. Van Bruijnsvoort M, Kok WTh, Tijssen R (2001) Hollow-fiber flow field-flow fractionation of synthetic polymers in organic solvents. Anal Chem 73:4736–4742
35. Roda B, Cioffi N, Ditaranto N, Zattoni A, Casolari S, Melucci D, Reschiglian P, Sabbatini L, Valentini A, Zambonin PG (2005) Biocompatible channels for field-flow fractionation of biological samples: correlation between surface composition and operating performance. Anal Bioanal Chem 381:639–646
36. Van Bruijnsvoort M, Tijssen R, Kok W (2001) Assessment of the diffusional behavior of polystyrene sulfonates in the dilute regime by hollow-fiber flow field flow fractionation. J Polym Sci Part B Polym Phys 39:1756–1765
37. Zhu R, Frankema W, Huo Y, Kok WTh (2005) Studying protein aggregation by programmed flow field-flow fractionation using ceramic hollow fibers. Anal Chem 77:4581–4586
38. Kang D, Moon MH (2005) Hollow fiber flow field-flow fractionation of proteins using a microbore channel. Anal Chem 77:4207–4212
39. Lee JY, Kim KH, Moon MH (2009) Evaluation of multiplexed hollow fiber flow field-flow fractionation for semi-preparative purposes. J Chromatogr A 1216:6539–6542
40. Chmelík J (2007) Applications of field-flow fractionation in proteomics. Presence and future. Proteomics 7:2719–2728
41. Reschiglian P, Moon MH (2008) Flow field-flow fractionation: a pre-analytical method for proteomics. J Proteomics 71:265–276
42. Roda A, Parisi D, Guardigli M, Zattoni A, Reschiglian P (2006) Combined approach to the analysis of recombinant protein drugs using hollow-fiber flow field-flow fractionation, mass spectrometry, and chemiluminescence detection. Anal Chem 78:1085–1092
43. Zattoni A, Rambaldi DC, Roda B, Parisi D, Roda A, Moon MH, Reschiglian P (2008) Hollow-fiber flow field-flow fractionation of whole blood serum. J Chromatogr A 1183:135–142

44. Kim KH, Kang D, Koo H, Moon MH (2008) Molecular mass sorting of proteome using hollow fiber flow field-flow fractionation for proteomics. J Proteomics 71:123–131
45. Kang D, Moon MH (2006) Development of non-gel based two-dimensional separation of intact proteins by an on-line hyphenation of capillary isoelectric focusing and hollow fiber flow field-flow fractionation. Anal Chem 78:123–131
46. Lee H, Williams SK, Wahl KL, Valentine NB (2003) Analysis of whole bacterial cells by flow field-flow fractionation and matrix-assisted laser desorption/ionization time-of-flight mass spectrometry. Anal Chem 75:2746–2752
47. Caldwell KD, Cheng ZQ, Hradecky P, Giddings JC (1984) Cell Biophys 6:233–251
48. Caldwell KD (2000) In: Schimpf ME, Caldwell KD, Giddings JC (eds) Field-flow fractionation handbook. Wiley-Interscience, New York, Chapter 5
49. Lucas A, Lepage F, Cardot P (2000) In: Schimpf ME, Caldwell KD, Giddings JC (eds) Field-flow fractionation handbook. Wiley-Interscience, New York, Chapter 29
50. Reschiglian P, Zattoni A, Roda B, Casolari S, Moon MH, Lee J, Jung J, Rodmalm K, Cenacchi G (2002) Bacteria sorting by field-flow fractionation. Application to whole-cell Escherichia coli vaccine strains. Anal Chem 74:4895–4904
51. Jones JJ, Stump MJ, Fleming RC, Lay JO Jr, Wilkins CL (2003) Investigation of MALDI-TOF and FT-MS techniques for analysis of escherichia coli whole cells. Anal Chem 75:1340–1347
52. Bernard A, Bories C, Loiseau PM, Cardot PJP (1995) Selective elution and purification of living Trichomonas vaginalis using gravitational field-flow fractionation. J Chromatogr B 664:444–448

Mass Spectrometry and Tandem Mass Spectrometry for Protein Biomarker Discovery and Bacterial Speciation

Alvin Fox and Karen Fox

Abstract After culture, MALDI-MS protein profiling, for species characterization, is widely used. DNA-based identification of bacterial species (with or without prior culture) often involves PCR and/or sequencing. 16S rRNA sequence cataloging is the gold standard but discrimination is often only at the genus level. This chapter discusses protein marker discovery and chemotaxonomy for threat agents using MS and MS/MS. Characterization of small acid soluble proteins (SASPs) of *Bacillus anthracis* and related species are used for illustrative purposes. The ultimate goal of our studies is universal applicability with species-level certainty in these identifications including biodetection without culture.

1   Introduction

Medical microbiology focuses on specific bacterial species (e.g. *Bacillus anthracis*) as the causative agents of infectious diseases (e.g. anthrax). Many molecular biology techniques for bacterial discrimination first target a specific gene for amplification (most commonly polymerase chain amplificationPCR) prior to analysis. Commonly, DNA regions are selected that are known to contain conserved regions (that can be targeted by appropriate PCR primers) flanking variable regions that provide sequence-specific information. 16S rRNA sequence is the gold standard that serves well as a cataloging technique. Although 16S rRNA sequencing discriminates at the genus level, there are limitations in its species-level identification capabilities.

In a recent study of 158 clinical isolates, 64.0% were identified to the genus and species level, 26.6% were identified to the genus level only and there was no correlation with conventional methods for 9.5% of isolates.[1] One explanation for the disparity between genus level and species level identification is that most 16S

A. Fox (✉) and K. Fox
Department of Pathology, Microbiology and Immunology, University of South Carolina,
School of Medicine, Columbia, SC 29223
e-mail: alvin.fox@uscmed.sc.edu

J. Banoub (ed.), *Detection of Biological Agents for the Prevention of Bioterrorism*,
NATO Science for Peace and Security Series A: Chemistry and Biology,
DOI 10.1007/978-90-481-9815-3_3, © Springer Science+Business Media B.V. 2011

rRNA sequences are highly conserved DNA regions which can be identical in some closely-related species. Bacterial discrimination can also become considerably more complicated when sampling includes environmental isolates that may be genetically closely related to disease-causing species. The environmental isolates may not have had their 16S rRNA sequences catalogued since these isolates are generally not encountered in clinical microbiology laboratories. This situation was observed in a study of urban dust where 16S rRNA arrays indicated the presence of 8,935 taxa, none of which were identified to the species level.[2]

Due to the limitations of 16S rRNA sequencing, the discovery of other biomarkers that can be used to differentiate between bacterial species is vital. Whole genome sequencing of multiple representatives of species, followed by bioinformatic comparison of these huge DNA sequence sets can allow biomarker discovery. Nonetheless, it is not always desirable to take such an expensive and time-consuming route to biomarker discovery.[3,4]

Similarly to gene amplification approaches, there are proteomic techniques involving protein amplification prior to MALDI-TOF-MS analysis. In one technique, a specific bacteriophage is amplified by multiplication within an infected bacterial host and the phage proteins are targeted for detection.[5] Another technique consists of incubating exotoxin-producing bacteria with a synthetic peptide substrate which upon enzymatic cleavage generates peptide products. In both instances, the peptide products are readily detected by MALDI-TOF-MS with exquisite sensitivity. In a clinical microbiology laboratory, such methods can be an excellent choice.[6] However, amplification of DNA or proteins targets adds biological complexity to a readily automatable analytical chemical system necessary for biodetection. Furthermore, such focused methods are not optimal for biomarker discovery.

The most widely used mass spectrometry technique for identification of bacterial species involves protein profiling using MALDI-TOF-MS, without prior amplification or protein fractionation.[7-10] As is widely-known, isolated bacterial colonies are sampled from culture plates, dried on a MALDI plate with the ionization matrix and subjected to MALDI-TOF-MS analysis. The whole process is completed in a few minutes and is simple to perform. A plot of abundance versus mass with a characteristic pattern is generated by the analysis and can be used to define a specific bacterial species in many instances. Mass profiling is a well-established approach which is commercially available (e.g. the Bruker Biotyper profiling software and mass spectral library).[11] The power of this approach is its universal applicability along with its simplicity. In spite of this, mass profiling still has some limitations.

In MALDI-TOF-MS analysis, there can be considerable variation between profiles for replicate cultures of the same strain and between different strains, particularly if grown on different media.[12-14] Thus individual masses can't be readily assigned as biomarkers. It is necessary to have sophisticated pattern recognition software to compensate for these variations. Another complication is the possibility that representatives of species are not in the database, making species identification impossible.

Counter-terrorism efforts are concerned with detecting bacterial species in as close to real-time as possible. The organism must be detected without culture from an environmental or clinical sample. Many samples collected for the purpose of biodetection are usually gathered through air sampling. The problem with this type of sampling is that many bacterial species can be present in these samples creating a large amount of background during analysis. The background species along with other components of the mixture will all contribute to a mixture mass spectrum with the potentially important peaks buried in the background. This is the reason why a mass profile cannot be used for the analysis of a complex sample without some type of fractionation. One solution to this problem is analysis of single particles by MALDI MS analysis. Unfortunately, this instrumentation is not available to the wider microbiology community and awaits exhaustive evaluation.

An additional limitation of MALDI-TOF-MS profiling is the preferential ionization of low mass peptides (<10,000 Da). This can prevent higher mass proteins (~50,000–100,000 Da) from being detected in a mass profile. These proteins may be potential biomarkers, making their detection essential in some cases. Prior separation, including the removal of competing low mass peptides and proteins, can allow the ionization of higher mass proteins.

With MALDI-TOF-TOF-MS/MS, the available commercial instruments are often limited to analyzing peptides in the 1,000–3,000 Da mass range. However, many of the proteins in a mass profile will be higher in mass, often greater than 10,000 Da, and fall outside the range of current MALDI-TOF-TOF-MS/MS In order to examine these higher mass proteins, they must be converted into lower mass peptides. This conversion is usually performed by tryptic digestion. Since there is no prior separation in mass profiling, a tryptic digest would consist of a mixture of peptides derived from all the original protein components. Determining which intact protein the different tryptic peptides are derived from can be difficult to deduce, further complicating the identification of biomarkers.

In the following sections, we shall describe some of our work on defining protein markers using MS/MS. We will also discuss efforts to build upon these efforts involving prior protein separation. The eventual aim is universal mass spectrometric biodetection without prior culture.

2   Identification of Markers by MALDI-MS and MS/MS

The *Bacillus cereus* group includes *Bacillus anthracis* and the closely related species, *Bacillus cereus* and *Bacillus thuringiensis* (the ACT group). *Bacillus mycoides*, *Bacillus pseudomycoides* and *Bacillus weihenstephanensis* are generally considered to be more distantly related members of the *B. cereus* group. However *B. mycoides* and *B. pseudomycoides are* still difficult to distinguish from ACT. Only *B. anthracis* and *B. cereus* commonly cause human disease. Thus the other species are less well-studied. The *B. subtilis* group (which includes numerous species including *B. subtilis* and *B. atrophaeusB. subtilis var. niger*) is the next most closely related group of bacilli. In military research, *B. atrophaeus* (sometimes incorrectly referred to as *B. globigii*) is often used a stimulant for *B. anthracis*, but since it is not a member of the *B. cereus* group this is not an optimal choice. Bacilli are soil organisms and are ubiquitous in airborne dust.[2]

Small acid-soluble proteins (SASPs) are a group of proteins present in large amounts in the core region of spores. Mass spectrometric analysis of SASPs generally involves their extraction from spores with a concentrated solution of trifluoroacetic acid prior to MALDI-TOF-MS.[15] The MW of SASPs can be determined using either MALDI-TOF-MS or ESI-MS.[16]

Since the mass of SASPs is in the range of 7,000, trypsin digestion is required prior to MALDI TOF-TOF MS/MS analysis. Demirev et al. first used mass spectrometry to identify SASPs using MALDI MS/MS to sequence the C-terminal fragment generated upon tryptic digestion.[17] They also noted that N-terminal methionine is commonly lost through post-translational modification in bacterial proteins, which accounted for the mass of the intact SASPs differing by 131 Da from that predicted from the genomic sequence.

SASPs have also been analyzed as intact proteins using ESI MS/MS.[16] For example, *B. cereus* strains studied fall into two clusters, one close genetically, as determined by DNA-DNA hybridization and one more distantly related to *B. anthracis*.[18,19] The closely related cluster is characterized by a β SASP with a single amino acid substitution, localized either close to the C terminus (phenylalanine to tyrosine, 16 mass change) or close to the N terminus (serine to alanine, also 16 mass change.[19] The more distantly related cluster displayed both amino acid substitutions (32 mass change).[20] See Figs. 1, 2 and 3, that provide illustrations of the discussion above.

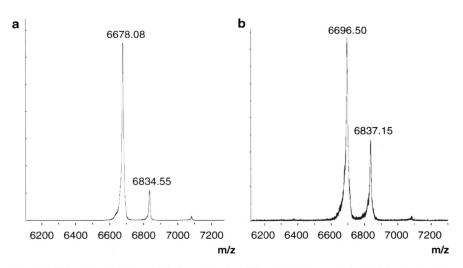

Fig. 1 M.W. of the SASPs of (**a**) *B. anthracis* ANR and (**b**) *B. mycoides* 6A14 measured by MALDI MS. The more prominent peak in each case is the β SASP and there is a single amino acid substitution. The other prominent peak is an α SASP, whose fixed mass serves as a natural internal standard

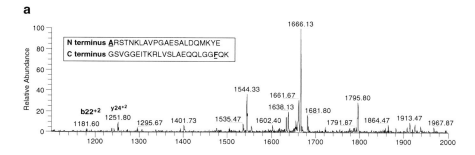

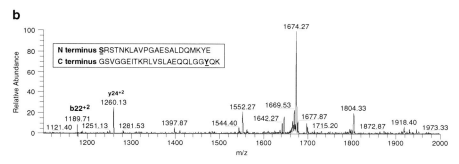

Fig. 2 N and C terminus sequence variation of intact SASPS distinguished using ESI MS/MS (**a**) *B. anthracis* and (**b**) *B. cereus*. The b22 ion (+2 charge, N terminus) and the y24 (+2, C terminus) each differ by an m/z, of 8 (16 mass units); A to S and F to Y shift respectively

3 Correlation of Measured Mass and Predicted Mass from Genomic Sequence

The predicted MWs measured with MALDI-TOF-MS, for representative strains of each of the species of the *B. cereus* group, have been shown to agree with the predicted MWs of the peptide sequence derived from deposited genomic sequences.[21] This suggests that modifications to protein structure, at least for SASPs, do not occur on MALDI-MS analysis (Table 1).[21]

4 Identification of Bacterial Peptides by Sequence

There are two broadly used approaches to computer assisted peptide identification that use MS/MS data: database methods and de novo sequencing. The database search methods are the workhorse for peptide identification. The wide use of commercial systems based on database search algorithms (primarily MASCOT and SEQUEST) indicates the maturity of this approach. The experimental spectrum is

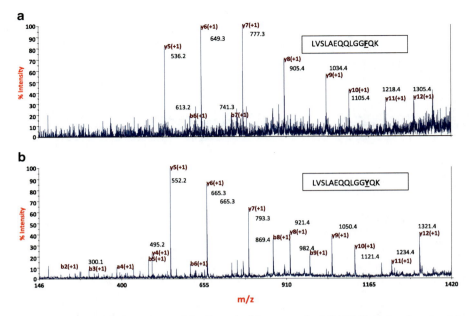

Fig. 3 MALDI MS/MS spectrum of the C terminal fragment of a β SASP. Note that there is a 16 mass unit shift between the corresponding marked masses between (**a**) (*B. anthracis* [parent mass 1518]) and (**b**) (*B. cereus* [parent mass 1534]) reflecting an F to Y shift in the sequence

Table 1 Comparison of common measured MWs (using MALDI-TOF-MS) and calculated MWs (from genomic sequences) of representative SASPs of species of the *B. cereus* group [21]

	Measured	Calculated	Measured	Calculated
Species	SASP α	SASP α	SASP β	SASP β
B. anthracis	6834.55	6834.59	6678.08	6678.48
B. thuringiensis,	6834.72	6834.59	6694.01	6694.48
B. cereus,				
B. mycoides				
B. pseudomycoides	6636.92	6636.41	6536.10	6536.28
B. weihenstephanensis	6874.18	6874.65	6524.37	6527.27

treated as a fingerprint and is compared with theoretical spectra computed for peptides in the database. The proteomic sequence database is largely generated from determination of a whole genome (by high-throughput DNA sequencing) followed by prediction of potentially expressed proteins based on predicted genes. However, database approaches are of limited value when trying to identify peptides from organisms with un-sequenced genomes. A major weakness of all database approaches is that they are unable to identify peptides that are not contained in the database. This is a particularly important limitation in the case of microbial peptides. It is well known that only a small fraction of microbes found in the environment are currently culturable. Indeed, even among culturable organisms many remain uncharacterized due to extreme diversity. Even in the case where a microbial genome has been sequenced, it is quite possible that the strain under investigation exhibits amino acid mutations relative to the peptides in the database.

Since *de novo* sequencing does not depend on a database of known peptides, it offers the possibility of identifying novel peptides. It also offers the possibility of studying a proteome before the genome has been sequenced. The key problems involved in de novo sequencing are those of identifying the subset of peaks in the spectrum that specify an ion ladder and then determining the sequence of amino acids that are most consistent with the ion ladder

One important piece of information that is often missing from current probabilistic and cross-correlation scoring function is the prior distribution of amino acid usage. This distribution describes the percentage of each amino acid as well as the probability of combinations of amino acids in peptide sequences. It captures the mutual information present in adjacent residues in the protein sequences from which the distribution was derived. By leaving this information out, one is effectively using a flat prior that treats all combination of amino acids as equally likely. Amino acid usage can vary widely from organism to organism. Typically, it is similar between closely related taxa but can be quite different when taxa are distantly related.

5   Ongoing Work and Conclusions

As illustrated above, in MS analysis, the experimental mass of peptides correlates with the predicted mass from the DNA sequence. If confirmatory results are obtained from two or more product ion spectra of different peptides, this often is sufficient for identification of a protein. Unfortunately, closely related bacterial species often synthesize proteins (like SASPs) that differ in sequence by a few amino acid substitutions, deletions or insertions. Thus regions of sequence variation within proteins must be the focus for identification of bacterial species.

Developments are needed in analytical microbiology methodology with the necessary simplicity, speed, sensitivity, and selectivity for biodetection. There should be a significant focus on discovery of protein markers for threat agents. Universal applicability with species-level certainty in these identifications is vital.

Proteins can be separated in liquid as a mixture after tryptic digestion or as intact proteins.[22–24] If tryptic digestion is performed prior to LC separation, the MW of a potential biomarker protein is not generally determined and it can sometimes be difficult to home in on relevant peptides that contain species-specific sequences present in the complex mixture using LC-ESI-MS/MS. Alternatively if separation of intact proteins is performed on a gel, (followed by *in situ* digestion and MALDI-MS/MS) analysis is time consuming but it is simpler to focus on individual peptides derived from the parent protein that can serve as peptide sequence markers for bacterial discrimination.

Variations in automated separations prior to MALDI-MS/MS and ESI-MS/MS are thus both viable options for biomarker discovery.[25,26] Biodetection would be a logical follow-up provided sensitivity of the analysis is adequate. In its simplest form, it would be desirable to analyze proteins without tryptic digestion using MS/MS analysis. However, in some applications, automated and rapid tryptic digestion may be feasible as demonstrated with model proteins including mammalian myoglobin and bovine serum albumin.[25] It is recognized that there is the still need for substantial developments in mass spectrometry instrumentation and software for microbial identification. It is our contention that these developments would occur most effectively (as relates to biodetection) if the microbiology and analytical chemical methods developments occurred in a coordinated and focused manner with improvements in one discipline stimulating directed changes in the other.

Acknowledgements Support for this work was provided by the Sloan Foundation (Indoor Air Program).

References

1. Morgan M et al (2009) Comparison of the Biolog OmniLog identification system and 16S ribosomal RNA gene sequencing for accuracy in identification of atypical bacteria of clinical origin. J Microbiol Meth 79:333–343
2. Brodie E et al (2007) Urban aerosols harbor diverse and dynamic bacterial populations. PNAS 104:299–304
3. Read T, Salzberg S, Pop M, Shumway M, Umayam L, Jiang L, Holzapple E, Busch J, Smith K, Schupp J, Solomon D, Keim P, Frazer C (2002) Comparative genome sequencing for discovery of novel polymorphisms in *Bacillus anthracis*. Science 296:2028–2033
4. Ivanova A et al (2003) Genome sequence of *Bacillus cereus* and comparative analysis with *Bacillus anthracis*. Nature 423:87–89
5. Madonna A, Cuyk S, Voorhees K (2003) Detection of *Escherichia coli* using immunomagnetic separation and bacteriophage amplification coupled with matrix-assisted laser desorption time-of-flight mass spectrometry. Rapid Commun Mass Spectrom 17:257–263
6. Boyer A et al (2009) Kinetics of lethal factor and Poly-D-glutamic acid antigenemia during inhalation anthrax in rhesus macaques. Infect Immun 77:3432–3441
7. Cain T, Lubman D, Weber J (1994) Differentiation of bacteria using protein profiles from matrix assisted laser desorption/ionization time of flight mass spectrometry. Rapid Commun Mass Spectrom 8:1026–1030
8. Claydon M, Davey S, Edwards-Jones V, Gordon D (1996) The rapid identification of intact microorganisms using mass spectrometry. Nat Biotechnol 14:1584–1586
9. Holland R, Wilkes J, Rafii F, Sutherland J, Persons C, Voorhees K, Lay J (1996) Rapid identification of intact whole bacteria based on spectral patterns using matrix assisted laser desorption/ionization with time-of-flight mass spectrometry. Rapid Commun Mass Spectrom 10:1227–1232
10. Krishnamurphy T, Ross P, Rajamani U (1996) Detection of pathogenic and non-pathogenic bacteria by matrix assisted laser desorption/ionization time-of-flight mass spectrometry. Rapid Method Commun Mass Spectrom 10:883–888
11. Fox K, Fox A, Elßner T, Feigley C, Salzberg D (2010) MALDI TOF mass spectrometry speciation of staphylococci and their discrimination from micrococci isolated from indoor air of school rooms. J Environ Monit 12:917–923
12. Arnold A, Karty J, Ellington A, Reilly J (1999) Monitoring the growth of a bacteria culture by MALDI-MS of whole cells. Anal Chem 71:1990–1996
13. Walker J, Fox AJ, Edwards-Jones V, Gordon D (2002) Intact cell mass spectrometry (ICMS) used to type methicillin-resistant *Staphylococcus aureus*: media effects and inter-laboratory reproducibility. J Microbiol Meth 48:117–126

14. Valentine N et al (2005) Effect of culture conditions on microorganism identification by matrix-assisted laser desorption ionization mass spectrometry. Appl Environ Microbiol 71:58–64

15. Hathout Y et al (1999) Identification of *Bacillus* spores by matrix-assisted laser desorption ionization-mass spectrometry. Appl Environ Microbiol 65:4313–4319

16. Castanha E, Fox A, Fox K (2006) Rapid discrimination of *Bacillus anthracis* from other members of the *B. cereus* group by mass and sequence of "intact" small acid soluble proteins (SASPs) using mass spectrometry. J Microbiol Meth 67:230–240

17. Demirev P, Ramirez J, Fenselau C (2001) Tandem mass spectrometry of intact proteins for characterization of biomarkers from *Bacillus cereus* T spores. Anal Chem 73:5725–5731

18. La Duc, Satomi M, Agata N, Venkateswaran K (2004) *gyrB* as a phylogenetic discriminator for members of the *Bacillus anthracis-cereus-thuringiensis* group. J Microbiol Meth 56:383–394

19. Castanha E et al (2007) *Bacillus cereus* strains fall into two clusters (one closely and one more distantly related) to *Bacillus anthracis* according to amino acid substitutions in small acid-soluble proteins as determined by tandem mass spectrometry. Mol Cell Probes 21:190–201

20. Callahan C, Castanha E, Fox K, Fox A (2008) The *B. cereus* containing sub-branch most closely related to *B. anthracis*, have single amino acid substitutions in small acid soluble proteins, but remaining sub-branches are more variable. Mol Cell Probes 22:207–211

21. Callahan C, Fox K, Fox A (2009) The small acid soluble proteins of *Bacillus weihenstephanensis and Bacillus mycoides* group 2 are the most distinct among the *Bacillus cereus* group. Mol Cell Probes 23:291–297

22. Lee S et al (2002) Direct mass spectrometric analysis of intact proteins of the yeast large ribosomal subunit using capillary LC-FTICR. PNAS 99:5942–5947

23. Everley R, Mott T, Toney D, Croley T (2009) Characterization of *Clostridium* species utilizing liquid chromatography/mass spectrometry of intact proteins. J Microbiol Meth 77:152–158

24. Shevchenko A et al (1996) Large-scale identification of yeast proteins from two dimensional gels. Proc Natl Acad Sci USA 93:14440–14445

25. Carol J et al (2005) Determination of denatured proteins and biotoxins by on-line size-exclusion chromatography–digestion–liquid chromatography–electrospray mass spectrometry. Anal Biochem 346:150–157

26. Hattan S, Vestal M (2008) Novel three-dimensional MALDI plate for interfacing high-capacity LC Separations with MALDI-TOF. Anal Chem 80:9115–9123

Intact Cell/Spore Mass Spectrometry of *Fusarium* Macro Conidia for Fast Isolate and Species Differentiation

Hongjuan Dong, Martina Marchetti-Deschmann, Wolfgang Winkler, Hans Lohninger, and Guenter Allmaier

Abstract The focus of this paper is the development of an approach called intact cell mass spectrometry (ICMS) or intact spore mass spectrometry (ISMS) based on the technique matrix-assisted laser desorption/ionization time-of-flight mass spectrometry (MALDI-TOF-MS) for the rapid differentiation and identification of *Fusarium* species. Several parameters, which are known to affect the quality of IC mass spectra, have been investigated in detail by varying the MALDI matrix as well as the solvent system, in which the matrix has been dissolved, the solvent system for sample purification and the type of sample/MALDI matrix deposition technique. In the end characteristic as well as highly reproducible IC or IS mass spectra or peptide/protein fingerprints of three *Fusarium* species (*F. cerealis*, *F. graminearum* and *F. poae*) including 16 *Fusarium* isolates derived from different hosts and geographical locations have been obtained. Unscaled hierarchical cluster analysis based on ICMS data of eight selected *Fusarium* isolates of two species *F. graminearum* and *F. poae* revealed significant difference among the peptide/protein pattern of them. The results of the applied cluster analysis proved that, ICMS is a powerful approach for the rapid differentiation of *Fusarium* species. In addition, an *on-target* tryptic digestion was applied to *Fusarium* macro conidia spores to identify proteins using MALDI post source decay (PSD) fragment ion analysis. Two kinds of trypsin, namely bead-immobilized–to favor cleavage of surface-associated proteins–and non-immobilized trypsin were applied and compared. The results showed that the latter is more suitable for generating sequence tags by PSD fragment ion analysis.

Keywords MALDI • Mass spectrometry • ICMS • Species differentiation • Fungi • *Fusarium* • *On-target* digestion

H. Dong, M. Marchetti-Deschmann, W. Winkler, H. Lohninger, and G. Allmaier (✉)
Institute of Chemical Technologies and Analytics, Vienna University of Technology, Getreidemarkt 9/164-IAC, A-1060, Vienna, Austria
e-mail: guenter.allmaier@tuwien.ac.at

J. Banoub (ed.), *Detection of Biological Agents for the Prevention of Bioterrorism*,
NATO Science for Peace and Security Series A: Chemistry and Biology,
DOI 10.1007/978-90-481-9815-3_4, © Springer Science+Business Media B.V. 2011

1   Introduction

To many microbiologists in medicine, homeland security and food sciences, rapid and reliable differentiation as well as identification of microorganisms without extensive manipulation has become of major interest as well as represent an analytical challenge. Direct profiling/imaging of the surface of intact microorganisms such as bacteria or fungi by intact cell mass spectrometry (ICMS) or intact spore mass spectrometry (ISMS) based on matrix-assisted laser desorption/ionization time-of-flight mass spectrometry (MALDI-TOF-MS) has emerged as a valuable research tool which is capable of fulfilling several tasks.[1-4] MALDI-TOF-MS is the technique of choice providing high tolerance against the presence of salts and detergents as well as the possibility of automation, making itself suitable for direct and rapid analysis of microorganisms. The sample preparation, after often time-consuming development, in ICMS/ISMS can be carried out within minutes in a straightforward manner, because the samples are analyzed "as a whole" with minimal direct sample pretreatment after collection and purification from a cell culture suspension.[5] Briefly, intact vegetative cells or spores are mixed with MALDI-MS matrix solution and deposited directly onto the MALDI-MS target to co-crystallize before transferred into the high vacuum ion source of the MALDI mass spectrometer for analysis. The matrix absorbs the laser energy and releases it into the solid preparation of intact cells/matrix molecules. Then a surface-associated peptides as well as small proteins derived from intact cells or spores are desorbed and ionized to form mostly singly charged molecules which can then be mass analyzed by the linear TOF mass analyzer. The mass spectra obtained by ICMS/ISMS are usually considered as mass spectral fingerprints (even without understanding which specific components were desorbed/ionized from the intact microorganisms) and the identification as well as differentiation was to be based on comparison with those from the others or mass spectrometric database.[6-9] Several algorithms.[10,11] based on m/z and intensity values as well as numerical analysis such as cluster analysis[12,13] have also been developed to compare and estimate the similarity between two or more mass spectra. All these techniques allow accurate typing (differentiation and identification, latter if a reference database is available) of microorganisms at genus, species, and even at the strain level.

The relative simplicity in sample preparation, sensitivity, broad capability, tolerance to contaminants and speed of the method indicates that ICMS/ISMS has great potential for routine use in differentiation, identification and classification of microorganisms. Most research groups have focused their research on identification of intact bacterial cells[1,8,14-16] or bacterial spores[17-19] allowing the use of well-established experimental methods and to a certain extent of mass spectral databases. So far the ICMS/ISMS technique has also applied successfully to only few intact fungal cells[20,21] and fungal spores.[2,22-26] Fungi are the fifth important kingdom of eukaryotes[27,28] and compared to bacterial cells, fungi are typically larger and are surrounded by relative rigid cell walls which are generally composed

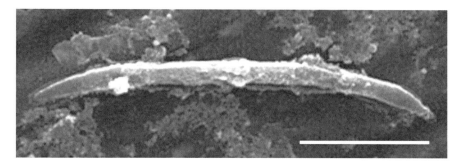

Fig. 1 Electron microscope image of *Fusarium* macro conidia spore (white bar corresponds to 10 μm)

by up to 90% polysaccharides. Proteins, lipids, polyphosphates and inorganic ions are also present in fungal cells[12,22] (Figure 1 showed the electron microscopic image of a macro conidia spore of the fungus *Fusarium*). Members of the ascomycetous genus *Fusarium* is a large genus of filamentous fungi which are distributed on plants and in soil, representing an important group of fungal plant pathogens and may cause various infections in humans, too.[29,30] Due to the latter facts it is of great importance to identify and differentiate them rapidly, for example by ICMS/ISMS in an early stage. In one recent paper,[31] 62 *Fusarium* isolates were identified by MALDI-TOF-MS based on the extracts of hyphae/ spores, however, not with intact spores. In previous studies on ICMS of fungal spores,[2,22–26] various parameters has been reported to influence mass spectra data such as the purification of intact spores, choice of matrix compounds as well as their dissolved solvents and the MALDI sample deposition techniques. Small variations during sample preparation will affect the quality and reproducibility of mass spectrum significantly. Another quite common problem is that, during the culturing of *Fusarium* as well as many other filamentous fungi, they often produce colored pigments of different types and intensities.[32–34] Especially for the *Fusarium* species of our interest, slightly orange to red-brown colored macro conidia spores were produced.[24] The presence of this colored components resulted in quite poor quality of IC/IS mass spectra. Therefore the deep-colored spores need to be pre-treated properly.

In this paper we reported on the differentiation of intact *Fusarium* macro conidia spores by MALDI TOF mass spectrometry. We optimized the experimental parameters for ISMS, showed that differentiation is feasible by cluster analysis and started to build a mass spectral database on *Fusarium* species as well as strains. In addition, identification of proteins from *Fusarium* spores preparation (*F. graminearum* CPK 2985) has been achieved by *on-target* digestion technique in combination with MALDI post source decay (PSD) fragment ion analysis.

2 Experimental

2.1 Sample Generation of Macro Conidia Spores

All *Fusarium* isolates (collection of the Institute of Chemical Engineering (Vienna University of Technology, Vienna, Austria)) were first vitalized on SNA-Plates (Synthetischer Nährstoffarme Agar) and the subsequently macro conidia spore generation was carried out at 28 °C in a shaker incubator at 160 rpm with mungbean soup (20 g mungbean in 1 L of water were heated for 20 min and directly used after filtration) as nutritional medium. The spores were collected by filtration through a sterile glass funnel containing glass wool and centrifugation of the filtered liquid suspension containing spores at 8,000 rpm and 4 °C for 10 min. The obtained supernatant was discarded while the pellet was mixed with an aqueous solution containing 20% glycerol (w/v) and stored at -20 °C if the MALDI-MS analysis was not performed immediately. The concentration of the spore solution was determined by spore counting in a light-optical microscope (Nikon Instruments Europe, Amstelveen, The Netherlands).

2.2 Sample Preparation

The major amount of glycerol and other contaminants in spore suspension has to be removed prior to MALDI-MS analysis. Several solvent systems were used to wash both colorless, light- and deep-colored *Fusarium* spores for removal of the colorizing surface components and were evaluated in terms of mass spectrometric performance (detailed description see reference 24). The spore suspension was washed three times with various solvent systems at $19,500 \times g$ for 10 min using Nanosep™ (Pall, Ann Arbor, MI, USA) centrifugal devices (MWCO 10 kDa). Then the spore pellet was resuspended in pure water with a final concentration of three million spores/μL and in case of *on-target* digest in 25 mM ammonium bicarbonate solution with a conidia spore concentration of 1.5 million/μL, respectively.

12 MALDI-MS matrix compounds in different concentrations and 15 different matrix solvent mixtures were evaluated (detailed description see reference 25). Five different sample deposition techniques comprise of dried droplet, mixed volume, thin layer, sandwich, two-layer volume technique were evaluated based on mass spectrometric reproducibility, peaks numbers and intensities.[24-26]

For *on-target* digestion, 1 μL *Fusarium* conidia spores suspension in 25 mM ammonium bicarbonate solution was placed onto stainless steel MALDI-MS target. After drying the spores solution at room temperature (RT), 0.5 μL of bead-immobilized TPCK trypsin solution[35,36] isolated from bovine pancreas (Pierce Biotechnology, No. 20230, Rockford, IL; Before using 20 μL bead-immobilized TPCK trypsin was washed following the instruction provided by the company and finally 4 times diluted (in terms of volume) with 25 mM ammonium bicarbonate solution) or 0.5 μL of 0.1 μg/μL non-immobilized trypsin isolated from bovine pancreas (Roche Diagnostics, Mannheim, Germany, Cat. No. 11418025001) in 25 mM ammonium bicarbonate solution were added onto each sample spot, respectively. Then conidia spore samples prepared on the MALDI-MS target plate were incubated in a humidity-controlled chamber at RT to prevent spot drying. *On-target* digestion was stopped after 25 min by allowing the samples to dry at RT. Finally, 0.5 µL of 10 mg/mL α-cyano-4-hydroxycinnamic acid (CHCA) in acetonitrile/0.1% aqueous trifluoroacetic acid (70/30, v/v) was added for MALDI-MS analysis and dried at RT.

2.3   MALDI-TOF-MS

Positive ion MALDI mass spectra were acquired on Axmia-CFR^{plus} instrument (Shimadzu Biotech Kratos Analytical, Manchester, UK) equipped with a nitrogen laser (337 nm, 3 ns pulse width) and a curved field reflector. The instrument was operated at an acceleration voltage of 20 kV. All the mass spectra were accumulated (up to 2,500 single unselected laser shots) across the whole matrix/analyte spot automatically in the rastering mode.

For IC/ISMS analysis, the instrument was operated in linear mode with delayed extraction (optimized for m/z 5,000) in the m/z range of 1,000–15,000. The blanking gate was set at m/z 1,000 to remove the ions below this m/z value arising from matrix (their clusters as well as fragments) and other unknown contaminants exhibiting low molecular mass. An external three-point calibration was performed prior to every automatic measurement of each *Fusarium* species/strain with the protein cytochrome c (protonated molecule at m/z 12361.2 and double protonated molecule at m/z 6181.1) and the standard peptide ACTH 7–38 (protonated molecule at m/z 3657.9). The singly charged peptides from tryptic auto digestion products at monoisotopic m/z 659.38, 805.42, 2163.06 and 2273.16 (mass accuracy of ± 0.03 Da) were used for internal calibration of the digested samples. The PSD mode has been calibrated with the synthetic peptide $(P)_{14}K$. For PSD fragment ion analysis 2,500 single unselected laser shots were accumulated. The average m/z values of fragment ions, which were used for manual *de novo* sequencing, were automatically derived from smoothed PSD spectra (Shimadzu Biotech supplied Savitzky-Golay algorithm, smoothing filter width 20) with baseline subtraction (baseline filter width 60). Typically PSD spectra showing an average mass accuracy of ± 0.5 Da in the low m/z range (< m/z 1,500) and in the m/z range > m/z 1,500 of ± 2 Da were acquired.

2.4   Data Handling

The first step of data analysis consisted of spectral preprocessing using the software supplied by the instrument manufacturer (Launchpad 2.7.3) by applying baseline subtraction (filter width 200) and smoothing (Savitzky-Golay algorithm, 20 channels).

The processed mass spectra were exported in mzXML[37] format and re-imported into mMass[38] software (version 2.4) for further analysis. Peaks were manually picked in the m/z range of 2,400–15,000 omitting those peaks known to be caused by the matrix as well as sodium or potassium adducts of peptide/protein analytes. For each sample (i.e. isolate), peak lists containing m/z and intensity (normalized by the most intense peak in the m/z range 2,400–15,000) values of those peaks, which could be detected in at least 50% of all replicate measurements were generated. An alignment of all peak lists was generated considering peaks from different mass spectra to be identical if their m/z difference was less than $1 + m/z / 3,000$ (variable threshold of 1.8 Da at m/z = 2,400 extending to 6 Da at m/z = 15,000). Peaks, which were not detected in one mass spectrum, but were present in others, were added with an intensity of zero. This procedure resulted in a data matrix consisting of a total of 270 features (i.e. peaks) defined for 95 mass spectra, which was used for hierarchical clustering (distance measure: Euclidean distance, linkage type: Ward's method) using Datalab (version 2.4, http://www.lohninger.com/datalab/de_home.html, Epina, Austria).

3 Results and Discussion

3.1 Short Description of the Applied Strategy

In this study, we demonstrated a rapid screening approach that employs MALDI-TOF-MS of intact conidia spores (Fig. 1) for differentiation of *Fusarium* species.

Figure 2 showed the strategy of IC/ISMS of *Fusarium* macro conidia.

Experimentally, after spore generation and a simple washing procedure with organic/water solvent mixture by centrifugation, the intact *Fusarium* spores and matrix solution were pre-mixed in an polypropylene tube (Fig. 3) or mixed on MALDI-MS target directly (based on different sample deposition technique) for sufficient interaction which indicates an intimate contact between *Fusarium* conidia spores, MALDI-MS matrix compound and solvents.

Then the prepared sample/matrix spot was allowed to dry by evaporating of solvents at RT and subsequently was used for MALDI-TOF-MS analysis. The generated mass spectral profiles of macro conidia spores turned out to be sufficient "different" to differentiate various *Fusarium* species.

3.2 Different Types of MALDI-MS Sample Preparations

Sample preparation of *Fusarium* conidia spores for IC/ISMS includes purification and concentrating of spores as well as subsequently transfer of the spores onto the MALDI target and embedding into the MALDI matrix. During these steps several parameters have been evaluated for the optimization of IC/ISMS based on mass

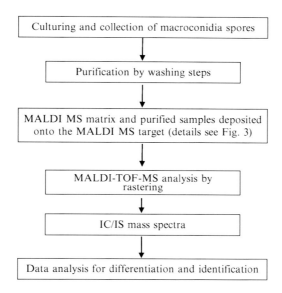

Fig. 2 Strategy of intact cell/spore mass spectrometry (IC/ISMS) of *Fusarium* macroconidia

spectrometric reproducibility, in terms of m/z values, as well as number and intensities of peaks. The washing solvent systems for purification, matrix compounds as well as solvents in which the matrix has been dissolved and sample deposition technique were included in the evaluation process.

A previous study[24] proved that the solvent system for washing *Fusarium* macro conidia spores especially for washing the deep-colored *Fusarium* spores is of vital importance for successful analysis. In fact, due to the rigid cell walls of fungal spores, the purpose of this procedure is not only the purification of spores but also for extraction of proteins/peptides (i.e. bringing them to the surface) in order to improve the signal-to-noise ratio. Several solvents composed of water, acetonitrile, methanol, ethanol, isopropanol, organic acids such as formic acid or trifluoroacetic acid have been investigated. They were either used in pure form or were mixed in a certain ratios (v/v). The result showed that the organic acid play an important role in this washing step. For example, the use of solvent system acetonitrile/water (7/3) for washing the *Fusarium* isolate CPK NO.2765 deep-colored spores generated a poor mass spectrum without any useful peaks. However, when aqueous 0.1% trifluoroacetic acid or aqueous 0.5% formic acid replaced pure water as washing system, the color of the intact spores became lighter and the quality of obtained MALDI mass spectra improved significantly (and for some spore preparations for the first time useful mass spectra were obtained). Many other researchers have also noted the utility of acid pre-treatment of other microorganisms for enhanced mass spectral data.[14,18,21,39] The use of organic acid changes the pH of the spore suspension and might modify the architecture of the spore cell wall resulting in a better release of proteins and peptides. Furthermore, the change of pH value of the solution might be an effective way of color fading (although not totally) of *Fusarium* macro conidia spores.

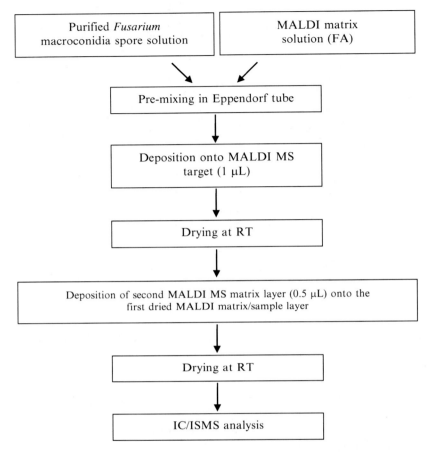

Fig. 3 IC/ISMS sample preparation with two-layer volume technique for *Fusarium* macroconidia

The MALDI mass spectra varied with the MALDI-MS matrix used significantly. For IC/ISMS analysis of *Fusarium* conidia spores, different matrix compounds in different concentrations and solvents, in which matrix has been dissolved, have been investigated in our previous study.[25] The commonly used acidic matrix compounds including CHCA, 2,5-dihydroxybenzoic acid (DHB), ferulic acid (FA), sinapinic acid (SA) and neutral ionic liquid matrix system composed of DHB/butyl amine and CHCA/butyl amine were tested. The solvent system for dissolving the matrix compounds is also a crucial factor. It has not only an impact on crystal structure but also on the degree of the spores (analyte) incorporated into the matrix layer. Various solvents namely acetonitrile, methanol, ethanol, isopropanol in pure form or combinations with pure water or TFA-acidified water were used for dissolving matrices. Through comparison based on the mass spectrometric performance (e.g. reproducibility or signal intensity) of the preparation resulted in the selection of the matrix FA dissolved in the traditional acetonitrile/0.1% aqueous trifluoroacetic acid (7/3, *v/v*) solvent system at a concentration of 10 mg/mL as most suitable for ISMS analysis of *Fusarium* conidia spores. In a study of ICMS of the other fungal genus *Aspergillus*, it was reported that the addition of organic acids in the MALDI-MS matrix solution provided sufficient dissolving ability to release bioactive compounds (e.g. small proteins) from the cell walls of intact spores.[23] Furthermore the organic acid in matrix/spore crystals might also provide a source of protons thus facilitating the protein/peptide ionization resulting in high quality MALDI mass spectra.

The optimal matrix/analyte deposition technique is one further key requirement for a successful MALDI-MS analysis. The morphology of the crystal layer of matrix and analyte varied with the change of deposition technique. Hence five types of sample deposition techniques in total were evaluated for IC/ISMS of *Fusarium* spores, namely thin layer, sandwich, dried droplet, mixed volume, two-layer volume technique.[24–26] At the beginning the first three deposition technique including thin layer, sandwich, and dried droplet were compared by Kemptner *et al.*[25] As a result, no significant peptide/protein profiles could be achieved with thin layer technique. The sandwich technique generated homogeneous crystal layers and in most cases a well-defined peptide/protein profile. However, the sample preparation reproducibility was not satisfactory. Direct mixing of *Fusarium* conidia spores and MALDI-MS matrix solution on the MALDI-MS steel target, called dried droplet technique, yielded highly reproducible as well as abundant peptide/protein profiles compared to the first two techniques. The reproducibility can be further improved by pre-mixing matrix solution and *Fusarium* conidia spore suspension in a tube prior to application onto the MALDI target, referred to as mixed volumes method.[26] But in later investigations, both dried droplet and mixed volume techniques were found not to work properly for the deep-colored *Fusarium* conidia spores, which is a serious drawback. So dried droplet, mixed volume and two-layer volume technique were compared in parallel for the development of a preparation technique applicable to deep-colored *Fusarium* conidia spores.[24] Comparing with dried droplet technique, the plain mixed volume technique resulted in a homogeneous matrix/spore crystal layer and generated an improvement of the MALDI mass spectrometric reproducibility. But the two-layer volume technique was found to be the most suitable technique with the highest number and abundances of peaks. The essential character of the two-layer volume technique is the application of second MALDI-MS matrix layer, which is connected to a dissolving and re-crystallization step, and resulted in a better incorporation of the spores into the matrix layer and maybe kind of purification. The flow chart of steps of the two-layer volume technique for IC/ISMS of *Fusarium* macro conidia spores is outlined in Fig. 3. Because two-layer volume technique is associated with the exposure of the spores to organic as well as acidic solvents in the MALDI-MS matrix solution, the volume ratio and time of matrix/spore solutions interaction were also investigated.[24,26] Finally for both, light-colored and deep-colored *Fusarium* conidia spores, a matrix/spore solution ratio of 1/1 (v/v) and the duration of mixing time of at least 3 min were recommended to use.

3.3   Verification of Peptides/Proteins Desorbed/Ionized from Conidia Spore Surface

In this work, MALDI PSD fragment ion analysis of protonated tryptic peptides generated via *on-target* tryptic digestion from the surface of *F. graminearum* macro conidia spores was tested for its applicability. Generally after spore purification, i.e. washing with acetonitrile/aqueous 0.5 % formic acid, the intact *Fusarium* spores were suspended in 25 mM ammonium bicarbonate and applied onto the MALDI target for 25 min tryptic *on-target* digestion (details please see experimental part 2.2). The generated tryptic peptides were analyzed directly by MALDI RTOF MS without any further sample cleanup generating a complex peptide mass fingerprint (PMF). Then the MALDI PSD fragment ion analysis was applied to the selected tryptic peptides with high abundance. In addition, the effectiveness of *on-target* digestion using both bead-immobilized and non-immobilized trypsin as enzyme has been compared and the results proved that the latter is more efficient for generating a good PMF. This strategy is rapid, easy to operate and the obtained positive ion PSD spectra were complex but still interpretable allowing partial peptide sequencing followed by database search. Figure 4 presents a typically a PSD spectrum of the abundant precursor peptide ion at m/z 1888.90. With manual *de novo* peptide sequencing from such PSD spectra and combined with database homology search, protein identification based on the shown PSD spectrum and another PSD spectrum from the precursor peptide ion at m/z 673.43 (data not shown) could be accomplished.

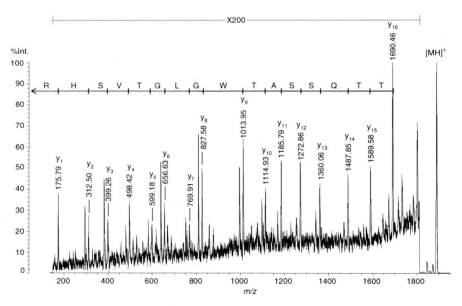

Fig. 4   MALDI PSD spectrum obtained from the precursor ion at m/z 1888.9 after on-target tryptic digestion of *Fusarium* macroconidia spores

Two other hypothetical proteins FG04915.1 and FG07774.1 that contribute precursor peptides at m/z 924.56 and m/z 2258.11 separately are also identified (data were not shown). However, due to the limitation of *Fusarium* database, the proteins identified in this study are all hypothetical proteins. But they can be tentatively explained based on their superfamily annotation.

3.4 Differentiation of IC/IS Mass Spectra of Different *Fusarium* Species and Isolates

Two previous studies[25,26] performed in our laboratory presented that the different reference *Fusarium* species can be differentiated based on their evident difference in their protein/peptide peak patterns. Here in this study three *Fusarium* species including 16 unique isolates which were derived from various hosts and geographical locations were included (Table 1) in such a differentiation study.

Figure 5a–c exhibit the IC/IS mass spectra of the species *F. cerealis*, *F. graminearum* and *F. poae* as well as from different isolates. The IC/IS mass spectra of *Fusarium* conidia spores were measured in the mass range from m/z 1,000 to 15,000. But most of the protein/peptide signals observed in the mass spectra occurred between the m/z values 2,000–10,000.

We found significant differences in the peak pattern of the three different *Fusarium* species. Furthermore the protein/peptide pattern of B isolates of the same species show common peaks in terms of m/z values but some peaks could be found with quite different relative abundance to each other (e.g. see Fig. 5a (a–d)). To evaluate quantitatively whether IC/ISMS could distinguish isolates of different species, we performed a hierarchical cluster analysis of selected 8

Table 1 Investigated *Fusarium* species and strains grown on different hosts in different locations of Austria

Species	Strains CPK No.	Host	Geographic origin	Label
F. cerealis	2739	maize	Wieselsdorf	a
F. cerealis	2740	maize	Wieselsdorf	b
F. cerealis	2741	maize	Mogersdorf	c
F. cerealis	2743	barley	Lambach	d
F. graminearum	2761	maize	Wieselsdorf	e
F. graminearum	2763	maize	Wieselsdorf	f
F. graminearum	2764	maize	Wieselsdorf	g
F. graminearum	2765	barley	Probstdorf	h
F. graminearum	2766	wheat	unknown	i
F. graminearum	1122	unknown	unknown	j
F. poae	2772	Maize	Tulln	k
F. poae	2774	Maize	Tulln	l
F. poae	2775	Barley	Probstdorf	m
F. poae	2781	Barley	Probstdorf	n
F. poae	2793	Barley	Lambach	o
F. poae	2794	Barley	Lambach	p

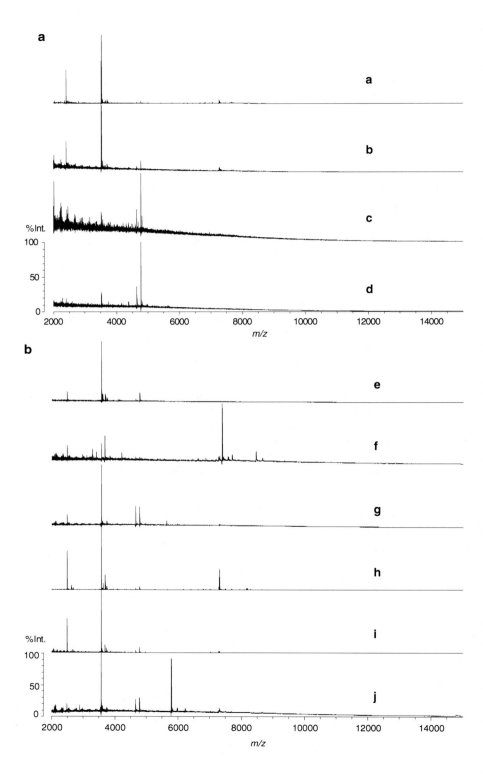

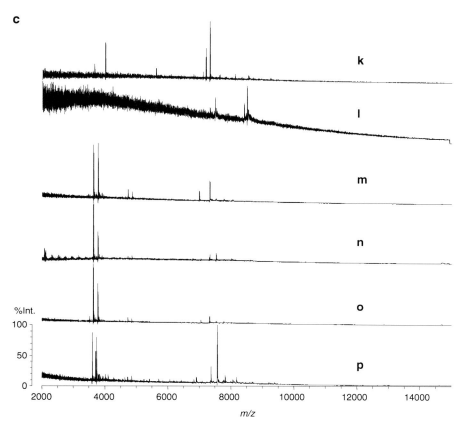

Fig. 5 (continued) (**a**) IC/IS mass spectra of four (a–d) strains of *F. cerealis*. (**b**) IC/IS mass spectra of six (e–j) strains of *F. graminearum*. (**c**) IC/IS mass spectra of six (k–p) strains of *F. poae*

isolates of two *Fusarium* species from various hosts and geographical locations (*F. graminearum* isolates CPK No. 2763, 2764, 2765, 1122 and *F. poae* isolates CPK NO. 2772, 2774, 2775, 2781) as well as analyzed them several times. To exclude peaks potentially derived from the MALDI-MS matrix as well as other contaminants, only peaks with m/z values between 2,400 and 15,000 in the IC/IS-mass spectra were used for the cluster analysis. The dendrogram is constructed using the degree of association and unscaled hierarchical clustering techniques that help visualize similarities between different protein/peptide fingerprints (Fig. 6).

Conceptually, the two fingerprints with the highest degree of association are joined together first, followed by those with the next highest degree of association, and so on. Figure 6 shows the derived dendrogram indicating clearly the differences. In the dendrogram, the two species *F. graminearum* and *F. poae* grouped separately from each other were strikingly dissimilar indicating that

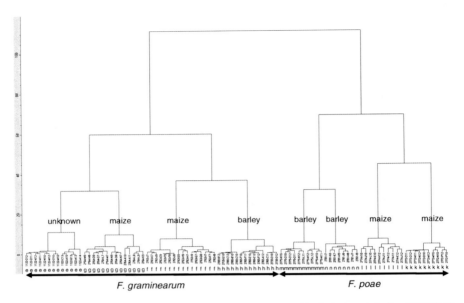

Fig. 6 Dendrogram representing the result of unscaled hierarchical cluster analysis of IC/ISMS data of *Fusarium* species/strains from different hosts (maize, barley and unknown) and locations

differentiation of isolates at species level is possible. Replicate analyses of single isolates showed a very high degree of association and therefore grouped tightly together. Although the isolates within a species are closely related in terms of their IC/IS mass spectra, some differences between each of them other were also observed. Until now it is difficult or impossible to differentiate the isolates derived from different hosts (e.g. corn, barley and unknowns) and geographical locations within eastern regions of Austria. The number of well-defined isolates which must be analyzed by IC/ISMS has to be increased significantly to present a final statement to the last topic.

4   Conclusions

We have described successfully the use of MALDI-TOF-MS approach for differentiation of macro conidia spores of *Fusarium* species. Sample preparation of intact *Fusarium* spores (stored for example at −80 ° C) from light- to deep-colored samples for IC/ISMS analysis do not need to employ any prior tedious separation or extraction procedure and can be finished in a few minutes. MALDI mass spectrometric measurement can be made quickly, with a relative high tolerance to contaminants, and can be automated. The described method is specific, reproducible and accurately identifies differences in large sets of closely related isolates of *Fusarium* species. It represents a fast approach for *Fusarium* macro conidia species differentiation and has the potential for other fungal microorganisms differentiation and meets the requirements for high-throughput screenings. Thus in the future it is easy to expect that IC/ISMS will continue to be widespread used for such demanding applications and the development as well as growth of IC/IS mass spectra databases will further foster the described method. In addition, *on-target* tryptic digestion of *Fusarium* spores is readily accomplished and has the capability to identify the proteins found in the IC/IS mass spectra of fungal spores by PSD fragment ion analysis TOF/RTOF or QqRTOF experiments in case of available databases.

Acknowledgements The investigation was supported financially in part by the Austrian Federal Ministry of Agriculture, Forestry, Environment and Water Management (grant number 100053). Furthermore the authors thank Christian P. Kubicek (Institute of Chemical Engineering, Vienna University of Technology, Vienna, Austria) for the supply of macro conidia spores during all the investigations.

References

1. Holland RD, Wilkes JG, Rafii F, Sutherland JB, Persons CC, Voorhees KJ, Lay JO (1996) Rapid identification of intact whole bacteria based on spectral patterns using matrix-assisted laser desorption/ionization with time-of-flight mass spectrometry. Rapid Commun Mass Spectrom 10(10):1227–1232
2. Chen HY, Chen YC (2005) Characterization of intact *Penicillium* spores by matrix-assisted laser desorption/ionization mass spectrometry. Rapid Commun Mass Spectrom 19(23):3564–3568
3. Dieckmann R, Graeber I, Kaesler I, Szewzyk U, von Döhren H (2005) Rapid screening and dereplication of bacterial isolates from marine sponges of the Sula Ridge by intact-cell-MALDI-TOF mass spectrometry (ICM-MS). Appl Microbiol Biotechnol 67(4):539–548
4. Fenselau C, Demirev PA (2001) Characterization of intact microorganisms by MALDI mass spectrometry. Mass Spectrom Rev 20(4):157–171
5. Lasch P, Nattermann H, Erhard M, Stämmler M, Grunow R, Bannert N, Appel B, Naumann D (2008) MALDI-TOF mass spectrometry compatible inactivation method for highly pathogenic microbial cells and spores. Anal Chem 80(6):2026–2034
6. Arnold RJ, Reilly JP (1998) Fingerprint matching of *E.coli* strains with matrix-assisted laser desorption/ionization time-of-flight mass spectrometry of whole cells using a modified correction approach. Rapid Commun Mass Spectrom 12(10):630–636
7. Nilsson CL (1999) Fingerprinting of *Helicobacter pylori* strains by matrix-assisted laser desorption/ionization mass spectrometric analysis. Rapid Commun Mass Spectrom 13(11):1067–1071
8. Jackson KA, Edwards-Jones V, Sutton CW, Fox AJ (2005) Optimisation of intact cell MALDI method for fingerprinting of methicillin-resistant *Staphylococcus aureus*. J Microbiol Methods 62(3):273–284
9. Vargha M, Takáts Z, Konopka A, Nakatsu CH (2006) Optimization of MALDI-TOF MS for strain level differentiation of *Arthrobacter* isolates. J Microbiol Methods 66(3):399–409
10. Jarman KH, Daly DS, Petersen CE, Saezn AJ, Valentine NB, Wahl KL (1999) Extracting and visualizing matrix-assisted laser desorption/ionization time-of-flight mass spectral fingerprints. Rapid Commun Mass Spectrom 13(15):1586–1594
11. Jarman KH, Cebula ST, Saenz AJ, Petersen CE, Valentine NB, Kingsley MT, Wahl KL (2000) An algorithm for automated bacterial identification using matrix-assisted laser desorption/ionization mass spectrometry. Anal Chem 72(6):1217–1223

62   H. Dong et al.

12. Vanlaere E, Sergeant K, Dawyndt P, Kallow W, Erhard M, Sutton H, Dare D, Devreese B, Samyn B, Vandamme P (2008) Matrix-assisted laser desorption ionisation-time-of of-flight mass spectrometry of intact cells allows rapid identification of *Brukholderia cepacia* complex. J Microbiol Methods 75(2):279–286

13. Hettick JM, Green BJ, Buskirk AD, Kashon ML, Slaven JE, Janotka E, Blachere FM, Schmechel D, Beezhold DH (2008) Discrimination of *Aspergillus* isolates at the species and strain level by matrix-assisted laser desorption/ionization time-of-flight mass spectrometry fingerprinting. Anal Biochem 380(2):276–281

14. Wang Z, Russon L, Li L, Roser DC, Long SR (1998) Investigation of spectral reproducibility in direct analysis of bacteria proteins by matrix-assisted laser desorption/ionization time-of-flight mass spectrometry. Rapid Commun Mass Spectrom 12(8):456–464

15. Evason DJ, Claydon MA, Gordon DB (2000) Effects of ion mode and matrix additives in the identification of bacteria by intact cell mass spectrometry. Rapid Commun Mass Spectrom 14(8):669–672

16. Ryzhov V, Fenselau C (2001) Characterization of the protein subset desorbed by MALDI from whole bacterial cells. Anal Chem 73(4):746–750

17. Hathout Y, Demirev PA, Ho YP, Bundy JL, Ryzhov V, Sapp L, Stutler J, Jackman J, Fenselau C (1999) Identification of *Bacillus* spores by matrix-assisted laser desorption ionization – mass specttometry. Appl Environ Microbiol 65(10):4313–4319

18. Ryzhov V, Hathout Y, Fenselau C (2000) Rapid characterization of spores of *Bacillus cereus* group bacteria by matrix-assisted laser desorption/ionization time-of-flight mass spectrometry. Appl Environ Microbiol 66(9):3828–3834

19. Ullom JN, Frank M, Gard EE, Horn JM, Labov SE, Langry K, Magnotta F, Stanion KA, Hack CA, Benner WH (2001) Discrimination between bacterial spore types using time-of-flight mass spectrometry and matrix free infrared laser desorption and ionization. Anal Chem 73(10):2331–2337

20. Valentine NB, Wahl JH, Kingsely MT, Wahl KL (2002) Direct surface analysis of fungal species by matrix-assisted laser desorption/ionization mass spectrometry. Rapid Commun Mass Spectrom 16(14):1352–1357

21. Amiri-Eliasi B, Fenselau C (2001) Characterization of protein biomarkers desorbed by MALDI from whole fungal cells. Anal Chem 73(21):5228–5231

22. Welham KJ, Domin MA, Johnson K, Jones L, Ashton DS (2000) Characterization of fungal spores by laser desorption/ionization time-of-flight mass spectrometry. Rapid Commun Mass Spectrom 14(5):307–310

23. Li TY, Liu BH, Chen YC (2000) Characterization of *Aspergillus* spores by matrix-assisted laser desorption/ionization time-of-flight mass spectrometry. Rapid Commun Mass Spectrom 14(24):2393–2400

24. Dong H, Kemptner J, Marchetti-Deschmann M, Kubicek CP, Allmaier G (2009) Development of a MALDI two-layer volume sample preparation technique for analysis of colored conidia spores of *Fusarium* by MALDI linear TOF mass spectrometry. Anal Bioanal Chem 359(5):1373–1383

25. Kemptner J, Marchetti-Deschmann M, Mach R, Druzhinina IS, Kubicek CP, Allmaier G (2009) Evaluation of matrix-assisted laser desorption/ionization (MALDI) preparation techniques for surface characterization of intact *Fusarium* spores by MALDI linear time-of-flight mass spectrometry. Rapid Commun Mass Spectrom 23(6):877–884

26. Kemptner J, Marchetti-Deschmann M, Kubicek CP, Allmaier G (2009) Mixed volume sample preparation method for intact cell mass spectrometry of *Fusarium* spores. J Mass Spectrom 44(11):1622–1624

27. Burnett J (2003) Populations and species. Oxford University Press, New York

28. Kavanagh K (2005) Fungi – biology and applications. Wiley, Chichester, UK

29. Phalip V, Delalande F, Carapito C, Goubet F, Hatsh D, Leize-Wagner E, Dupree P, Dorsselaer AV, Jeltsch JM (2005) Diversity of the exoproteome of *Fusarium graminearum* grown on plant cell wall. Curr Genet 48(6):366–379

30. Varga M, Bartók T, Mesterházy Á (2006) Determination of ergosterol in *Fusarium*-infected wheat by liquid chromatography-atmospheric pressure photoionization mass spectrometry. J Chromatogr A 1103(2):278–283

31. Marinach-Patrice C, Lethuillier A, Marly A, Brossas JY, Gené J, Symoens F, Datry A, Guarro J, Mazier D, Hennequin C (2009) Use of mass spectrometry to identify clinical *Fusarium* isolates. Clin Microbiol Infect 15(7):634–642

32. Malz S, Grell MN, Thrane C, Maier FJ, Rosager P, Felk A, Albertsen KS, Salomon S, Bohn L, Schäfer W, Giese H (2005) Identification of a gene cluster responsible for the biosynthesis of aurofusarin in the *Fusarium graminearum* species complex. Fungal Genet Biol 42(5):420–433

33. Bhardwaj S, Shukla A, Mukherjee S, Sharma S, Guptasarma P, Chakraborti AK, Chakrabarti A (2007) Putative structure and characteristics of a red water-soluble pigment secreted by *Penicillium marneffei*. Med Mycol 45(5):419–427

34. Medentsev AG, Arinbasarova AY, Akimenko VK (2005) Biosynthesis of naphthoquinone pigments by fungi of the genus *Fusarium*. Appl Biochem Microbiol 41(5):503–507

35. Warscheid B, Jackson K, Sutton C, Fenselau C (2003) MALDI analysis of *Bacilli* in spore mixtures by applying a quadrupole ion trap time-of-flight tandem mass spectrometry. Anal Chem 75(20):5608–5617

36. Warscheid B, Fenselau C (2003) Characterization of *Bacillus* spore species and their mixtures using postsource decay with a curved-field reflectron. Anal Chem 75(20):5618–5627

37. Pedrioli PG, Eng JK, Hubley R, Vogelzang M, Deutsch EW, Raught B, Pratt B, Nilsson E, Angeletti RH, Apweiler R, Cheung K, Costello CE, Hermjakob H, Huang S, Julian RK, Kapp E, McComb ME, Oliver SG, Omenn G, Paton NW, Simpson R, Smith R, Taylor CF, Zhu W, Aebersold R (2004) A common open representation of mass spectrometry data and its application to proteomics research. Nat Biotechnol 22(11):1459–1466

38. Strohalm M, Hassman M, Kosata B, Kodicek M (2008) mMass data miner: an open source alternative for mass spectrometric data analysis. Rapid Commun Mass Spectrom 22(6):905–908

39. Madonna AJ, Basile F, Ferrer I, Meetani MA, Rees JC, Voorhees KJ (2000) On-probe sample pretreatment for the detection of proteins above 15 kDa from whole cell bacteria by matrix-assisted laser desorption/ionization time-of-flight mass spectrometry. Rapid Commun Mass Spectrom 14(23):2220–2229

Bacteriophage Amplification-Coupled Detection and Identification of Bacterial Pathogens

Christopher R. Cox and Kent J. Voorhees

Abstract Current methods of species-specific bacterial detection and identification are complex, time-consuming, and often require expensive specialized equipment and highly trained personnel. Numerous biochemical and genotypic identification methods have been applied to bacterial characterization, but all rely on tedious microbiological culturing practices and/or costly sequencing protocols which render them impractical for deployment as rapid, cost-effective point-of-care or field detection and identification methods. With a view towards addressing these shortcomings, we have exploited the evolutionarily conserved interactions between a bacteriophage (phage) and its bacterial host to develop species-specific detection methods. Phage amplification-coupled matrix assisted laser desorption time-of-flight mass spectrometry (MALDI-TOF-MS) was utilized to rapidly detect phage propagation resulting from species-specific in vitro bacterial infection. This novel signal amplification method allowed for bacterial detection and identification in as little as 2 h, and when combined with disulfide bond reduction methods developed in our laboratory to enhance MALDI-TOF-MS resolution, was observed to lower the limit of detection by several orders of magnitude over conventional spectroscopy and phage typing methods. Phage amplification has been combined with lateral flow immunochromatography (LFI) to develop rapid, easy-to-operate, portable, species-specific point-of-care (POC) detection devices. Prototype LFI detectors have been developed and characterized for *Yersinia pestis* and *Bacillus anthracis*, the etiologic agents of plague and anthrax, respectively. Comparable sensitivity and rapidity was observed when phage amplification was adapted to a species-specific handheld LFI detector, thus allowing for rapid, simple, POC bacterial detection and identification while eliminating the need for bacterial culturing or DNA isolation and amplification techniques.

C.R. Cox and K.J. Voorhees (✉)
Department of Chemistry and Geochemistry, Colorado School of Mines, Golden, CO 80401
USA
e-mail: kvoorhee@mines.edu

J. Banoub (ed.), *Detection of Biological Agents for the Prevention of Bioterrorism*,
NATO Science for Peace and Security Series A: Chemistry and Biology,
DOI 10.1007/978-90-481-9815-3_5, © Springer Science+Business Media B.V. 2011

Keywords Bacterial detection • Bacterial identification • Bacteriophage amplification • Lateral flow immunochromatography • Matrix assisted laser desorption time-of-flight mass spectroscopy

1 Introduction

The Advanced Biodetection Technology Laboratory at the Colorado School of Mines has been involved in biodetection research for over 23 years. Several major accomplishments in bacterial biodetection have been realized in that time. Some of these contributions include:

- Adaptation of tandem mass spectrometry to identification of biomarkers in bacterial pyrolysates.[1]
- Development of pyrolysis chemistry for the U.S. Army Chemical Biological Mass Spectrometer (CBMS).[2-4].
- Development of a lipid in situ derivatization process for the CBMS version II.[5]
- Development of matrix assisted laser desorption ionization – time of flight mass spectrometry (MALDI-TOF-MS) for analysis of whole cell bacteria.[6,7]
- Development of immunological techniques for separation of mixed species bacterial cultures prior to MALDI-TOF-MS analysis.[8]
- Development and analysis of bacterial detection and identification methods incorporating phage amplification-coupled MALDI-TOF-MS.[9-11]
- Development of phage amplification methods for rapid determination of bacterial antibiotic resistance.[12]
- Coupling of phage amplification with real-time PCR for rapid detection and identification of viable *Bacillus anthracis*[13].
- Development of phage amplification-coupled lateral flow immunochromatography for rapid, POC bacterial detection and identification.[14]
- Development of new MALDI-TOF-MS sample pretreatment methods for enhanced resolution of viral capsid protein structures.[15]
- Development of novel phage absorption-based bacterial detection and identification methods.[16]
- Application of phage amplification-coupled surface enhanced Raman spectroscopy for bacterial detection and identification.[16]

Most common methods of species-specific bacterial detection and identification including those developed in our lab and others are complex, time-consuming, and often require expensive specialized equipment and highly trained personnel. Numerous biochemical and genotypic identification methods have been applied by others to bacterial characterization, but all rely on tedious microbiological culturing practices and/or costly sequencing protocols which render them impractical for deployment as rapid, cost-effective POC or field detection and identification methods. With a view towards addressing these shortcomings, we are presently exploiting the evolutionarily conserved interactions between a bacteriophage (phage) and its bacterial host to develop species-specific detection methods. Phage amplification, when coupled to contemporary detection platforms is a novel approach for rapid detection and identification of bacterial agents. In this process, a bacterium is infected with a species-specific bacteriophage and allowed to progress through the infection, resulting in the release of many progeny phage. With each progeny phage produced, there is an amplification of both protein and nucleic acid content, which can be used as indirect indicators of the presence of a specific bacterial host.

Phage amplification-coupled MALDI-TOF-MS was shown to rapidly detect phage propagation resulting from species-specific in vitro bacterial infection. This novel signal amplification method allowed for bacterial detection and identification in as little as 2 h, and when combined with disulfide bond reduction methods developed in our laboratory to enhance MALDI-TOF-MS resolution, was observed to lower the limit of detection by several orders of magnitude over conventional spectroscopy and phage typing methods. Phage amplification was also combined with lateral flow immunochromatography (LFI) to develop rapid, easy-to-operate, portable, species-specific point-of-care detection devices. Prototype LFI detectors have been developed and characterized for *Yersinia pestis* and *Bacillus anthracis*, the etiologic agents of plague and anthrax, respectively. Comparable sensitivity and rapidity to that of MALDI-TOF-MS were observed when phage amplification was adapted to a species-specific handheld LFI detector, thus allowing for rapid, simple, point-of-care bacterial detection and identification while eliminating the need for bacterial culturing or DNA isolation and amplification techniques.

2   Contemporary Bacterial Detection Methods

Numerous biochemical and genotypic identification methods have been applied to bacterial detection, identification, and characterization with varied levels of success. 16S-23S rRNA gene sequencing-based methods have become the gold standard of genotypic bacterial identification.[17–19] Other genotypic identification methods such as multilocus enzyme electrophoresis (MLEE),[20–22] multilocus sequence typing (MLST),[23–25] pulsed-field gel electrophoresis (PFGE),[26,27] amplified fragment length polymorphism (AFLP),[28,29] repetitive sequence-based PCR (REP-PCR),[28,30] and Raman Spectroscopy-based systems[31,32] have also been used, but all rely on tedious microbiological culturing practices and/or costly and time consuming sequencing protocols utilizing highly specialized equipment which render them impractical for deployment as rapid, cost-effective POC or field detection and identification methods.

While these methods have been used effectively as tools for laboratory detection and identification of bacterial pathogens, there is an interest among military, clinical, and academic institutions for much simpler methods of rapid bacterial detection and identification, which are much more cost-effective, do not rely on complex equipment, and allow for positive pathogen identification without the need for highly trained personnel. The use of phage amplification techniques combined with commonly used lateral flow diagnostic technologies addresses all of the shortcomings of current detection and identification methodologies.

3 Bacteriophage Biology

Since the independent discovery of bacteriophages by Frederick Twort in 1915[33] and Felix d'Herelle in 1917,[34] the elucidation of the phage infection cycle has contributed significantly to the general bodies of microbiology, molecular biology and genetics. Chief among the knowledge gained from phage biology in regards to bacterial detection and identification, is the concept of phage-based bacterial species typing. Traditional phage-typing assays consist of bacterial host infection with an appropriate species-specific phage, followed by plating onto a lawn of the bacterial host on nutrient agar.[35,36] If the bacterial host is present, phage infection and replication are initiated as illustrated in Fig. 1.

Phage first adsorb to the bacterial cell surface via a number of possible interactions, including attachment to Gram-negative outer membrane porins,[37] transporter proteins,[38] or by attachment to bacterial cell wall components such as lipopolysaccharide[39] and Gram-positive peptidoglycan.[40,41] Phage genetic material is then transferred into the host by a number of possible mechanisms depending on the phage[42,43] where it is incorporated by recombination into the genome resulting in transcription and translation of phage genes by host cellular machinery.[44] The resulting outcome is the production of many new progeny phage. As intracellular phage replication and assembly progress, a lysis event or burst is triggered releasing progeny phage into the surrounding milieu, followed by continued infection of remaining uninfected bacteria in the vicinity.[44] Numerous infectious cycles lead to the formation of visible lysis zones

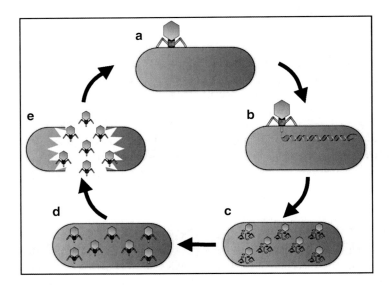

Fig. 1 (**a**) Species-specific phage attachment to bacterial host is immediately followed by insertion of phage genetic material into host (**b**), which reprograms bacterial replication machinery to produce numerous progeny phage components (**c**). Progeny phage are then assembled intracellularly (**d**) followed by eventual lysis of the host bacterium (**e**) releasing new phages for subsequent infection

Table 1 Glossary of phage terminology

Terminology	Definition
Adsorption	The extracellular attachment of a phage to its host bacterium
Burst size	The number of progeny phage released from a lysis event
Burst time	The average amount of time required for release of progeny phage following bacterial infection
Multiplicity of infection (MOI)	The ratio of infectious phages to targeted host bacteria
Lytic phage	A phage that infects a host bacterium, produces progeny phage, and lyses that host to release progeny phage
Lysis	The event of host cell destruction following phage infection and production of progeny phage
Lysis from without	The phenomenon of bacterial lysis resulting not from infection, but from the adsorption of multiple phages to a single bacterium
Progeny phage	A phage that has been produced as the result of a bacterial infection

(plaques) on a bacterial lawn. Any resulting plaque formation is indicative of the presence of a specific bacterial host and can be used in combination with serial dilution techniques to quantify phage as well.[45] Using this traditional plaque assay method, many human bacterial pathogens have been routinely typed, including *Bacillus anthracis*,[46,47] *Escherichia coli*,[48] *Staphylococcus aureus*,[49] *Pseudomonas aeruginosa*,[50] *Camphylobacter jejuni*,[51] *Listeria monocytogenes*,[52] and *Salmonella typhimurium*,[53] By combining a defined multiplicity of infection (MOI), which is the ratio of infecting phage to target bacteria, a phage burst of known size and time can be reproducibly triggered. This can then be coupled with modern detection instrumentation to exploit the evolutionarily conserved phage infection process and produce an amplified detectable signal that is directly attributable to the presence of the targeted bacteria[10] (Table 1).

4   MALDI-TOF-MS

MALDI-TOF-MS is an effective tool for the analysis and identification of a large range of peptides, proteins, and intact bacteria and viruses.[6,54,55] Protein profiling, or mass fingerprinting using MALDI-TOF-MS has emerged as a rapid method of bacterial detection and identification.[6,9] The basic concept of MALDI-TOF-MS centers on the rapid volatilization of target peptides or proteins by laser ablation of samples embedded in a UV-absorbing matrix, combined with time-of-flight mass spectrum analysis. Typical MALDI-TOF-MS applications involve mixing a protein-containing sample (biological fluids such as serum, urine, tissue extracts, or whole bacterial or viral preparations) with an organic matrix solution that catalyzes the co-crystallization of target proteins with matrix on a chargeable target grid. Samples are subjected to a laser pulse under vacuum resulting in the liberation of protein ions from the matrix. Liberated ions are then accelerated by an electrical field towards a detector. Smaller proteins are accelerated to higher velocities than heavier proteins and thus reach the detector sooner. The time of flight for a given target protein is then proportional to its relative mass to charge ratio (m/z). The charge (z) of an ionized protein, if given a value of 1, thus makes the m/z value equal to the mass of the protein. The output of each detection event can then be compiled with a series of readings to generate a spectral profile for each protein present in a sample. When plotted graphically each protein present within the detection limits of the device is described by its mass to charge ratio along the horizontal axis, and by a peak-intensity value on the vertical axis to arrive at a spectral protein profile.

5 Phage Amplification-Coupled MALDI-TOF-MS Bacterial Detection

The coupling of MALDI-TOF-MS protein profiling with bacteriophage amplification (Fig. 2), by virtue of the production of many copies of phage from each bacterial infection, further extends the capability of MALDI-TOF-MS for bacterial

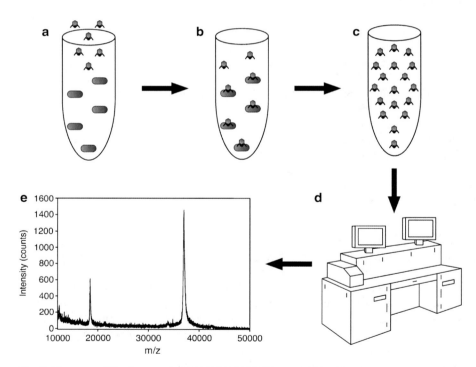

Fig. 2 Phage amplification-coupled MALDI-TOF-MS bacterial identification. (**a**) Application of species-specific phage to a sample suspected to contain target bacteria leads to phage infection (**b**), followed by propagation and amplification of phage (**c**). MALDI-TOF-MS detection of phage-specific capsid proteins (**d**) results in the production of phage-specific MCP profile (37.8 kDa peak) (**e**), indicating the presence of target bacterial species

detection by effectively lowering the amount of bacteria necessary to elicit a detectable signal. This concept was used to develop tractable MALDI-TOF-MS detection methods in our laboratory for the detection of *E. coli*[9,10,12] and *Salmonella enterica* (subsp. *enterica* serovar *typhimurium*).[11] Madonna et al. determined that a typical MALDI-TOF-MS limit of detection (LOD) for *E. coli* of 1.0×10^5 CFU/mL could be decreased by two orders of magnitude by the application of MS2 phage amplification techniques[9] while Rees and Voorhees extended this work to demonstrate the simultaneous detection of *E. coli* and *S. enterica* with MS2 and MPSS1 phage amplification.[11]

Much of our research has focused on the development of rapid, highly species-specific detection and identification methods for bacterial pathogens. With a view towards the evolution of the basic phage plaque assay into a more rapid, species-specific bacterial detection method, we have characterized a number of phage amplification-based MALDI-TOF-MS, LFI, and real-time PCR detection techniques for bacterial pathogens including enterohemorrhagic *E. coli*, *S. typhimurium*, *B. anthracis* and *Y. pestis*.[9,11,13,56] Representative MALDI-TOF-MS spectra resulting from the infection of an *E. coli* culture with MS2 phage are shown in Fig. 3. A pure, high CFU/mL culture of *E. coli* yields a MALDI- TOF mass spectrum showing multiple bacterial proteins (Fig. 3a). In contrast, purified, high titer MS2 phage yields a mass spectrum consisting of one main peak representative

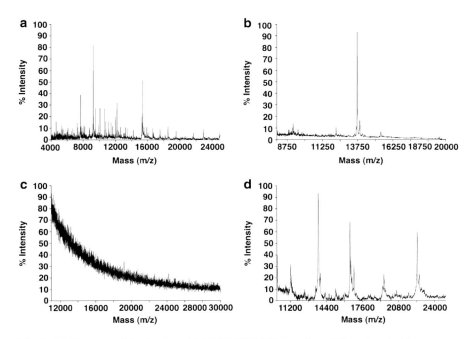

Fig. 3 (**a**) Phage amplification-based MALDI-TOF-MS detection of *B. anthracis*. Mass spectra are shown for (**a**) *B. anthracis*, (**b**) γ phage, (**c**) *B. anthracis*-γ phage mixture before phage amplification (bacterial and phage concentrations below the MALDI-TOF-MS LOD), and (**d**) the same mixture after 2 h of phage amplification

of a previously described 13.7 kDa capsid assembly protein[57] (Fig. 3b). Immediately following infection of *E. coli* with MS2 phage at concentrations below the limit of MALDI-TOF-MS detection, mass spectra show no discernible bacterial or viral peaks (Fig. 3c). However, after a 2-h incubation at room temperature, the same infection reveals the presence of the MS2 capsid protein among the peaks generated by lysed bacterial cells, effectively demonstrating the use of phage amplification as an indirect indicator of the presence of *E. coli* (Fig. 3d).

Similar research conducted in our laboratory demonstrated that the species-specific anthrax typing phage Gamma (γ)[46,58] could be utilized for amplification-coupled MALDI-TOF-MS for rapid, highly accurate *Bacillus anthracis* detection (Fig. 4).[14,59] A pure, high CFU/mL culture of *B. anthracis* yields a MALDI-TOF mass spectrum showing multiple bacterial proteins (Fig. 4a). Purified, high titer γ phage yields a mass spectrum consisting of two primary peaks representative of previously described 22 and 32.0 kDa minor and major capsid structural proteins, respectively (Fig. 4b).[60,61] A 16 kDa doubly charged ion of the major capsid protein is also observed. Initial infection of *B. anthracis* with γ phage at bacterial and phage concentrations below the limit of MALDI-TOF-MS detection yields mass spectra that lack bacterial or viral peaks (Fig. 4c). After 2 h of incubation at room temperature,

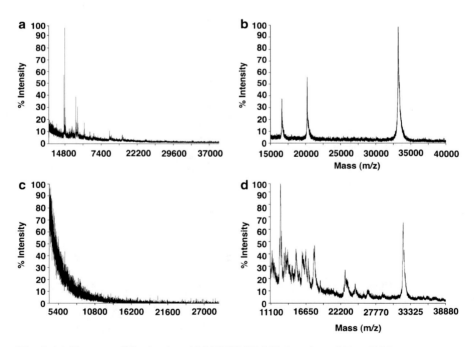

Fig. 4 (**a**) Phage amplification-based MALDI-TOF-MS detection of *E. coli*. Mass spectra are shown for (**a**) *E. coli*, (**b**) MS2 phage, (**c**) *E. coli*-MS2 phage mixture before phage amplification (bacterial and phage concentrations below the MALDI-TOF-MS LOD), and (**d**) the same mixture after 3 h of phage amplification

the same infection reveals the presence of the γ phage capsid proteins among the peaks generated by lysed bacterial cells, demonstrating the use of phage amplification as an indirect indicator of the presence of *B. anthracis* (Fig. 4d).

6   Enhanced Phage Amplification-Coupled MALDI-TOF-MS

Phage amplification-coupled MALDI-TOF-MS relies on the liberation of phage structural proteins, primarily major capsid protein monomers (MCPs), to produce a mass spectrum. While intact phages are preferentially used for MALDI-TOF-MS detection, the inherently large mass of most phage capsid assemblies requires some level of capsid disruption to derive a signal within the operational size capabilities of most mass spectrophotometers. To that end, previous reports have demonstrated successful use of various acid pretreatment methods to achieve sufficient viral disassembly for mass analysis.[9,11,57] It should be noted, however, that observations on the use of common acidic MALDI-TOF-MS matrices combined with laser ablation suggest these methods are unreliable for reproducible phage capsid disassembly.[14] As a result of these findings and based on knowledge gleaned from available phage genomic sequencing data,[62–64] it was recently hypothesized that of the limited number of possible molecular interactions controlling tertiary structure in phage capsid complexes, disulfide bond formation was likely contributing to the formation of such large structures.[15] It follows then, that disruption of these disulfide bonds should liberate MCP monomers and effectively enhance MALDI-TOF-MS phage detection. To test this hypothesis, several disulfide bond reducing agents were applied in the pretreatment of six different phage types and evaluated by MALDI-TOF-MS.[15]

Comparative MALDI-TOF-MS spectra produced by pretreatment of *Y. pestis*-specific phage φA1122 illustrating the requirement of disulfide bond reduction for resolution of the φA1122 MCP are shown in Fig. 5. Figure 5a shows a representative spectrum obtained by conventional MALDI-TOF-MS sample preparation. Figure 5b shows a representative spectrum obtained by acetic acid pretreatment, and Fig. 5c illustrates results obtained by 2-mercaptoethanol (βME) pretreatment. Based on previously published genome sequencing data,[62] the 15.8 kDa peak in Fig. 5a was identified as a φA1122 *gene 13*-encoded scaffolding protein involved in viral capsid assembly. During virion assembly the scaffolding protein serves as a template around which MCP monomers are assembled.[62] The expected molecular weight for the φA1122 major capsid protein is between 35 and 40 kDa[14] which agrees with its previously described molecular weight of 36.588 kDa for φA1122.[62] The function of the 11 kDa peak in Fig. 5a and b is unknown, but is likely encoded by φA1122 gene 5.5. Based on homology to gene 5.5 of phage T7, this open reading frame has been putatively identified as an abundantly expressed nucleoid protein H-NS-binding protein.[62,65] Spectra obtained after βME pretreatment (Fig. 5c) reveal the appearance of a strong peak in the 36.3 kDa range indicative of the φA1122 MCP. A doubly charged ion of this protein is seen at 18.1 kDa.

Fig. 5 MALDI-TOF-MS
spectra of *Y. pestis* phage
φA1122 with (**a**) conventional sample preparation,
(**b**) refluxed with 90% (v/v)
acetic acid in water, and
(**c**) pretreated with βME.
Only capsid assembly pro-
teins (15.7 kDa) are visual-
ized by conventional and acid
pretreatments (**a** and **b**),
while the 36.3 kDa MCP is
resolved with disulfide bond
reduction using βME (**c**).

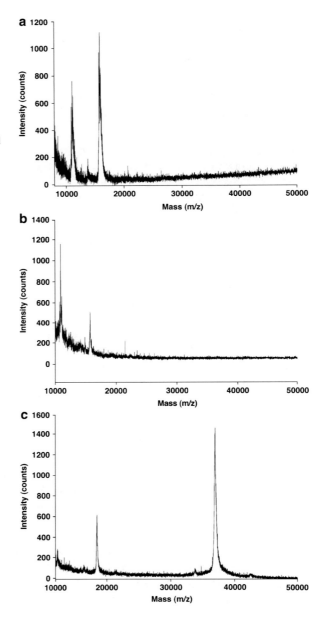

The elaboration of a 36 kDa peak by disulfide bond reduction, which is not
observed by conventional MALDI-TOF-MS sample preparation, suggests that
high-copy multimeric protein complexes such as the virion capsid assembly present in many phage types must be robustly disassembled prior to analysis in order

to resolve their monomeric constituents. When properly resolved, this data indicates, due to the large number of MCP monomers present, a significant enhancement in detectable MALDI-TOF-MS signal.

7   Lateral Flow Immunochromatography

MALDI-TOF-MS is considered by many to be the gold standard for protein identification. However, as a matter of practicality, an ideal biodetection system should be considerably less complex, far less expensive, and amenable to mobile applications. It is for this reason that we have initiated a movement towards easy-to-operate, portable detection platforms with similar, if not superior, detection capabilities when compared to current mass spectroscopic methods of bacterial detection and identification. LFI detection has been adapted in many formats for detection of a large number of analytes, but the basic concept remains constant (Fig. 6). A liquid sample is applied to a test strip and an analyte of interest moves across a polymeric membrane by capillary action passing various zones containing reporters or other molecules specifically designed to interact with the analyte.

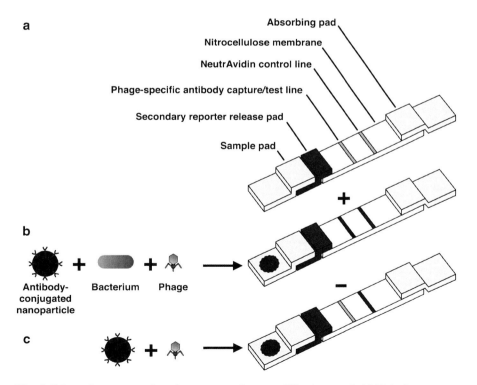

Fig. 6 Schematic representation of a prototype phage amplification-coupled LFI device

If the given analyte is present, it is arrested in combination with a myriad of possible colored, fluorescent, or luminescent reporter conjugates in a detection zone often involving analyte-specific antibodies. As a product of the concentration of the analyte-reporter complex along the detection zone, a visible or otherwise detectable line is formed indicating the presence of the target analyte. In the case of phage amplification-coupled bacterial detection, a patient sample (urine, serum, skin swab, etc.), hospital surface swab, or other sample suspected of containing a target bacterial species is mixed with a previously prepared species-specific phage and incubated to allow for phage infection and amplification. Following incubation, a primary reporter consisting of colored polystyrene nanoparticles conjugated with phage-specific monoclonal antibodies is added to the infection reaction. An aliquot of this is applied to a test strip where the liquid sample wicks from the sample pad into a secondary control reporter conjugate release pad (in this case containing a second color of polystyrene nanoparticles conjugated with biotin). The secondary reporter and any phage-primary reporter complexes are carried onto a nitrocellulose detection membrane striped with a detection line composed of immobilized, phage-specific monoclonal antibodies. In the case of a positive test, phage-primary reporter complexes concentrate along this line resulting in the formation of a visible colored line. At the same time, the secondary control reporter is carried across a second detection zone consisting of a stripe of NeutrAvidin. The biotin-conjugated secondary reporter concentrates along this line, resulting in the formation of a secondary colored line, indicating that the sample was completely transported across the detection zone. A positive test is thus indicated by the formation of two separate colored lines – one at the detection line, and one at the control line (Fig. 6b). It follows that a negative test is indicated by the formation of a single control line only (Fig. 6c).

In addition to bacterial and viral detection platforms, LFI devices have been reported for a wide variety of biomedical, veterinary, agricultural, and environmental applications.[66] First described in 1980 as a pregnancy test, the technique was originally termed the "sol particle immunoassay".[67] Since then, LFI devices have been utilized with various levels of success for detection of human bacterial pathogens, including *B. anthracis*,[68] *Brucella abortus*,[69] *E. coli*,[70] *Helicobacter pylori*,[71] *S. aureus*,[72] *Streptococcus pneumoniae*,[73] and *Treponema pallidum*.[74] LFI has also been successfully applied to detection of a number of human,[75–78] plant,[79] and animal viral pathogens.[80–82]

Figure 7 depicts a prototype phage amplification-coupled *B. anthracis* detection device. Figure 7a shows a positive result following the addition of a *B. anthracis*-specific phage to a liquid bacterial sample. A bacterial culture was infected with an optimized phage concentration in order to maximize burst size and prevent lysis from without (the non-specific lysis of host bacteria resulting from numerous phage adsorption events to a single cell without the production of progeny phage). After a short incubation, an aliquot of the phage infection reaction was applied to the sample-loading port of a handheld cassette (S), and allowed to wick across the test window. A positive result is indicated by the formation of two test lines – one at the test line (T), and another at the control line (C). In order to verify that this positive test was the result of phage amplification, accompanying control tests were run at

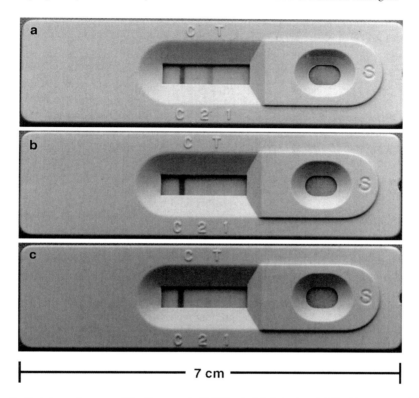

7 cm

Fig. 7 Prototype phage amplification-coupled LFI bacterial detection. (**a**) Positive test resulting from phage amplification in presence of target bacterial species as indicated by development of two lines. (**b**) Phage-free water control showing a single control line, which indicates a negative result. (**c**) Phage-free bacterial culture control showing a single negative control line. Control line (C), test line (T), sample applicator pad (S)

the same time. Figure 7b shows the result of prototype tests run without phage or a bacterial target, verifying the absence of a false positive as indicated by the formation of only a single control line. The test run in Fig. 7c was conducted with the same bacterial sample used in Fig. 7a, but without the addition of phage. A single test line in the absence of phage was observed only at the control line.

As these preliminary results indicate, a handheld prototype LFI device such as that shown in Fig. 7 allows for the rapid, user-friendly, species-specific detection of any number of phage-bacterial host pairs. In doing so, this platform bypasses the need for laborious microbial culturing and/or costly DNA amplification methods, or complex detection instrumentation.

In conclusion, experience in biodetection research at CSM has defined a set of problems associated with most biodetectors. Research in our laboratory has focused on overcoming these problems. Careful application of the evolutionarily conserved relationship between phages and their bacterial hosts, and the resulting amplification associated with phage infection has allowed for significant advances in bacterial detection without many of the common pitfalls of other contemporary detection methods. Phage amplification-coupled MALDI-TOF-MS has allowed for instrumental advances in bacterial detection via rapid, species-specific phage protein profiling. This novel signal amplification method allowed for bacterial detection and identification in as little as 2 h, and when combined with disulfide bond reduction methods developed in our laboratory to enhance MALDI-TOF-MS resolution, was observed to lower the limit of detection by several orders of magnitude over conventional spectroscopy and phage typing methods. With the goal of bypassing the need for complex instrumentation and highly trained personnel, methods for the use of phage amplification-coupled LFI have been explored and found to provide for a combination of rapid, sensitive bacterial detection in an inexpensive, portable configuration, which is ideal for POC detection and identification of bacterial pathogens.

References

1. Voorhees KJ, Durfee SL, Holtzclaw JR, Enke C, Bauer M (1988) Pyrolysis-tandem mass spectrometry of bacteria. J Anal Appl Pyrol 14:7
2. DeLuca S, Sarver EW, Harrington PD, Voorhees KJ (1990) Direct analysis of bacterial fatty acids by Curie-point pyrolysis tandem mass spectrometry. Anal Chem 62:1465
3. DeLuca SJ, Sarver EW, Voorhees KJ (1992) Direct analysis of bacterial glycerides by Curie-point Py-MS. J Anal Appl Pyrol 23:1
4. Beverly MB, Basile F, Voorhees KJ, Hadfield TL (1996) A rapid approach for the detection of dipicolinic acid in bacterial spores using pyrolysis/mass spectrometry. Rapid Commun Mass Spectrom 10:455
5. Basile F, Beverly MB, Abbas-Hawks C, Mowry CD, Voorhees KJ, Hadfield TL (1998) Direct mass spectrometric analysis of in situ thermally hydrolyzed and methylated lipids from whole bacterial cells. Anal Chem 70:1555
6. Holland RD, Wilkes JG, Rafii F, Sutherland JB, Persons CC, Voorhees KJ, Lay JO Jr (1996) Rapid identification of intact whole bacteria based on spectral patterns using matrix- assisted laser desorption/ionization with time-of-flight mass spectrometry. Rapid Commun Mass Spectrom 10:1227
7. Holland RD et al (1999) Identification of bacterial proteins observed in MALDI TOF mass spectra from whole cells. Anal Chem 71:3226
8. Madonna AJ, Basile F, Furlong E, Voorhees KJ (2001) Detection of bacteria from biological mixtures using immunomagnetic separation combined with matrix-assisted laser desorption/ionization time-of-flight mass spectrometry. Rapid Commun Mass Spectrom 15:1068
9. Madonna AJ, Van Cuyk S, Voorhees KJ (2003) Detection of *Escherichia coli* using immunomagnetic separation and bacteriophage amplification coupled with matrix-assisted laser desorption/ionization time-of-flight mass spectrometry. Rapid Commun Mass Spectrom 17:257
10. Madonna AJ, Voorhees KJ, Rees JC (2007) Method for detection of low concentrations of a target bacterium that uses phages to infect target bacterial cells, United States Patent and Trademark Office. Colorado School of Mines, Golden, CO
11. Rees JC, Voorhees KJ (2005) Simultaneous detection of two bacterial pathogens using bacteriophage amplification coupled with matrix-assisted laser desorption/ionization time-of-flight mass spectrometry. Rapid Commun Mass Spectrom 19:2757
12. Majors LK, Doan LG, Rees JC, Voorhees KJ (2004) A rapid method to determine minimum inhibitory concentration of antibiotics in *Staphylococcus aureus* by bacteriophage amplification MALDI-TOF-MS. In: 52nd American Society of Mass Spectrometry Meeting, Nashville, TN, May 2004

13. Reiman RW, Atchley DH, Voorhees KJ (Mar 2007) Indirect detection of *Bacillus anthracis* using real-time PCR to detect amplified gamma phage DNA. J Microbiol Methods 68:651
14. Rees JC (2005) Detection of bacteria using MALDI-TOF-MS and immunodiagnostics with emphasis on bacteriophage amplification detection. PhD thesis, Colorado School of Mines, Golden, CO
15. McAlpin CR, Cox CR, Matyi SA, Voorhees KJ (2010) Enhanced matrix-assisted laser desorption/ionization time-of- flight mass spectrometric analysis of bacteriophage major capsid proteins with beta-mercaptoethanol pretreatment. Rapid Commun Mass Spectrom 24:11
16. Kim Y, Boyanapalli R, Mondesire R, Voorhees KJ (2010) Indirect detection of bacteria using MS2-bacteriophage amplificaiton- coupled with a lateral flow immunochromatography detection device. In: 239th American Chemical Society National Meeting and Exposition San Francisco, CA
17. Bottger EC (1989) Rapid determination of bacterial ribosomal RNA sequences by direct sequencing of enzymatically amplified DNA. FEMS Microbiol Lett 53:171
18. Clarridge JE 3rd (2004) Impact of 16S rRNA gene sequence analysis for identification of bacteria on clinical microbiology and infectious diseases. Clin Microbiol Rev 17:840
19. Fredericks DN, Relman DA (1996) Sequence-based identification of microbial pathogens: a reconsideration of Koch's postulates. Clin Microbiol Rev 9:18
20. Selander RK, Caugant DA, Ochman H, Musser JM, Gilmour MN, Whittman TS (1986) Methods of multilocus enzyme electrophoresis for bacterial population genetics and systematics. Appl Environ Microbiol 51:873
21. Duan G, Liu Y, Qi G (1991) Use of multilocus enzyme electrophoresis for bacterial population genetics, classification and molecular epidemiology. Zhonghua Liu Xing Bing Xue Za Zhi 12:177
22. Seltmann G, Beer W, Claus H, Seifert H (1995) Comparative classification of *Acinetobacter baumannii* strains using seven different typing methods. Zentralbl Bakteriol 282:372
23. Maiden MC, Bygraves JA, Fil E, Morelli G, Russell JE, Urwin R, Zhang Q, Zhou J, Zurth K, Caugant DA, Feavers IM, Achtman M, Spratt BG (1998) Multilocus sequence typing: a portable approach to the identification of clones within populations of pathogenic microorganisms. Proc Natl Acad Sci USA 95:3140
24. Maiden MC (2006) Multilocus sequence typing of bacteria. Annu Rev Microbiol 60:561
25. Naser SM, Thompson FL, Hoste B, Gevers D, Dawyndt P, Vancanneyt M, Swigs J (2005) Application of multilocus sequence analysis (MLSA) for rapid identification of *Enterococcus* species based on rpoA and pheS genes. Microbiology 151:2141
26. Durmaz R, Otlu B, Hosoglu F, Ozturk R, Ersoy Y, Aktas E, Gursoy NC, Calkiskan A (2009) The optimization of a rapid pulsed- field gel electrophoresis protocol for the typing of *Acinetobacter baumannii*, *Escherichia coli* and *Klebsiella* spp. Jpn J Infect Dis 62:372
27. Nasonova ES (2008) Pulsed field gel electrophoresis: theory, instruments and applications. Tsitologiia 50:927
28. Jayarao BM, Dore JJ Jr, Oliver SP (1992) Restriction fragment length polymorphism analysis of 16S ribosomal DNA of *Streptococcus* and *Enterococcus* species of bovine origin, J Clin. Microbiology 30:2235
29. Koeleman JG, Stoof J, Biesmans DJ, Savelkoul PH, Vandenbroucke-Grauls CM (1998) Comparison of amplified ribosomal DNA restriction analysis, random amplified polymorphic DNA analysis, and amplified fragment length polymorphism fingerprinting for identification of *Acinetobacter* genomic species and typing of *Acinetobacter baumannii*. J Clin Microbiol 36:2522
30. Grisold AJ, Zarfel G, Strenger V, Feieri G, Leitner E, Masoud L, Hoenigl M, Raggam RB, Dosch V, Marth E (2010) Use of automated repetitive-sequence-based PCR for rapid laboratory confirmation of nosocomial outbreaks. J Infect 60:44
31. Maquelin K, Dijkshoorn L, van der Reijden TJ, Puppels GJ (2006) Rapid epidemiological analysis of *Acinetobacter* strains by Raman spectroscopy. J Microbiol Methods 64:126
32. Patel IS, Premasiri WR, Moir DT, Ziegler LD (2008) Barcoding bacterial cells: a SERS based methodology for pathogen identification. J Raman Spectrosc 39:1660

33. Twort FW (1915) An investigation on the nature of ultra- microscopic viruses. Lancet 2:1241
34. d'Herelle F (1917) Sur un microbe invisible an- tagonistic des bacilles dysenterique. C R Acad Sci Paris 165:373
35. Cherry WB, Davis BR, Edwards PR, Hogan RB (1954) A simple procedure for the identification of the genus *Salmonella* by means of a specific bacteriophage. J Lab Clin Med 44:51
36. Stewart GS, Jassim SA, Denyer SP, Newby P, Linley K, Dhir VK (1998) The specific and sensitive detection of bacterial pathogens within 4 h using bacteriophage amplification. J Appl Microbiol 84:777
37. Hashemolhosseini S, Montag D, Kramer L, Henning U (1994) Determinants of receptor specificity of coliphages of the T4 family. A chaperone alters the host range. J Mol Biol 241:524
38. Hantke K (1978) Major outer membrane proteins of *E. coli* K12 serve as receptors for the phages T2 (protein Ia) and 434 (protein Ib). Mol Gen Genet 164:131
39. Heller KJ (1992) Molecular interaction between bacteriophage and the gram-negative cell envelope. Arch Microbiol 158:235
40. Duplessis M, Moineau S (2001) Identification of a genetic determinant responsible for host specificity in *Streptococcus thermophilus* bacteriophages. Mol Microbiol 41:325
41. Valyasevi R, Sandine WE, Geller BL (1990) The bacteriophage kh receptor of *Lactococcus lactis* subsp. *cremoris* KH is the rhamnose of the extracellular wall polysaccharide. Appl Environ Microbiol 56:1882
42. Letellier L, Plancon L, Bonhivers M, Boulanger P (1999) Phage DNA transport across membranes. Res Microbiol 150:499
43. Perez GL, Huynh B, Slater M, Maloy S (2009) Transport of phage P22 DNA across the cytoplasmic membrane. J Bacteriol 191:135
44. Kutter E, Raya R, Carlson K (2005) In: Kutter E, Sulakvelidze A (eds) Bacteriophages biology and applications. CRC Press, New York, pp 165–222
45. Carlson K (2005) In: Kutter E, Sulakvelidze A (eds) Bacteriophages biology and applications. CRC Press, New York, pp 437–494
46. Abshire TG, Brown JE, Ezzell JW (2005) Production and validation of the use of gamma phage for identification of *Bacillus anthracis*. J Clin Microbiol 43:4780
47. Thal E, Nordberg BK (1968) On the diagnostic of *Bacillus anthracis* with bacteriophages. Berl Münch Tierärztl Wochenschr 81:11
48. Nicolle P, Le Minor L, Buttiaux R, Ducrest P (1952) Phage typing of *Escherichia coli* isolated from cases of infantile gastroenteritis. II. Relative frequency of types in different areas and the epidemiological value of the method. Bull Acad Natl Med 136:483
49. Wallmark G, Laurell G (1952) Phage typing of *Staphylococcus aureus* some bacteriological and clinical observations. Acta Pathol Microbiol Scand 30:109
50. Postic B, Finland M (1961) Observations on bacteriophage typing of *Pseudomonas aeruginosa*. J Clin Invest 40:2064
51. Grajewski BA, Kusek JW, Gelfand HM (1985) Development of a bacteriophage typing system for *Campylobacter jejuni* and *Campylobacter coli*. J Clin Microbiol 22:13
52. Shcheglova MK, Neidbailik IN (1968) Experience in phage typing of *Listeria*. Veterinariia 45:102
53. Felix A (1956) Phage typing of *Salmonella typhimurium*: its place in epidemiological and epizootiological investigations. J Gen Microbiol 14:208
54. Karas M, Hillenkamp F (1988) Laser desorption ionization of proteins with molecular masses exceeding 10, 000 daltons. Anal Chem 60:2299
55. Tanaka K, Waki H, Ido Y, Akita S, Yoshida Y, Yoshida T (1988) Protein and polymer analyses up to m/z 100,000 by laser ionization time-of-flight mass spectrometry. Rapid Commun Mass Spectrom 2:151
56. Luna LG (2006) Development and evaluation of prototype bacteriophage amplification immunochromatography strips for the identificaiton of *Yersinia pestis*. PhD thesis, Colorado School of Mines, Golden, CO

57. Thomas JJ, Falk B, Fenselau C, Jackman J, Ezzell J (1998) Viral characterization by direct analysis of capsid proteins. Anal Chem 70:3863
58. Brown ER, Cherry WB (1955) Specific identification of *Bacillus anthracis* by means of a variant bacteriophage. J Infect Dis 96:34
59. Reiman RW (2007) Indirect detection of *Bacillus anthracis* (anthrax) using amplified gamma phage-based assays. PhD thesis, Colorado School of Mines, Golden, CO
60. Fouts DE, Rasko DA, Cer RZ, Jiang L, Fedorova NB, Shvartsbeyn A, Vamathevan JJ, Tallon L, Althoff R, Arbogast TS, Fadrosh DW, Read TD, Gill SR (2006) Sequencing *Bacillus anthracis* typing phages gamma and cherry reveals a common ancestry. J Bacteriol 188:3402
61. Watanabe T, Morimoto A, Shiomi T (1975) The fine structure and the protein composition of gamma phage of *Bacillus anthracis*. Can J Microbiol 21:1889
62. Garcia E, Elliott JM, Ramanculov E, Chain PS, Chu MC, Molineux IJ (2003) The genome sequence of *Yersinia pestis* bacteriophage phiA1122 reveals an intimate history with the coliphage T3 and T7 genomes. J Bacteriol 185:5248
63. Dunn JJ, Studier FW (1983) Complete nucleotide sequence of bacteriophage T7 DNA and the locations of T7 genetic elements. J Mol Biol 166:477
64. Pajunen MI, Elizondo MR, Skurnik M, Kieleczawa J, Molineux IJ (2002) Complete nucleotide sequence and likely recombinatorial origin of bacteriophage T3. J Mol Biol 319:1115
65. Cerritelli ME, Studier FW (1996) Assembly of T7 capsids from independently expressed and purified head protein and scaffolding protein. J Mol Biol 258:286
66. Posthuma-Trumpie GA, Korf J, van Amerongen A (2009) Lateral flow (immuno)assay: its strengths, weaknesses, opportunities and threats. A literature survey. Anal Bioanal Chem 393:569
67. Leuvering JH, Thal PJ, van der Waart M, Schuurs AH (1980) Sol particle immunoassay (SPIA). J Immunoassay 1:77
68. Carter DJ, Cary RB (2007) Lateral flow microarrays: a novel platform for rapid nucleic acid detection based on miniaturized lateral flow chromatography. Nucleic Acids Res 35:e74
69. Clavijo E, Diaz R, Anguita A, Garica A, Pinedo A, Smits HL (2003) Comparison of a dipstick assay for detection of *Brucella*-specific immunoglobulin M antibodies with other tests for serodiagnosis of human brucellosis. Clin Diagn Lab Immunol 10:612
70. Aldus CF, Van Amerongen A, Ariens RM, Peck MW, Wichers JH, Wyatt GM (2003) Principles of some novel rapid dipstick methods for detection and characterization of vero-toxigenic *Escherichia coli*. J Appl Microbiol 95:380
71. Kato S, Ozawa K, Okuda M, Nakayama Y, Yoshimura N, Konno M, Minoura T, Linuma K (2004) Multicenter comparison of rapid lateral flow stool antigen immunoassay and stool antigen enzyme immunoassay for the diagnosis of *Helicobacter pylori* infection in children. Helicobacter 9:669
72. Fong WK, Modrusan Z, McNevin JP, Marostenmaki J, Zin B, Bekkaoui F (2000) Rapid solid-phase immunoassay for detection of methicillin-resistant *Staphylococcus aureus* using cycling probe technology. J Clin Microbiol 38:2525
73. Zuiderwijk M, Tanke HJ, Sam Niedbala R, Corstjens PL (2003) An amplification-free hybridization-based DNA assay to detect *Streptococcus pneumoniae* utilizing the up-converting phosphor technology. Clin Biochem 36:401
74. Oku Y et al (2001) Development of oligonucleotide lateral-flow immunoassay for multi-parameter detection. J Immunol Methods 258:73
75. Al-Yousif Y, Anderson J, Chard-Bergstrom C, Kapil S (2002) Development, evaluation, and application of lateral-flow immunoassay (immunochromatography) for detection of rotavirus in bovine fecal samples. Clin Diagn Lab Immunol 9:723
76. Chaiyaratana W, Chansumrit A, Pongthanapisth V, Tangnararatchakit K, Lertwongrath S, Yoksan S (2009) Evaluation of dengue nonstructural protein 1 antigen strip for the rapid diagnosis of patients with dengue infection. Diagn Microbiol Infect Dis 64:83
77. Kikuta H, Sakata C, Gamo R, Ishizaka A, Koga Y, Konno M, Ogasawara Y, Sawada H, Taguchi Y, Takahashi Y, Yasuda K, Ishiguro N, Hayashi A, Ishiko H, Kobayashi K (2008) Comparison of a lateral-flow immunochromatography assay with real-time reverse transcription-PCR for detection of human metapneumovirus. J Clin Microbiol 46:928

78. Li L, Zhou L, Yu Y, Zhu Z, Lin C, Lu C, Yang R (2009) Development of up-converting phosphor technology-based lateral-flow assay for rapidly quantitative detection of hepatitis B surface antibody. Diagn Microbiol Infect Dis 63:165
79. Kusano N, Hirashima K, Kuwahara M, Narahara K, Imamura T, Mimori T, Nakahira K, Torii K (2007) Immunochromatographic assay for simple and rapid detection of Satsuma dwarf virus and related viruses using monoclonal antibodies. J Gen Plant Pathol 73:66
80. Joon Tam Y, Mohd Lila MA, Bahaman AR (2004) Development of solid-based paper strips for rapid diagnosis of Pseudorabies infection. Trop Biomed 21:121
81. Lyoo YS, Kleiboeker SB, Jang KY, Shin NK, Kang JM, Kim CH, Lee SJ, Sur JH (2005) A simple and rapid chromatographic strip test for detection of antibody to porcine reproductive and respiratory syndrome virus. J Vet Diagn Invest 17:469
82. Sithigorngul W, Rukpratanporn S, Sittidilokratna N, Pecharaburanin N, Longyant S, Chaivisuthangkura P, Sithigorngul P (2007) A convenient immunochromatographic test strip for rapid diagnosis of yellow head virus infection in shrimp. J Virol Methods 140:193

Matrix Assisted Laser Desorption Ionization Mass Spectrometric Analysis of *Bacillus anthracis*: From Fingerprint Analysis of the Bacterium to Quantification of its Toxins in Clinical Samples

Adrian R. Woolfitt, Anne E. Boyer, Conrad P. Quinn, Alex R. Hoffmaster, Thomas R. Kozel, Barun K. De, Maribel Gallegos, Hercules Moura, James L. Pirkle, and John R. Barr

Abstract A range of mass spectrometry-based techniques have been used to identify, characterize and differentiate *Bacillus anthracis*, both in culture for forensic applications and for diagnosis during infection. This range of techniques could usefully be considered to exist as a continuum, based on the degrees of specificity involved. We show two examples here, a whole-organism fingerprinting method and a high-specificity assay for one unique protein, anthrax lethal factor.

Keywords *Bacillus anthracis* • Anthrax • Anthrax toxin • Anthrax lethal factor • MALDI MS • Fingerprinting • Statistical analysis • Quantification

1 Introduction

Bacillus anthracis, the causative agent of anthrax, is a gram-positive, spore-forming bacterium. The three distinct routes of human infection are dermal, ingestion and inhalation.[1] Each route leads to a different form of the disease. The spores are very resistant to environmental stress and have been weaponized by several countries to facilitate aerosol distribution. Today, naturally-occurring cases of anthrax are relatively rare in the US and Western Europe, but inhalation anthrax has a subtle and rapid onset with a high mortality rate. During the 2001 anthrax incidents in the

A.R. Woolfitt, A.E. Boyer, M. Gallegos,
H. Moura, J.L. Pirkle, and J.R. Barr (✉)
Centers for Disease Control and Prevention, 4770 Buford Highway, Atlanta, GA 30341, GA, USA
e-mail: ahw9@cdc.gov

C.P. Quinn, A.R. Hoffmaster, and B.K. De
Centers for Disease Control and Prevention, 1600 Clitton Rd, Atlanta, GA 30333 USA

T.R. Kozel
University of Nevada School of Medicine, Reno, NV 89557, USA

J. Banoub (ed.), *Detection of Biological Agents for the Prevention of Bioterrorism*,
NATO Science for Peace and Security Series A: Chemistry and Biology,
DOI 10.1007/978-90-481-9815-3_6, © Springer Science+Business Media B.V. 2011

United States, the case: fatality ratio was 45% despite the administering of antibiotics and aggressive supportive care.[2,3] Vegetative cells circumvent the host defenses, by producing both a poly-D-glutamic acid (PGA) capsule to evade phagocytosis, and two binary toxins which disrupt immune signaling.[1] As infection progresses, vegetative cells enter the bloodstream, often resulting in systemic bacteremia, sepsis-induced shock, respiratory distress, extensive hemorrhage, and death. *B. anthracis* produces two binary toxins including lethal toxin (LTx) which is a complex of lethal factor (LF, 93 kDa) and protective antigen (PA, 63 kDa in the activated form), and edema toxin which is a complex of edema factor (EF) and PA.[1] The genes for the three toxin proteins are carried on plasmid pX01, while the genes directing the PGA capsule synthesis are carried on plasmid pX02. The modes of action for both toxins have been well characterized. While PA targets the cell surface transporting LF and EF into the cell, both LF and EF have enzymatic activities. LF is a zinc-dependent endoproteinase known to cleave specific peptide bonds in the family of mitogen-activated protein kinase kinase (MAPKK) response regulators and EF is a calmodulin dependent adenylate cyclase.

Diagnosis and characterization of anthrax is achieved using classical microbiological methods that are based upon identification of genotypic and phenotypic differences between *B. anthracis* and related organisms in the *B. cereus* group. However, microbiological methods that require culturing techniques rely on viable organisms and thus may yield negative results in the presence of antibiotic treatment. In addition most of these take more than 24 h for culture incubation. There is a need for appropriate tools for rapid diagnosis of anthrax in clinical samples following a bioterrorism incident.[4] Following the 2001 anthrax attacks however, a range of molecular biochemical methods were successfully used, including multiple-locus variable number tandem repeat analysis (MLVA), sequencing of the PA gene (*pagA*) used for sub-typing,[5] and real-time polymerase chain reaction (PCR).[4] Immunohistochemical analysis of tissue biopsies and serological analyses have also featured prominently in confirmatory diagnosis of anthrax.[6-8] To complement and enhance our capabilities for rapid, high sensitivity and high specificity diagnosis of anthrax, a range of mass spectrometry (MS)-based methods have been developed to identify *B. anthracis* and diagnose anthrax. Many of the MS-based methods can be assigned to one of three broad classes, listed in order of increasing specificity:

1. Fingerprinting methods
2. Proteomic approaches
3. Highly targeted analyses

The fingerprinting methods often use matrix-assisted laser-desorption ionization (MALDI) to generate a mass spectral pattern or fingerprint of the proteins or other compounds from *B. anthracis* spores or vegetative cells.[9,10] Typically, statistical analyses are used to aid in correct identification of the organism and/or strain, and this becomes increasingly important as larger numbers of organisms and strains are included in the studies.[11,12] An important distinction between the MS fingerprinting methods and the other classes of methods is that specific identification of individual proteins is not required. However, it is common that certain of the mass spectral peaks will be tentatively or definitively assigned to specific proteins, generally by using other techniques or sources of information for confirmation. For example, tentative assignments can be obtained by comparison with the protein databases.[12,13] If such a protein is identified, it can then be investigated as a candidate biomarker for that organism or strain.[14,15]

MS-based proteomic approaches for the analysis of *B. anthracis* and related organisms rely upon the identification of specific peptides and proteins by matching of MS and tandem MS (MS/MS) data with information in the genetic and protein databases. The presence of proteins unique to a given organism or strain can provide an unambiguous identification of that organism; in that sense, proteomic approaches are more specific than fingerprinting methods. Currently, the rapid increase in the number of fully-sequenced microorganisms listed in the databases makes these proteomic approaches increasingly applicable. Many researchers favor strategies which include tryptic digestion of extracted proteins, followed by MALDI MS/MS or liquid chromatography-MS/MS (LC-MS/MS).[16] Alternatively, mass spectral peaks from the intact small acid-soluble proteins (α, β, α-β and γ SASPs; around 6.5–10 kDa) from *B. anthracis* and related organisms in conjunction with the protein database entries have been used for unequivocal identification.[14,15]

Highly-targeted MS methods can be characterized by the use of one or more highly specific target compounds for unequivocal identification of the organism or strain. The choice of compounds ranges from small molecules through lipids,[17] carbohydrates or glycoproteins,[18] to proteins such as specific cell-surface markers or extracellular secreted toxins such as anthrax LF.[19] In many cases, the biomarkers used in targeted MS-based methods have first been identified and characterized using other biochemical techniques, which may have involved decades of research. This is true for example for anthrax LF.[19] It is relatively uncommon that a targeted method can quickly or easily be developed based solely upon a compound initally discovered in an MS fingerprinting study.

This report presents an example of an MS-based fingerprinting method for discrimination of *B. anthracis* strains and of a highly targeted method for quantitative analysis of anthrax lethal factor that exploits the enzyme activity of this protein. These methods utilize sample preparation equipment and MS instruments found in a well-equipped laboratory.

2   MALDI-MS Fingerprint Analysis of *B. anthracis* Strains

The method described here is intended for the analysis of pure cultures of *B. anthracis* and related organisms. Although the method is not fast since it relies upon culture (a minimum of 2 days plus 2–3 days to ensure that materials are rendered safe), it can potentially be used to identify microorganisms at the strain level and may assist in forensic or epidemiologic analyses.

2.1   Sample Preparation

Seven well-characterized strains of *B. anthracis* were chosen to represent genotypic and origin diversity, along with five closely related bacilli chosen to represent differential diagnostic challenge, Table 1. Three separate harvestings of vegetative cells and spores were obtained, to include the variability resulting from the growth conditions. Organisms were inactivated using gamma irradiation (10^6 rads, 4 h), and absence of viability was confirmed by 2- to 3-day culture. This part of the work was carried out using approved BSL-3 facilities, while observing appropriate safety precautions. Four other organisms were also cultured for use as statistical outliers.

2.2   Maldi Analysis

Cultures (around 10^6 organisms/mL) were mixed with an equal volume of 10 mg/mL sinnapinic acid in 70:30 (v/v) water:acetonitrile plus 0.1% trifluoroacetic acid. Samples (0.5 µL) were plated on stainless steel MALDI targets, and were analyzed in positive-ion linear mode on an Applied Biosystems (AB) 4700 Proteomics Analyzer. At least ten replicate mass spectra were acquired, over the *m/z* range

Table 1 *Bacilli* and other organisms selected for MALDI fingerprint analysis. Fully-numeric identification codes refer to organisms in the CDC collection. The vegetative and spore codes are used as abbreviations on other figures

Organism	Identification	Plasmids; genotype	Vegetative code	Spore code
B. anthracis	ATCC 4229; Pasteur strain	pX01−, pX02+	V1	S1
B. anthracis	Sterne strain	pX01−, pX02+	V2	S2
B. anthracis	2000031661	pX01+, pX02+	V3	S3
B. anthracis	2000031653	pX01+, pX02+	V4	S4
B. anthracis	2000031658	pX01+, pX02+	V5	S5
B. anthracis	2000032777	pX01−, pX02−	V6	S6
B. anthracis	2000031136	pX01+, pX02+	V7	S7
B. cereus	ATCC 14579	pBClin15	V8	S8
B. megaterium	ATCC 14581	Theta-pBMs	V9	S9
B. mycoides	2000032765	Cryptic plasmids	V10	S10
B. subtilis	2000031480	palT1, palT2	V11	S11
B. thuringiensis	ATCC 13367	Multiple plasmids	V12	S12
Escherichia coli	ATCC 25922	−	X13	−
Enterococcus faecalis	ATCC 29212	−	X14	−
Streptococcus pyogenes	ATCC 700294	−	X15	−
Staphylococcus aureus	ATCC 29213	−	X16	−

2,000–14,000, from each of three MALDI wells so as to include mass spectral variability in the data sets.

2.3 Statistical Analysis

Profile mass spectra were exported from the MS data system in text format, and were subjected to a custom sequence of preprocessing steps designed to apply normalization, summing, background subtraction and smoothing,[13,20] and finally standardizing and denoising,[21] Fig. 1. The preprocessing yielded data sets comprising profile mass spectra containing only statistically significant peaks with a background of zero intensities at all other m/z values. This was advantageous because such data can relatively easily be subjected to a range of statistical algorithms, such as hierarchical clustering analysis (HCA),[13] partial least squares discriminant analysis (PLS-DA)[20] and random forest (RF) analysis.[21] Here we used the PAST program[22] for HCA, and Fortran code for RF.[23,24]

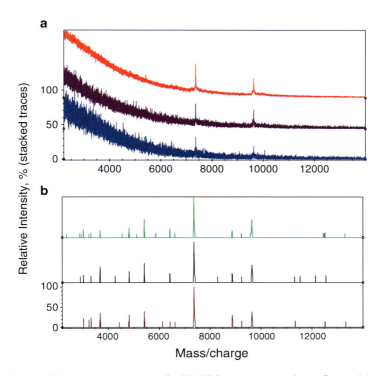

Fig. 1 Details of the preprocessing steps for MALDI mass spectra prior to fingerprint analysis. (**a**) Three individual raw spectra, from *B. anthracis* Pasteur spores; (**b**) Three summed spectra, each comprising ten raw spectra acquired from each of three MALDI wells, after background subtraction, smoothing, standardizing and denoising, This preprocessing sequence gives near-optimal Random Forest classification

2.4   Fingerprinting Results

Figures 2 and 3 show MALDI spectra of the vegetative and spore forms of seven strains of *B. anthracis*, along with five other bacilli and four non-bacillus organisms. Most *B. anthracis* strains yielded good-quality spectra. Most spectra of the spore forms contained a pair of peaks in the *m/z* 6,500–7,000 region, although in several cases the intensities were low. We presume that these are the *B. anthracis* α- and β-SASPs at 6,836 and 6,679 Da respectively, that have been characterized by others.[14,15] Interestingly, many of the vegetative forms of *B. anthracis* also yielded similar peaks. Spectra of the other bacilli analyzed here did not contain the β-SASP peak at around *m/z* 6,683, which is thought to be unique to *B. anthracis*.[15]

HCA was able to group most harvestings of each organism and growth form correctly. This suggests that the variability across replicate harvestings is less than that between organisms and growth forms. However, the inter-relationships suggested by the protein fingerprint similarities do not easily match the expected phylogeny based

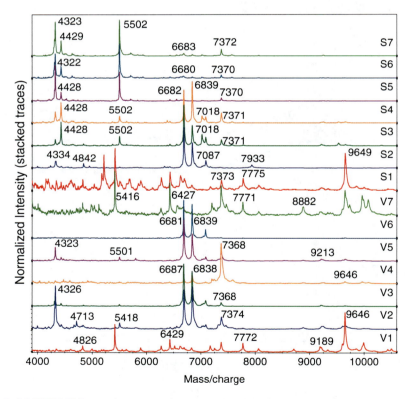

Fig. 2 MALDI MS fingerprint spectra of seven strains of vegetative forms and spores of *Bacillus anthracis*. Spectra are identified using the short codes as listed in Table 1; V1–V7 are the vegetative and S1–S7 are the spore forms of organisms 1 through 7 respectively. Traces for corresponding vegetative and spore forms also share the same colors

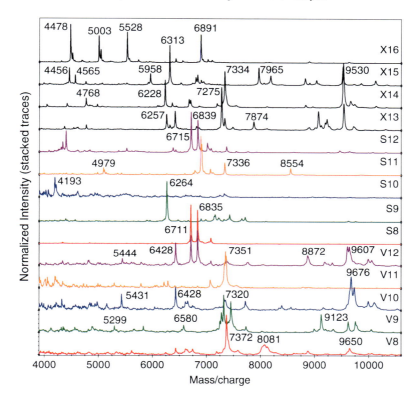

Fig. 3 MALDI MS fingerprint spectra of strains of non- *Bacillus anthracis* organisms. Spectra are identified using the short codes as listed in Table 1; V8–V12 are the vegetative and S8–S12 are the spore forms of organisms 6 through 12 respectively. Traces for corresponding vegetative and spore forms also share the same colors. Traces of four non-*Bacillus* organisms, coded X13–X16 and shown in *black*, were included as outgroups for statistical purposes

on genetic data. In our experience RF has a higher discriminatory power[21] than both HCA and PLS-DA, and a richer information output than HCA. RF typically gave around 95–99% accuracy in the classification of MALDI spectra from the three harvestings, the only errors involving the incorrect assignment of some of the poorest-quality spectra, which had been obtained from *B. anthracis* Pasteur spores (Fig. 2). Figure 4 shows an RF proximity plot, in which the densest grays represent the highest similarities. In general, individual spectra are classified by RF as being most similar to other spectra from the same organism and growth form, which generates the diagonal line of dense squares. RF was always able to correctly identify organisms and growth forms, and could successfully classify all strains of *B. anthracis* in their vegetative forms. In addition, four of the spore forms of *B. anthracis* were always correctly classified. However, three of the spore forms could not be reliably distinguished from each other. Proximity plots indicated that the similarity is more diffuse for these three spore forms of *B. anthracis*, which can account for the misclassification in this region.

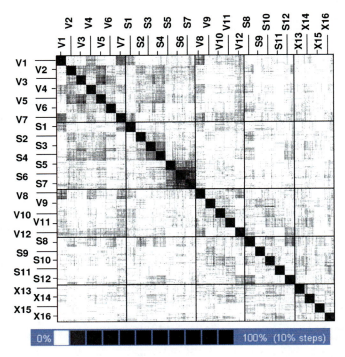

Fig. 4 Random forest proximity matrices, for 504 observed spectra classified as 28 classes. Increasing spectral similarity is represented by increasing image density. V1–V7 and S1–S7, *B. anthracis* vegetative and spore forms respectively, as coded as in Table 1. V8–V12 and S8–S12, other *Bacillus* vegetative and spore forms respectively. X13–X16, outgroups (*E. coli*, etc.)

It is now widely accepted that MALDI protein fingerprinting should include an appropriate statistical analysis, although there is no firm consensus on the best method. Some researchers favor 'pure' statistical algorithms such as PLS-DA,[20] RF[21,23] or artificial neural networks,[12] while others combine statistical analyses with protein database searching.[25]

3   Quantitative MALDI-MS Analysis of Anthrax Lethal Factor Activity

The method described below supports clinical diagnostics through the analysis of anthrax LF levels in human serum or plasma, following infection with *B. anthracis*. It has been developed based on animal studies using rabbits and rhesus macaques exposed via the inhalation route, and its performance was compared to a range of other assays for clinical diagnostic markers of anthrax.[19,26]

3.1 Animal Study

Most of the work described here involved analysis of serum and plasma samples from rhesus macaques that had been exposed to known amounts of aerosolized *B. anthracis* Ames strain spores (around 2×10^7 spores, equivalent to 300–400 times the 50% lethal dose), using appropriate safety precautions and approved humane methods in BSL-3 facilities.[26] Serum was drawn pre-exposure and at regular intervals post-exposure.

3.2 Sample Preparation

Briefly, the method incorporates a monoclonal antibody (mAb) capture step using a non-neutralizing mAb bound to magnetic beads, designed to extract the anthrax LF from serum or plasma.[19] The beads are washed to remove serum proteins, and are transferred to an appropriate buffer system at pH 7.3 which includes an optimized peptide substrate that mimics the primate MAPKK target sequences. The mAb-bead bound LF cleaves the substrate peptide at a specific location, releasing N-terminal (NT) and C-terminal (CT) product peptides, Fig. 5. We use two formats for the assay, involving 20 or 200 μL of plasma and 2 or 20 h incubations at 37°C, to cover both high and low levels of LF and different sensitivity ranges. Each analytical run includes a set of recombinant LF-spiked standards, blanks and quality control (QC) samples prepared in human serum along with the unknown samples to be quantified. All sample preparation was carried out in BSL-2 facilities and using appropriate personal protective equipment.

3.3 Maldi Analysis

After the LF reaction is complete, supernatant is mixed with MALDI matrix (α-cyano-4-hydroxycinnamic acid) and two isotopically-labeled internal standard peptides, which are identical in sequence but 7 Da higher in mass than the native NT and CT products. Samples are spotted on stainless steel 384 spot plates and are analyzed by MALDI-MS using an AB 4800 Proteomics Analyzer operated in positive-ion reflectron mode, Fig. 5. Samples are plated in quadruplicate, and four spectra (one per MALDI well) are acquired for each sample, in order to incorporate the variability from the MALDI analysis. In the absence of LF the predominant peaks correspond to the singly and doubly-charged substrate peptide, Fig. 5. However, when LF is present the substrate peaks are diminished and two peaks that correspond to the LF-cleaved peptide products are present with intensities that are related to a given amount of LF in the original sample.

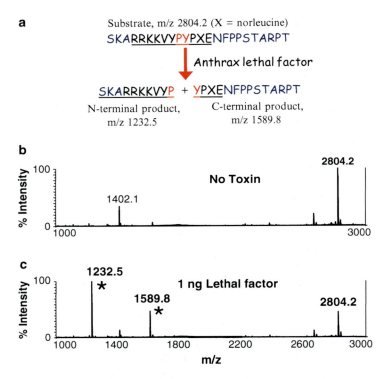

Fig. 5 Detection of anthrax lethal factor activity. (**a**) Detail of the enzymatic reaction catalyzed by lethal factor. Red amino acids (AAs) indicate the cleavage site and underlined AAs indicate the MAPKK consensus region. (**b** and **c**) MALDI mass spectra of the reaction mixture incubated without (**b**), and with (**c**) lethal factor, respectively. Asterisks indicate the two product peptides generated in the presence of lethal factor

3.4 Quantitative Analysis

MALDI mass spectra are processed using a custom software application which extracts isotopic peak areas of the native NT and CT products, the internal standards, and the substrate. Isotope-dilution quantification is performed based on the peak area ratios of the pairs of native and internal standard peptides, which is designed to correct for variability in the MALDI signals. The CT product was the primary target for quantification of LF, and the NT product was used for confirmation. Curve-fitting is achieved using a sliding-fit algorithm which carries out linear regression over narrow ranges, usually involving three or four adjacent levels of standards. Quantitative data from analytical runs are only accepted when the QC samples are in-control.

3.5 Comparison with Other Methods for Anthrax in Clinical Samples

The MALDI method for anthrax LF in plasma is highly specific and sensitive; limits of detection (LODs) are around 0.025 and 0.005 ng/mL of LF (93 kDa) in 200 μL of plasma, with 2 and 20-h incubations respectively. Quantitative results from the 2-h incubations could be reported within about 4-h of receipt of the sample. Coefficients of variation (CVs) for the QC materials are within 8–15%.

The MALDI method for LF was compared with culture (bacteremia), *pagA* PCR, PA ELISA, and PGA ELISA results for five inhalation-infected rhesus macaques. Culture results were negative until 48-h post-challenge, Fig. 6. In four out of five animals, culture reversed from a positive on day 2 to negative or low positive on day 3 post-challenge. This indicated that its use for clinical diagnosis may be limited. In contrast, the LF method gave positive results 24 h after exposure. The LF method also gave positive results in every case in which culture results had reversed. PA ELISA for the five animals was only positive in late infection (LOD 4.8 ng/mL), while PGA ELISA and *pagA* PCR were positive from 48 h onward

Diagnostic	Animal ID	24h	48h	72h	96h	120h
PA	A	−	−	−	147*	
	B	−	−	−	−	3,153*
	C	−	−	−	−	
	D	−	−	−	−	−
	E	−	−	−	19,434*	
LF	A	0.006	45.2	10.3	103*	
	B	−	25.9	19.0	63.6	1,236*
	C	0.200	51.7	10.6	34.1*	
	D	0.018	38.1	8.1	4.5	27.7
	E	−	42.0	25.1	3,886*	
PGA	A	−	10,394	20	32,000*	
	B	−	48	17	681	1.1e6*
	C	−	7,061	16	3,541*	
	D	−	1,523	7	3	32
	E	−	6,527	18	1.9e6*	
pagA PCR	A	−	+	+	+*	
	B	−	−	+	NS	+*
	C	−	+	+	+*	
	D	−	+	−	NS	NS
	E	−	+	+	+*	
Bacter-emia Status	A	−	+	−	+*	
	B	−	+	±	NS	+*
	C	−	+	−	+*	
	D	−	+	−	NS	NS
	E	−	+	+	+*	

Fig. 6 Results of five diagnostic tests carried out on five rhesus macaques between 24 and 120 h following inhalational exposure to anthrax. PA, protective antigen ELISA. LF, the lethal factor mass spectrometric method. PGA, poly-D-glutamic acid ELISA. These three diagnostics were quantitative, and amounts are recorded as nanograms per millilitre of PA, LF or PGA. Two qualitative diagnostics were also used: *pagA* PCR, PCR of the PA gene, and culture status (bacteremia). Coding: (−), negative; (+), positive; (±), low positive; (NS), no sample. *Asterisks* indicate results at time of euthenasia

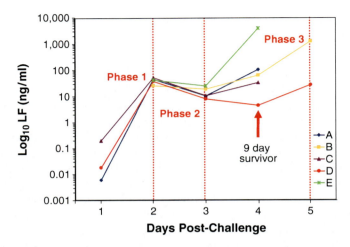

Fig. 7 Triphasic kinetics of toxemia in five rhesus macaques exposed to inhalation anthrax (*Bacillus anthracis* Ames strain). Data are for anthrax Lethal Factor measured in the serum using the mass spectrometric method, for individual animals coded A through E as described in Fig. 6. Three animals died on the fourth day post-challenge, one died on the fifth day, and one survived until the ninth day

(LOD 2.25 ng/mL). In two instances, *pagA* PCR was negative when LF was positive, but compensated for two of the negative culture results at 72 h with positive results. Overall, the LF method out-performed the other four methods, enabling both earlier and more reliable diagnosis.

The LF method also revealed triphasic kinetics of toxemia, in which LF levels first increase, then briefly decrease, and then increase again prior to death. Figure 7 shows LF levels for five animals, four of which died on days 4 or 5 and one on day 9. The triphasic pattern was also observed for LF in studies with New Zealand White rabbits with inhalation anthrax (unpublished results), although the time periods were shorter than in the macaque model. The triphasic kinetics described here help to explain the clinical description of anthrax, which initially resembles a mild flu-like illness and is sometimes followed by a period in which the patient feels better. However, after that point, death follows rapidly.[27] Finally, it should be noted that the method has been successfully applied to measure LF levels in a single human case of anthrax.[28]

4   Conclusions

From our point of view, a continuum of approaches for MALDI MS-based analyses of *B. anthracis* exists, and no single MS-based method is likely to suit all purposes in the event of an intentional release of anthrax. We have shown examples from the extremes of this continuum, ranging from a fingerprinting method requiring relatively little information about the organism, to a highly specific method for the enzyme activity of one protein unique to this organism, lethal factor. Although the quantitative method for LF is in many ways superior to the fingerprinting method, it is interesting to note that a strain of *B. cereus* that carries the anthrax toxin genes and causes pneumonia has been characterized.[29] Such *B. cereus* strains have been shown to be fatal[30] and although the LF method might misdiagnose this as a potential *B. anthracis* infection, it would alert clinicicians to the presence of potentially deadly anthrax-related toxins. These possibilities reiterate the importance of incorporating multiple diagnostic tests for confirmation of anthrax. To cover both toxins for a comprehensive clinical picture, we are currently working on an LC-MS/MS based method for edema factor activity which appears to be as sensitive as the LF method. The LF method is designed for clinical use only and therefore does not provide information about the strain involved. For forensic and epidemiologic applications and strain identification the MALDI MS fingerprinting or proteomics-based methods are most suitable.

Acknowledgements References in this article to any specific commercial products, process, service, manufacturer, or company do not constitute an endorsement or a recommendation by the US government or the Centers for Disease Control and Prevention. The findings and conclusions in this report are those of the authors and do not necessarily represent the views of CDC.

References

1. Mock M, Mignot T (2003) Anthrax toxins and the host: a story of intimacy. Cell Microbiol 5:15–23
2. Jernigan JA, Stephens DS, Ashford DA, Omenaca C, Topiel MS, Galbraith M et al (2001) Bioterrorism-related inhalational anthrax: the first 10 cases reported in the United States. Emerg Infect Dis 7:933–944
3. Jernigan DB, Raghunathan PL, Bell BP, Brechner R, Bresnitz EA, Butler JC et al (2002) Investigation of bioterrorism-related anthrax, United States, 2001: epidemiologic findings. Emerg Infect Dis 8:1019–1028
4. Hoffmaster AR, Meyer RF, Bowen MP, Marston CK, Weyant RS, Barnett GA, Sejvar JJ, Jernigan JA, Perkins BA, Popovic T (2002) Evaluation and validation of a real-time polymerase chain reaction assay for rapid identification of *Bacillus anthracis*. Emerg Infect Dis 8:1172–1182
5. Hoffmaster AR, Fitzgerald CC, Ribot E, Mayer LW, Popovic T (2002) Molecular subtyping of *Bacillus anthracis* and the 2001 bioterrorism-associated anthrax outbreak, United States. Emerg Infect Dis 8:1111–1116
6. Shieh WJ, Guarner J, Paddock C, Greer P, Tatti K, Fischer M, Layton M, Philips M, Bresnitz E, Quinn CP, Popovic T, Perkins BA, Zaki SR (2003) The critical role of pathology in the investigation of bioterrorism-related cutaneous anthrax. Am J Pathol 163:1901–1910
7. Quinn CP, Semenova VA, Elie CM, Romero-Steiner S, Greene C, Li H, Stamey K, Steward-Clark E, Schmidt DS, Mothershed E et al (2002) A specific, sensitive and quantitative enzyme linked immunosorbent assay for human immunoglobulin G antibodies to anthrax toxin protective antigen. Emerg Infect Dis 8:1103–1110
8. Quinn CP, Dull PM, Semenova V, Li H, Crotty S, Taylor TH, Steward-Clark E, Stamey KL, Schmidt DS, Wallace Stinson K, Freeman AE, Elie CM, Martin SK, Greene C, Aubert RD, Glidewell J, Perkins BA, Ahmed R, Stephens DS (2004) Immune responses to *Bacillus anthracis* protective antigen in individuals with bioterrorism-associated cutaneous and inhalation anthrax. J Infect Dis 190:1228–1236

9. Dickinson DN, La Duc MT, Haskins WE, Gornushkin I, Winefordner JD, Powell DH, Venkateswaran K (2004) Species differentiation of a diverse suite of Bacillus spores by mass spectrometry-based protein profiling. Appl Environ Microbiol 70:475–482

10. Claydon MA, Davey SN, Edwards-Jones V, Gordon DB (1996) The rapid identification of intact microorganisms using mass spectrometry. Nat Biotechnol 14:1584–1586

11. Jarman KH, Cebula ST, Saenz AJ, Petersen CE, Valentine NB, Kingsley MT, Wahl KL (2000) An algorithm for automated bacterial identification using matrix-assisted laser desorption/ionization mass spectrometry. Anal Chem 72:1217–1223

12. Lasch P, Beyer W, Nattermann H, Stämmler M, Siegbrecht E, Grunow R, Naumann D (2009) Identification of Bacillus anthracis using MALDI-ToF mass spectrometry and artificial neural networks. Appl Environ Microbiol 75:7229–7242

13. Moura H, Woolfitt AR, Carvalho MG, Pavlopoulos A, Teixeira LM, Satten GA, Barr JR (2008) MALDI-TOF mass spectrometry as a tool for differentiation of invasive and noninvasive Streptococcus pyogenes isolates. FEMS Immunol Med Microbiol 53:333–342

14. Hathout Y, Setlow B, Cabrera-Martinez RM, Fenselau C, Setlow P (2003) Small, acid-soluble proteins as biomarkers in mass spectrometry analysis of Bacillus spores. Appl Environ Microbiol 69:1100–1107

15. Castanha ER, Fox A, Fox KF (2006) Rapid discrimination of Bacillus anthracis from other members of the B. cereus group by mass and sequence of "intact" small acid soluble proteins (SASPs) using mass spectrometry. J Microbiol Methods 67:230–240

16. Warscheid B, Fenselau C (2004) A targeted proteomics approach to the rapid identification of bacterial cell mixtures by matrix-assisted laser desorption/ionization mass spectrometry. Proteomics 4:2877–2892

17. Xu M, Voorhees KJ, Hadfield TL (2003) Repeatability and pattern recognition of bacterial fatty acid profiles generated by direct mass spectrometric analysis of in situ thermal hydrolysis/methylation of whole cells. Talanta 59:577–589

18. Fox A, Stewart GC, Waller LN, Fox KF, Harley WM, Price RL (2003) Carbohydrates and glycoproteins of Bacillus anthracis and related bacilli: targets for biodetection. J Microbiol Methods 54:143–152

19. Boyer AE, Quinn CP, Woolfitt AR, Pirkle JL, McWilliams LG, Stamey KL, Bagarozzi DA, Hart JC Jr, Barr JR (2007) Detection and quantification of anthrax lethal factor in serum by mass spectrometry. Anal Chem 79:8463–8470

20. Pierce CY, Barr JR, Woolfitt AR, Moura H, Shaw EI, Thompson HA, Fernandez FM (2007) Strain and phase identification of the U.S. Category B agent Coxiella burnetii by matrix assistred laser desorbtion/ionization time-of-flight mass spectrometry and multivariate pattern recognition. Anal Chim Acta 583:23–31

21. Satten GA, Datta S, Moura H, Woolfitt AR, Carvalho MG, Carlone GM, De BK, Pavlopoulos A, Barr JR (2004) Standardization and denoising algorithms for mass spectra to classify whole-organism bacterial specimens. Bioinformatics 20:3128–3136

22. Hammer Ø, Harper DAT, Ryan PD (2001) PAST: Paleontological statistics software package for education and data analysis. Palaeontol Electron 4:9. http://palaeo-electronica.org/2001_1/past/issue1_01.htm

23. Breiman L (2001) Random forests. Mach Learn 45:5–32

24. Breiman L, Cutler A (2009) Random forests. http://www.stat.berkeley.edu/~breiman/RandomForests/cc_home.htm. Accessed October 2009

25. Pineda FJ, Antoine MD, Demirev PA, Feldman AB, Jackman J, Longenecker M, Lin JS (2003) Microorganism identification by matrix-assisted laser/desorption ionization mass spectrometry and model-derived ribosomal protein biomarkers. Anal Chem 75:3817–3822

26. Boyer AE, Quinn CP, Hoffmaster AR, Kozel TR, Saile E, Marston CK, Percival A, Plikaytis BD, Woolfitt AR, Gallegos M, Sabourin P, McWilliams LG, Pirkle JL, Barr JR (2009) Kinetics of lethal factor and poly-D-glutamic acid antigenemia during inhalation anthax in rhesus macaques. Infect Immun 77:3432–3441

27. Brachman PS (1980) Inhalation anthrax. Ann NY Acad Sci 353:83–93

28. Walsh JJ, Pesik N, Quinn CP, Urdaneta V, Dykewicz CA, Boyer AE, Guarner J, Wilkins P, Norville KJ, Barr JR, Zaki SR, Patel JB, Reagan SP, Pirkle JL, Treadwell TA, Messonnier NR, Rotz LD, Meyer RF, Stephens DS (2007) A case of naturally acquired inhalation anthrax: clinical care and analyses of anti-protective antigen immunoglobulin G and lethal factor. Clin Infect Dis 44:968–971
29. Hoffmaster AR, Ravel J, Rasko DA, Chapman GD, Chute MD, Marston CK, De BK, Sacchi CT, Fitzgerald C, Mayer LW, Maiden MC, Priest FG, Barker M, Jiang L, Cer RZ, Rilstone J, Peterson SN, Weyant RS, Galloway DR, Read TD, Popovic T, Fraser CM (2004) Identification of anthrax toxin genes in a *Bacillus cereus* associated with an illness resembling inhalation anthrax. Proc Natl Acad Sci U S A 101:8449–8454
30. Avashia SB, Riggins WS, Lindley C, Hoffmaster A, Drumgoole R, Nekomoto T, Jackson PJ, Hill KK, Williams K, Lehman L, Libal MC, Wilkins PP, Alexander J, Tvaryanas A, Betz T (2007) Fatal pneumonia among metalworkers due to inhalation exposure to *Bacillus cereus* containing *Bacillus anthracis* toxin genes. Clin Infect Dis 44:414–416

Microorganism Identification Based On MALDI-TOF-MS Fingerprints

Thomas Elssner, Markus Kostrzewa, Thomas Maier, and Gary Kruppa

Abstract Advances in MALDI-TOF mass spectrometry have enabled the development of a rapid, accurate and specific method for the identification of bacteria directly from colonies picked from culture plates, which we have named the MALDI Biotyper. The picked colonies are placed on a target plate, a drop of matrix solution is added, and a pattern of protein molecular weights and intensities, "the protein fingerprint" of the bacteria, is produced by the MALDI-TOF mass spectrometer. The obtained protein mass fingerprint representing a molecular signature of the microorganism is then matched against a database containing a library of previously measured protein mass fingerprints, and scores for the match to every library entry are produced. An ID is obtained if a score is returned over a pre-set threshold. The sensitivity of the techniques is such that only approximately 10^4 bacterial cells are needed, meaning that an overnight culture is sufficient, and the results are obtained in minutes after culture. The improvement in time to result over biochemical methods, and the capability to perform a non-targeted identification of bacteria and spores, potentially makes this method suitable for use in the detect-to-treat timeframe in a bioterrorism event. In the case of white-powder samples, the infectious spore is present in sufficient quantity in the powder so that the MALDI Biotyper result can be obtained directly from the white powder, without the need for culture. While spores produce very different patterns from the vegetative colonies of the corresponding bacteria, this problem is overcome by simply including protein fingerprints of the spores in the library. Results on spores can be returned within minutes, making the method suitable for use in the "detect-to-protect" timeframe.

T. Elssner(✉)
Bruker Daltonik GmbH, Permoser Strasse 15, 04318 Leipzig, Germany
e-mail: thomas.elssner@bdal.d

M. Kostrzewa and T. Maier
Bruker Daltonik GmbH, Fahrenheitstrasse 4, 28359 Bremen, Germany

G. Kruppa
Bruker Daltonis,Inc., 40 Manning Road, 01821, Billerica, MA, USA

J. Banoub (ed.), *Detection of Biological Agents for the Prevention of Bioterrorism*,
NATO Science for Peace and Security Series A: Chemistry and Biology,
DOI 10.1007/978-90-481-9815-3_7, © Springer Science+Business Media B.V. 2011

1   Introduction: MALDI-TOF Identification of Bacteria

Despite advances in PCR based methods for the rapid detection and identification
of biological warfare agents (BWA), mass spectrometry possesses several key characteristics that make it attractive as a complementary method to other techniques.[1]
PCR based methods are generally targeted, meaning that a PCR primer for each
suspected BW agent must be used in each test. Identification based on the molecular signatures from protein mass fingerprints is non-targeted and can identify as
many species as are present in a database to which the spectra are matched, and
previous studies have shown the technique to be very generally applicable to all
culturable microorganisms, including yeast, fungi and bacteria. MALDI-TOF mass
spectrometry has also been shown to be useful for the direct analysis and identification of *Bacillus* spores.[2,3] A further advantage of the MALDI Biotyper over other
methods is that limited consumables are required and the cost per sample is quite
low. While fast time to result, sensitivity and specificity are important for a detect-
to-treat method for bioterrorism events, these last advantages of cost and limited
consumables can also be very important in a suspected bioterrorism event. First
responder labs must stockpile the consumables needed to be able to run the large
number of samples that can be expected to be submitted in a short time. Anecdotal
reports about the recent H1N1 response in the US indicate that many labs faced
shortages of consumables and long re-stocking times due to shortages faced by
suppliers, because of the spike in demand. With the MALDI Biotyper all consumables needed to run thousands or tens of thousands of samples can be easily stored
in a small space and at low cost as will be discussed in more detail below.

For the method to work well in the clinical routine and in state health labs and
other field labs which may respond to a bioterrorism event, the method also must
have robust and simple sample preparation that can be carried out by microbiology
technicians. While the sample preparation can be as simple as smearing a picked
bacterial colony on a sample plate and pipetting a drop of matrix solution onto it,
alternative and equally simple sample preparation methods exist for different situations. If highly infectious organisms or spores are suspected in the sample, an
extraction protocol may be employed instead, in which the sample is inactivated by
harsh treatment with organic acid and/or boiling, and extracted protein material sent
for analysis. The inactivated extract samples are stable for days, so the analysis may
also be carried out at central labs; there is no need for culture or shipment of
putatively "live" samples to a central lab.

1.1   Prinicples of Microorganism Identification
by the MALDI Biotyper

In this section, the principles of microorganism ID by MALDI-TOF mass spectrometry will be discussed. While a full discussion of the development of MALDI-
TOF mass spectrometry for microorganism ID is beyond the scope of this chapter,

a brief discussion of previous work in the field is important and will be used to highlight aspects of the MALDI Biotyper that make it robust, easy to use and have led to its adoption by the clinical microbiology community. References to recent reviews and other work are given for those wishing to do further reading in this area. Next, the MALDI Biotyper workflow, including sample preparation, acquisition of data, and interpretation of the results will be presented.

1.2 Development of MALDI-TOF Mass Spectrometry for Microorganism Identification

The first efforts of t Anhalt and Feneslau to identify bacteria based on mass spectrometry date back to 1975[4] and several recent reviews and papers cover the development of the technique up to the present.[1,5–7] Many of the early efforts in mass spectrometry tried to mimic the biochemical techniques that were standard at the time, and remain as the most commonly used ID method today. These biochemical techniques rely on the fact that different bacteria produce different sets of metabolites, e.g. fatty acids, and tests were developed that could identify the species present by determining what subset of metabolites are present from a given cultured colony. Mass spectrometry can also be used to measure and identify small molecule metabolites, but the advantages of mass spectrometry for this type of measurement were limited. Furthermore, culture conditions can easily influence relative ratios of metabolites. The development of Electrospray Ionization (ESI) and Matrix Assisted Laser Desorption Ionization (MALDI) enabled the measurement of larger molecules, including proteins. The advantages in terms of reproducibility were noted in two early papers on the use of MALDI-TOF for identification of bacteria.[8,9] ESI will not be discussed further here, but a brief description of MALDI-TOF mass spectrometry is included as the rest of this work will focus on the use of this technique for the identification of microorganisms. The field of MALDI-TOF mass spectrometry is vast, and this summary is not meant to be an exhaustive introduction, but only a summary of the key features of the technique that are important for understanding why it is so useful for the identification of microorganisms.

While there are variations in the exact methodology, the first step in MALDI-TOF mass spectrometry involves mixing the sample with one to a few microliters of a matrix solution. For protein measurements, the most common matrix solutions are concentrated solutions of organic acids with strong absorption in the UV in the range of 320–350 nm. The concentration of the matrix is usually in the millimolar range, commonly in a mixture of acetonitrile and water. Ideally the amount of protein sample used is adjusted so that the molar excess of the matrix is about 10^4, though this ratio is not extremely critical, and the technique works in the presence of salts, low amounts of detergents, and other small molecules. This insensitivity to the ratio of sample to matrix, and to the presence of small molecule contaminants is one of the characteristics that makes MALDI-TOF very easy to use and suitable for situations where non-experts will be preparing the samples. A common technique is to apply a microliter of the protein sample to a target plate first, and then apply a microliter of the matrix solution from another pipet and mix the two by stirring or sucking the drop back into the pipet tip and pushing it back out a few times. The drop is allowed to dry, which results in the matrix crystallizing, with protein molecules dispersed in the crystalline matrix. This is the first key function of the matrix; it separates the protein molecules from each other, preventing aggregation of the protein molecules on the target. The target is then inserted into the MALDI-TOF mass spectrometer, in what is known as the ion source region. As mentioned above the matrices are chosen such that they have a strong absorption in the UV, and as shown in Fig. 1, a laser is fired at the crystals on the target, the matrix absorbs the energy of the laser and desorbs into the gas phase carrying the intact protein molecules with it. This is the second important function of the matrix, it absorbs the energy of the laser allowing intact proteins to be carried into the gas phase in the ion source region. Finally, the energy of the laser desorption event causes plasma to be formed, containing both positive and negative ions. The acidic nature of the matrix enables matrix ions to transfer protons to the basic sites in the protein molecules, producing positively charged proteins, this is the third important function of the matrix.

As shown in Fig. 1, the ions are formed in the source of a Time Of Flight (TOF) mass spectrometer, the target plate is at ground potential and is close to a set of electrodes with a high positive potential (in the MALDI Biotyper this potential is 20 kV). After traversing the source region between the target plate and the grid all of the protein ions have the same kinetic energy of 20 kV. Since $E = 1/2 (mv^2)$, the lighter ions will have a higher velocity and will traverse the field free region of the flight tube to the detector faster and hit the detector first. A plot of the intensity of protein ions hitting the detector as a function of time can be converted to a plot of

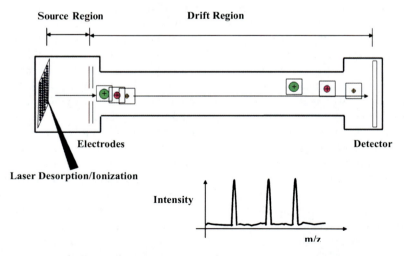

Fig. 1 A schematic of the MALDI-TOF mass measurement. The target plate with the dried crystalline matrix containing embedded proteins is placed in the source region of the mass spectrometer. The details of the measurement are described in the text

signal intensity as a function of protein mass, and this is the protein mass finger print, as shown schematically in Fig. 1.

Ryzhov and Feneslau observed that MALDI-TOF done directly on bacteria produces very reproducible patterns of proteins, primarily ribosomal proteins.[9] In the MALDI Biotyper, when a bacterial colony is placed on a MALDI target and the MALDI matrix solution is deposited on top of it, the high organic concentration of the matrix solution lyses the cell membranes, and soluble proteins are extracted from the bacteria into the matrix solution. It is these soluble proteins that are detected and produce the protein mass fingerprint after allowing the matrix solution to dry and carrying out the MALDI-TOF experiment as described above. Since no fractionation or other techniques are employed, the spectrum is dominated by the smaller, soluble and abundant proteins which are mostly ribosomal proteins in agreement with the work by Ryzhov and Feneslau.[9] Figure 2 shows a typical protein mass fingerprint of *E. coli* produced by the MALDI Biotyper, where many of the observed proteins have been identified by a combination of accurate mass measurements on higher performance MALDI-TOF instruments, and are mostly ribosomal proteins. Genetic techniques have been used to provide further evidence for these identifications, mutations in codons for the ribosomal proteins result in corresponding mass shifts in the observed peaks.[10] This observation is key to the specificity and reproducibility of the MALDI Biotyper. The ribosome is a complex of a large number of proteins and RNA, where the ratio of the proteins that make up the ribosome is constant. Ribosomes are key components of the cell and are produced by all living cells regardless of health state. Thus the protein mass fingerprints obtained from these ribosomal and other abundant proteins are easy to obtain and

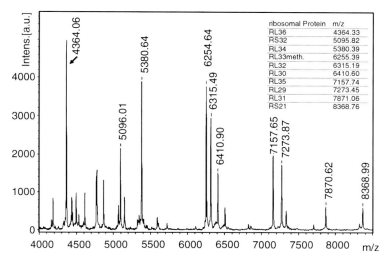

Fig. 2 A typical spectrum of an *E. coli* from the MALDI Biotyper. The labeled masses were identified as shown in the inset table, using accurate mass from a higher performance MALDI-TOF instrument and other evidence

highly reproducible regardless of culture conditions or state of the bacteria. One exception to this rule is spore forming bacteria, the spore state of these bacteria produce very different patterns from the vegetative state, but those patterns are also reproducible and can be used to identify the bacteria, as will be discussed separately. As an illustration of the independence of the protein mass fingerprint pattern from growth conditions, the protein mass fingerprint from a *Pseudomonas oleovorans* grown on a variety of common growth media, is shown in Fig. 3. From species to species there is sufficient variation in the masses of these common proteins that the protein mass fingerprints allow identification of bacterial genus and species with high specificity. Based on the fingerprints, identification on species level is possible. For this purpose, results obtained by usage of the MALDI Biotyper were evaluated in comparison to 16 S rRNA gene sequencing.[11]

1.2.1   The MALDI Biotyper Workflow

In this section the workflow for the identification of bacteria and bacterial spores using the MALDI Biotyper will be covered. The basic MALDI Biotyper workflow as used in the clinical routine is illustrated in Fig. 4.

Generally, the starting point for the analysis is a single colony from an agar plate, a few microliters of a liquid culture or a small amount of "white powder". In the simplest case, non-pathogenic biological material from an agar plate can be transferred directly onto the MALDI target plate and overlaid with MALDI matrix.

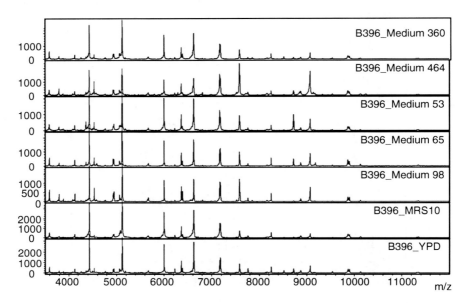

Fig. 3 *Pseudomonas oleovorans* grown on a variety of growth media (the codes for the media are labeled on the spectra)

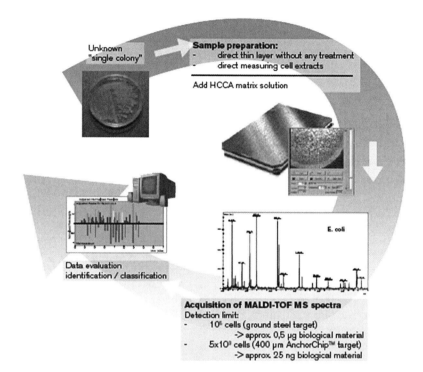

Fig. 4 The MALDI Biotyper workflow for microbial identification

Up to 96 samples can be prepared on a single target. Additionally, a commercially available calibration standard is placed on the target. However, in case of an unknown sample or white powder the biological material should be inactivated before analysis. Several extraction/inactivation protocols using ethanol/formic acid and/or trifluoroacetic acid (TFA) can be applied prior to target preparation. Especially, spores of *Bacillus* sp. have to be inactivated and disrupted by 80% TFA. The use of single cultured colony generally insures that only a single species is used for analysis, but mixed cultures can be analysed if the relative abundance ratio between the species in the mixture is not less than 1:10. If the ratio is too low, the less abundant microorganism may not be identified in the analysis of a mixture, but it will not lead to a wrong classification of the more abundant species.

For analysis by MALDI-TOF mass spectrometry the prepared sample is deposited onto a MALDI target plate. After drying, the same sample position on the target is covered with matrix solution (alpha-cyano-4-hydroxycinnamic acid). The target is then placed in the source of a MALDI-TOF mass spectrometer, and the instrument then automatically measures the protein mass fingerprint for each sample spot. Mass spectra are acquired between 2,000 and 20,000 Da, automatically controlled via dedicated easy-to-use software interface.

Microorganisms are then identified by comparison of their individual peak lists (MS fingerprint spectrum) via pattern matching with reference library entries.

Library spectra were generated by measurement of known bacterial species and strains. Pattern matching is accomplished through the calculation of a matching score. This score is calculated using a dedicated, proprietary algorithm on the basis of the number of matched peaks and the correlation of the overall intensity profile of the spectra. Matching peaks with each library spectrum are determined by a patented alignment procedure which can correct for mass measurement differences between the experimental and library peak lists. The scores are calculated on a log scale, with 3.0 being a perfect match, and 0.0 representing no correlation between the experimental result and the library spectrum. Identification results with high reliability are based on significant matching scores which are clearly separated from worse matches. Correct matches with highly probable species identification (score greater than 2.0) are displayed in green, whereas matches with scores so low that they are not matches are shown in red (scores < 1.7). Scores in between 1.7 and 2.0 are considered reliable enough to indicate the genus of the microorganism. The influence of peak intensities is reduced in the applied algorithm and identification is mainly based on accurate mass determination of the peaks. Therefore, the effect of instrument parameter settings is significantly reduced. This approach makes the identification exceptionally robust and accurate and enables inter-laboratory comparability of results and the creation of standard databases.

A key element for the reliability of the MALDI Biotyper is the database itself. The database has been developed by a group of microbiologists at Bruker Daltonics, in collaboration with microbiologists, clinicians and scientists at a number of hospitals, laboratories and institutions (e.g. German collection of microorganisms and cell cultures). There is a high proportion of reference and type strains in the library. Most reference spectra in the library were produced at Bruker Daltonics, either direct from fresh cultures, or from extracts sent to us by our collaborators. The current standard database includes 3,400 entries representing over 1,800 species. A separate security-related database contains entries for BWA species (e.g. *Bacillus anthracis*, *Yersinia pestis*, *Francisella tularensis*. *Brucella melitensis*...). The security-related database contains 110 database entries from 12 different species.

2 Studies Using the MALDI Biotyper: Reliability, Speed, Sensitivity, Specificity, and Cost

2.1 Studies of Inter-Laboratory Reproducibility

We made the claim in Section 1 that the protein mass fingerprints produced by the MALDI Biotyper are robust, insensitive to culture conditions and are easily reproduced between laboratories and even by non-expert users such as microbiology technicians in clinics. As a first example from our own experience, Fig. 5 shows two spectra of *Arthrobacter sufureus*. The top spectrum was obtained on the species grown from an strain in the library collection at the Helmholtz Center Institute

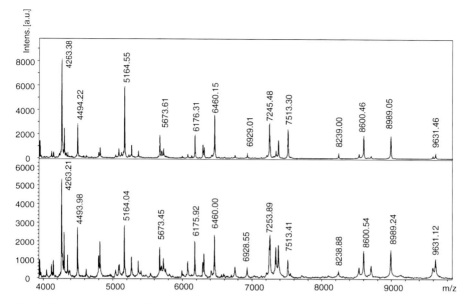

Fig. 5 Both spectra are MALDI Biotyper results for the species *Arthrobacter sulfureus*, strain DSM 20157. The top spectrum was acquired in the Bruker Daltonics applications laboratory, by Bruker scientists in 2005 on an autoflex MALDI-TOF instrument, with the strain sourced from the Helmholtz Center Institute for Environmental Resarch. The bottom spectrum was acquired at the laboratories of the DSMZ, the German collection of microorganisms and cell cultures in 2006 sourced from their own collection, and run by their technologists. Despite this degree of variability in the measurements, the observed protein mass finger prints are nearly identical

for Environmental Research, on an instrument known as an autoflex™, in the Bruker Daltonics R&D laboratory. The lower spectrum was obtained on the same species grown from a strain in the library of the German collection of microorganism and cell cultures (known as DSMZ from the German name), on a microflex™ instrument run by the DSMZ technologists in their own laboratories. While slight differences in the two spectra can be seen, they are trivial compared to the variation typically seen between species, and both spectra result in high scores indicating a species match. The microflex is the instrument that is most commonly packaged with the MALDI Biotyper software and databases for use in routine microbiology clinical labs as it has the advantage of being an entirely self-contained benchtop unit. This shows that despite tremendous variability in conditions, the results are consistent, and that the method can easily be transferred out of the R&D laboratory onto a routine benchtop instrument for use by clinical microbiology lab technicians.

A recently published and more detailed study of inter-laboratory reproducibility involved 60 samples sent with blind codes to eight laboratories worldwide.[12] The 60 samples consisted of 30 pure cultures of non-fermenting bacteria and 30 preprocessed ethanol cellular extracts of the same strains. The strains were a mix obtained from culture collections or isolated from the clinical routine at the Institute of

Hygiene in Muenster, Germany, and all strains were unambiguously identified by 16S rRNA gene sequencing prior to the study. As a further test of the method, the strains selected were all from species classified as non-fermenters, these species are difficult to identify using traditional biochemical methods, because they do not present great biochemical diversity. In addition to the samples, the participating laboratories also received a standard sample preparation guide, and all laboratories used the MALDI Biotyper 2.0 software release and the standard Bruker Daltonics micro-flex™ MALDI-TOF instrument. Of these 480 samples, 474 were correctly identified in all eight laboratories. Of the six that were misidentified, four were misidentified due to samples being incorrectly handled (samples were interchanged), one sample was contaminated in the lab, and one sample gave insufficient signal intensity. This represents an impressive 98.75% inter-laboratory reproducibility. The excellent reproducibility achieved on both the freshly cultivated samples and the extracts again shows that the MALDI Biotyper can be used in a central lab setting, with samples sent in from field labs after cultivation and extraction, and shows that the method is suitable by use by technicians in routine clinical microbiology labs.

2.2 Clinical Studies Demonstrating Speed, Specificity, Reliability and Cost Advantages

The MALDI Biotyper has already been installed at more than 100 clinical micro-biology sites in Europe as well as several in the US. There is a large and growing list of publications demonstrating the advantages of the technique with respect to: Advantages in performance on species that are slow growing and difficult to iden-tify using standard biochemical test methods; Speed, accuracy, and cost advantages when used for the identification of more routine isolates that do work well by bio-chemical testing; Advantages in cost compared to sequencing when used for spe-cies where sequencing has typically been used prior to the introduction of the MALDI Biotyper. Since many of these advantages are also important in a device to be used for BWA identification in detect to treat situations, many of these studies are discussed in more detail below.

The problem of non-fermenting bacteria was mentioned briefly above, and was the subject of a detailed study which showed that the MALDI Biotyper performs well for the identification of species from this grouping.[12] Non-fermenters are a group of spe-cies which cannot ferment sugars. Due to their limited biochemical reactivity and variable morphology, non-fermenters are often misidentified by traditional methods, including biochemical tests. In the study referenced here, the reference library was enhanced with the addition of reference spectra from 248 culture collection strains of non-fermenters. With this library, correct species identification of a clinical sampling of 80 non-fermenting bacteria were identified by the MALDI Biotyper, and correct identification to the species level was obtained on 85.2% of the samples and to the genus level on 95.2% of the samples. While species level identification is preferred, genus level identification is often clinically useful as well. Correct identifications were determined by comparison to results obtained by sequencing. These results were significantly better than results obtained by identification techniques relying on biochemical methods for similar sets of samples.

Another genus which has proven challenging to commonly used biochemical methods for identification, and for which genomic methods such as 16S RNA gene sequencing is often used, is *Clostridium*. In a study published in 2008, Grosse-Herrenthey and co-workers showed that 64 *Clostridal* strains representing 31 different species produced unique MALDI-TOF protein fingerprints, which were added to the MALDI Biotyper library. Using this library 25 clinical isolates were correctly identified in minutes after cultivation using the MALDI Biotyper.[13] The genus *Clostridia* is also anaerobic and contains spore forming species as well. Other biochemical methods for identifying these bacteria require closely controlled and precise culture conditions. For the work presented in this study, the bacteria were grown on standard agar media, under anaerobic conditions, and at a variety of culture times, and the MALDI-TOF protein fingerprint patterns were still highly reproducible and identifications could be obtained under all conditions studied. While the number of isolates identified in this study was limited, it shows the potential advantages of the MALDI Biotyper for working with bacteria that present challenges for identification by other methods.

Several more recent studies have looked at larger clinical sample sets for a direct comparison of the MALDI Biotyper against various biochemical identification systems and sequencing. One study compared identification of bacteria and yeast with the MALDI Biotyper to biochemical methods including the widely used API and Biomerieux Vitek II system.[14] In this study, 16S RNA gene sequencing was used as the "gold standard" to determine the correct identification in the case of discrepancies between the identification given by the MALDI Biotyper and the biochemical methods. For 980 clinical isolates of bacteria and yeast, the MALDI Biotyper gave 92.2% correct species identification, compared to 83.1% for the biochemical methods. Importantly, the MALDI Biotyper false positive rate is near zero, of the 980 clinical isolates, incorrect genus identifications were generated in only 0.1% of the cases compared with the MALDI Biotyper, compared to 1.6% false genus identifications from the biochemical methods. In fact, mis-identifications or lack of identification from the MALDI Biotyper was observed to be mainly due to missing or insufficient numbers of entries in the database for those species, so the already excellent performance of the MALDI Biotyper for identification is expected to improve as the reference spectra library is enlarged and improved. In another recently published study involving a large number of samples run in a routine clinical setting, the performance of MALDI Biotyper identification against the Vitek II and API biochemical methods was evaluated.[15] Again in this study, 16S RNA gene sequence was used to determine the correct identification when there was a discrepancy between the results obtained from the biochemical methods and the MALDI Biotyper. This study found that the MALDI Biotyper gave confident identifications for 94.4% of 720 clinical isolates, of which only 6 (0.9%) turned out to be incorrect, while 24 incorrect identifications (3.3%) were given by the biochemical methods. This study also contains a detailed analysis of time to result and costs showing advantages in both of these categories for the MALDI Biotyper and concludes that MALDI TOF based methods will be widely used in clinical microbiology in the future.

2.3 New Developments, Summary and Future Perspective on Clinical Applications

A new area of application for the MALDI Biotyper is identification direct from urine and liquid media. Two studies have been published to date about identification direct from positive blood culture bottles. Blood borne infections which can lead to sepsis are extremely dangerous, and treatment with antibiotics is often started as soon as a positive blood culture bottle is observed, as identification by traditional biochemical methods takes another 12–48 h. One recent study showed that correct identifications could be obtained in minutes directly from 76% of 584 positive blood culture bottles.[16] In a more recent study showed that correct identifications could be obtained from 80.2% of 212 positive blood culture bottles analyzed.[17] Both studies concluded that the more rapid results would have a positive impact on treatment regimens, as antibiotic treatment appropriate for the infectious organism could be chosen much sooner.

Also both studies pointed to improvements in the MALDI Biotyper library and software that will lead to increased yields of correct results direct from blood culture bottles, and work is also underway to reduce the amount of manual labor to isolate the bacterial pellets from aliquots taken from such bottles. Continuing improvements to the library will include the addition of new species, as well as more strains for species already in the library to increase coverage. The MALDI Biotyper library is extremely well curated, but with more than 3,400 entries it is inevitable that an a few errors will be found. We continuously monitor feedback from clinical users of the MALDI Biotyper to correct errors, as well as add annotations to warn users of the occasional species that cannot be differentiated by MALDI-TOF (e.g. *Shigella* species cannot be differentiated from closely related *E. coli* by the MALDI Biotyper workflow).

In addition to the library, developments are underway to even further simplify the use of the technique, and to make the instrument even more robust in a high throughput clinical setting. We have collaborators working to integrate the system in a fully automated microbiology lab for institutions that require extremely high throughput. The first studies on identification directly from blood culture employed somewhat laborious multiple centrifugation steps to isolate the bacteria from the blood culture. A simpler one step process is under development to simplify this procedure. These continuing improvements should only increase the already great clinical utility of the MALDI Biotyper.

2.4 Use for the Identification of Spore Forming Bacteria and Spores in White Powder

A special feature of a few bacterial genera is their ability to form spores under certain conditions. Sporulation is initiated mostly under stressful environmental conditions to ensure the survival of bacteria. Bacterial spores are characterized by a high resistance to physical and chemical agents. Regarding their macromolecular constituents spores consist of a germ cell, where the DNA is embedded in small, acid-soluble proteins (SASP) to protect the spore from UV radiation and heat, and a surrounding cortex and coat, which excludes large toxic molecules.

The most important spore-forming bacteria belong to the genera *Clostridium* and *Bacillus*.[18] Besides the importance as food-poisoning agents of e.g. *Clostridium botulinum*, and *Bacillus cereus* the identification of *B. anthracis* spores – as causative agent of anthrax – is of particular interest. Discrimination of *B. anthracis* from closely related *B. cereus*, *B. thuringiensis* and *B. mycoides* (*B. cereus* group) is a challenging task because of their very close phylogenetic relationship, reflected also by DNA homology.[19] SASPs were found to be biomarkers for spore differentiation/ identification by Mass Spectrometry developed a MALDI-TOF mass spectrometry[20,21] compatible inactivation method for highly pathogenic microbial cells and spores. SASPs could be inactivated and extracted from spores after treatment with 80% TFA followed by a filtration through a TFA resistant filter (pore size: 0.22 µm). Mass spectra of *B. anthracis* spores and *B. cereus* spores exhibit discriminating biomarkers due to the existence of these SASPs. Spore spectra were run against the regular Bruker MALDI Biotyper database as well, but no matches with common microorganisms exhibiting a reliable score could be observed. Calculation of a dendrogram comprising MALDI-TOF MS spectra of *Bacillus* sp. (vegetative cells and spores) belonging to the *Bacillus cereus* group (Fig. 6) confirmed that the pattern of spore spectra are completely different compared to spectra of corresponding vegetative cells. Moreover a clear separation of *B. anthracis* strains from the rest of the *B. cereus* group was apparent.

Furthermore, some "white powders" which have been reported to be potentially used as hoaxes/fakes in a terrorist attack were analyzed in the same way. In this context powders available in typical households or pharmacies (e.g. flour, coffee whitener, baby powder) were analyzed. However, only a few powders gave profile spectra (e.g. flour, sauce thickener, whole-wheat flour and white pepper). But, the analysis of these spectra using the MALDI Biotyper revealed no matches with bacteria or spore spectra of the reference database.

3   Conclusion

As noted in some recent publications, the MALDI Biotyper represents nothing short of a revolution in clinical microbiology.[6] As the technique is adopted by more clinical and public health labs the library and software will continue to be refined and the clinical utility will only increase. Many of the advantages of the MALDI Biotyper over commonly used genotypic and biochemical identification methods outlined here are not only useful for clinical applications, but also for BWA analysis in detect to treat situations. The section on spores demonstrates that the MALDI Biotyper also has applications in the detection of BWA in the detect to protect mode. We expect MALDI TOF identification of microorganisms to be even more

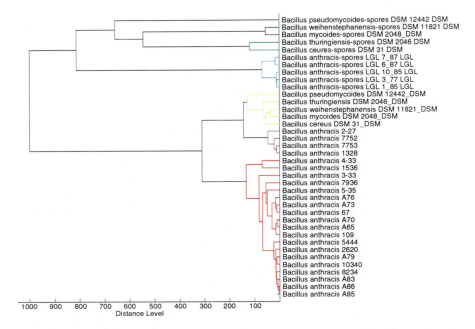

Fig. 6 Score oriented dendrogram calculated for MALDITOF MS spectra of *Bacillus* sp. (vegetative cells and spores) belonging or similar to the *Bacillus cereus* group

widely adopted by clinical and public health laboratories. Note: In some countries in Europe the MALDI Biotyper is available as an IVD-CE certified system for clinical use, while in the US and other regions of the world, it is currently for Research Use Only (RUO).

References

1. Demirev PA, Feneslau C (2008) Mass spectrometry for rapid characterization of microorganisms. Ann Rev Anal Chem 1:71–93 and Fenselau C, Demirev P (2001) Characterization of intact microorganisms by MALDI mass spectrometry. Mass Spectrom Rev 20:157–171
2. Demirev PA, Fenselau C (2008) Mass spectrometry in biodefense. J Mass Spectrom 43:1441–1457
3. Hathout Y, Demirev PA, Ho Y-P, Bundy JL, Ryzhov V, Sapp L, Stutler J, Jackman J, Fenselau C (1999) Identification of *Bacillus* spores by matrix-assisted laser desorption ionization-mass spectrometry. Appl Environ Microbiol 65(10):4313–4319
4. Anhalt JP, Fenselau C (1975) Identification of bacteria using mass spectrometry. Anal Chem 47:219–225
5. Eigner U, Holfelder M, Oberdorfer K, Betz-Wild U, Bertsch D, Fahr AM (2009) Performance of a matrix-assisted laser desorption ionization-time-of-flight mass spectrometry system for the identification of bacterial isolates in the clinical routine laboratory. Clin Lab 55(7–8):289–296

6. Seng P, Drancourt M, Gouriet F, La Scola B, Fournier PE, Rolain JM, Raoult D (2009) Ongoing revolution in bacteriology: routine identification of bacteria by matrix-assisted laser desorption ionization time-of-flight mass spectrometry. Clin Infect Dis 49(4):543–551
7. Lay JO Jr (2001) MALDI-TOF mass spectrometry of bacteria. Mass Spectrom Rev 20(4):172–194
8. Krishnamurthy T, Ross PL, Rajamani U (1996) Detection of pathogenic and non-pathogenic bacteria by matrix-assisted laser desorption/ionization time-of-flight mass spectrometry. Rapid Commun Mass Spectrom 10:883–888
9. Ryzhov V, Feneslau C (2001) Characterization of the protein subset desorbed by MALDI from whole bacterial cells. Anal Chem 73:746–750
10. Ilina EN, Borovskaya AD, Malakhova MM, Vereshchagin VA, Kubanova AA, Kruglov AN, Svistunova TS, Gazarian AO, Maier T, Kostrzewa M, Govorun VM (2009) Direct bacterial profiling by matrix-assisted laser desorption-ionization time-of-flight mass spectrometry for identification of pathogenic Neisseria. J Mol Diagn 11(1):75–86
11. Mellmann A, Cloud J, Maier T, Keckevoet U, Ramminger I, Iwen P, Dunn J, Hall G, Wilson D, Lasala P, Kostrzewa M, Harmsen D (2008) Evaluation of matrix-assisted laser desorption/ionization time-of-flight mass spectrometry (MALDI-TOF MS) in comparison to 16S rRNA gene sequencing for species identification of nonfermenting bacteria. J Clin Microbiol 46:1946–1954
12. Mellmann A, Bimet F, Bizet C, Borovskaya AD, Drake RR, Eigner U, Fahr AM, He Y, Ilina EN, Kostrzewa M, Maier T, Mancinelli L, Moussaoui W, Prévost G, Putignani L, Seachord CL, Tang YW, Harmsen D (2009) High interlaboratory reproducibility of matrix-assisted laser desorption ionization-time of flight mass spectrometry-based species identification of nonfermenting bacteria. J Clin Microbiol 47:3732–3734
13. Grosse-Herrenthey A, Maier T, Gessler F, Schaumann R, Böhnel H, Kostrzewa M, Krüger M (2008) Challenging the problem of Clostridial identification with matrix-assisted laser desorption/ionisation – time of flight mass spectrometry (MALDI – TOF MS). Anaerobe 14(4):242–249
14. van Veen SQ, Claas ECJ, Kuijper EJ (2010) High-throughput identification of bacteria and yeast by matrix-assisted laser desorption ionization mass spectrometry (MALDI-TOF MS) in routine medical microbiology laboratory. J Clin Microbiol 48(3):900–907
15. Cherkaoui A, Hibbs J, Emonet S, Tangomo M, Girard M, Francois P, Schrenzel J (2010) Comparison of two matrix-assisted laser desorption ionization-time of flight mass spectrometry methods with conventional phenotypic identification for routine bacterial speciation. J Clin Microbiol 48(4):1169–1175
16. La Scola B, Raoult D (2009) Direct identification of bacteria in positive blood culture bottles by matrix-assisted laser desorption ionisation time-of-flight mass spectrometry. PLoS One 25:e8041
17. Stevenson LG, Drake SK, Murray PR (2010) Rapid identification of bacteria in positive blood culture broths by MALDI-TOF mass spectrometry. J Clin Microbiol 48(2):444–447
18. Russel AD (1990) Bacterial spores and chemical sporicidal agents. Clin Microbiol Rev 3(2):99–119
19. Henderson I, Duggleby CJ, Turnbull PC (1994) Differentiation of Bacillus anthracis from other Bacillus cereus group bacteria with the PCR. Int J Syst Bacteriol 44:99–105
20. Hathout Y, Setlow B, Cabrera-Martinez RM, Fenselau C, Setlow P (2003) Small, acid-soluble proteins as biomarkers in mass spectrometry analysis of Bacillus spores. Appl Environ Microbiol 69(2):1100–1107
21. Lasch P, Nattermann H, Erhard M, Stämmler M, Grunow R, Bannert N, Appel B, Naumann D (2008) MALDI-TOF mass spectrometry compatible inactivation method for highly pathogenic microbial cells and spores. Anal Chem 80:2026–2034

Mass Spectrometric Detection of Botulinum Neurotoxin by Measuring its Activity in Serum and Milk

Suzanne R. Kalb, James L. Pirkle, and John R. Barr

Abstract Botulinum neurotoxins (BoNTs) are bacterial protein toxins which are considered likely agents for bioterrorism due to their extreme toxicity and high availability. A new mass spectrometry based assay called Endopep MS detects and defines the toxin serotype in clinical and food matrices via toxin activity upon a peptide substrate which mimics the toxin's natural target. Furthermore, the subtype of the toxin is differentiated by employing mass spectrometry based proteomic techniques on the same sample. The Endopep-MS assay selectively detects active BoNT and defines the serotype faster and with sensitivity greater than the mouse bioassay. One 96-well plate can be analyzed in under 7 h. On higher level or "hot" samples, the subtype can then be differentiated in less than 2 h with no need for DNA.

Keywords Botulinum neurotoxin • Mass spectrometry • MALDI • Antibody extraction

1 Introduction

Botulinum neurotoxins (BoNTs) are produced by some species of the genus *Clostridium*, particularly *Clostridium botulinum, C. butyricum, C. baratii, and C. argentinese*. BoNTs cause the disease known as botulism, which can be lethal if untreated. Rapid determination of exposure to BoNT is an important public health goal. By weight, BoNT is the most lethal substance known with an estimated oral LD_{50} in humans of approximately 70 μg for the average weight human.[1] This extreme toxicity has led to its current CDC designation as a category A agent, making it one of the most likely agents for bioterrorism.[1,2]

S.R. Kalb, J.L. Pirkle, and J.R. Barr (✉)
Centers for Disease Control and Prevention, 1600 Clifton Rd, 30333 Atlanta, GA, USA
e-mail: ssk7@cdc.gov

J. Banoub (ed.), *Detection of Biological Agents for the Prevention of Bioterrorism*, NATO Science for Peace and Security Series A: Chemistry and Biology, DOI 10.1007/978-90-481-9815-3_8, © Springer Science+Business Media B.V. 2011

Botulinum neurotoxins are currently classified into seven serotypes, labeled A-G, and serotypes A, B, E, and F are known to cause disease in humans. BoNTs are highly specific proteases which target neuronal proteins. BoNT/A, /C, and /E cleave SNAP-25 (synaptosomal-associated protein)[3-8] whereas BoNT/B, /D, /F, and /G cleave synaptobrevin−2 (also known as VAMP−2).[9-13] Only BoNT/C is known to cleave more than one protein as it also cleaves syntaxin.[3,14,15] Cleavage of any of these proteins which form the SNARE (**S**oluble **N**SF **A**ttachment Protein **R**eceptors) complex depicted in Fig. 1 results in an inability to form this complex and therefore stops nerve impulses and causes paralysis.

Our laboratory has developed an assay for BoNT termed the Endopep-MS method.[16-22] This method involves incubation of BoNT with a peptide substrate that mimics the toxin's natural target. Each BoNT cleaves the peptide substrate in a specific, toxin-dependent location. The reaction mixture then is introduced into a mass spectrometer, which detects any peptides within the mixture and accurately reports the mass of each. Detection of the peptide cleavage products corresponding to their specific toxin-dependent location indicates the presence of a particular BoNT serotype. If the peptide substrate either remains intact or is cleaved in a location other than the toxin-specific site, then that BoNT serotype is not present. We have demonstrated that this method can detect BoNT at levels comparable with or lower than levels detected with mouse bioassays, which historically have been the major means for BoNT detection.[23] Endopep-MS is made both more selective and more specific through the addition of an antibody-extraction step prior to the enzymatic reaction. Magnetic protein G beads coated with antibodies to a specific serotype of BoNT are incubated with the sample matrix which may or may not contain the toxin. If the toxin of that particular

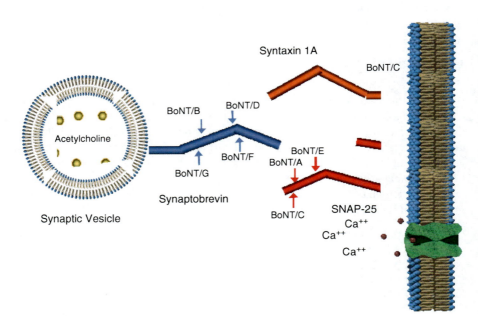

Fig. 1 Depiction of the SNARE complex and sites cleaved by BoNTs

serotype is present, it binds to the antibodies on the beads. Other proteins in the sample matrix are washed away, and the beads containing toxin are then incubated with the peptide substrate. Therefore, in its entirety, the Endopep-MS method has three layers of specificity; the first is antibody affinity, the second is enzymatic activity of the toxin upon the peptide substrate, and the third is an accurate mass measurement of the cleavage products of the peptide substrate. The combination of these techniques has led to the development of a highly specific and selective assay for a dangerous protein toxin considered to be a likely agent for bioterrorism. Additional information about the toxin can be obtained through tryptic digestion of the toxin and mass spectral analysis of the resultant tryptic peptides.

2 Detection of BoNT Activity in Clinical and Food Matrices

The detection of BoNT in any matrix is currently accomplished through antibody-extraction of the toxin from the matrix and subsequent incubation of the toxin with its peptide substrate.

2.1 Detection of BONT/A

The peptide substrate which mimics BoNT/A's natural target is Biotin-KGSNRTRIDQGNQRATRXLGG K-Biotin, where X is norleucine and the average molecular weight (MW) is 2,879.4 Da. Cleavage by BoNT/A produces two peptides with the sequence Biotin-KGSNRTRIDQGNQ (average MW = 1,699.9 Da) and RATR XLGGK-Biotin (average MW = 1,197.5 Da). These peaks are present in Fig. 2a, which is the mass spectrum of 100 mLD$_{50}$ of BoNT/A spiked into 0.5 mL of serum, extracted with antibody-coated beads, and analyzed via MALDI-TOF mass spectrometry. Peaks at m/z 1,197.7 and 1,700.0 (marked with asterisks) are indicative of the presence of BoNT A in this sample. Similar results are obtained from an antibody-extraction of 100 mouse LD$_{50}$ (mLD$_{50}$)of BoNT/A spiked into 0.5 mL of milk followed by incubation of the beads with the peptide substrate. Figure 2b is slightly different from Fig. 2a as it is the spectrum of 100 mLD$_{50}$ of BoNT/A spiked into 0.1 mL of stool extract. Stool extract is a more complicated matrix consisting of a greater number of proteases, some of which bind nonspecifically and cleave the peptide substrate, resulting in a more complex mass spectrum. However, the presence of peaks at m/z 1,197.7 and 1,700.0 indicates that BoNT/A can be detected in this sample.

2.2 Detection of BONT/B

The peptide LSELDDRADALQAGASQ FESSAAKLKRKYWWKNLK with an average molecular weight of 4,025.5 Da mimics the natural target of BoNT/B. Cleavage by BoNT/B yields two peptides with the sequence LSELDDRADALQAGASQ

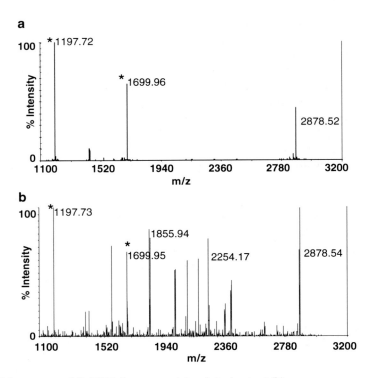

Fig. 2 Mass spectra of BoNT/A from serum (**a**) and stool extract (**b**)

(average MW = 1,759.9 Da) and FESSAAKLKRKYWWKNLK (average MW = 2,283.7 Da). Figure 3a is the mass spectrum of the reaction of 100 mLD$_{50}$ of BoNT/B spiked into 0.5 mL of serum, extracted with antibody-coated beads, and analyzed via MALDI-TOF mass spectrometry. Peaks at m/z 1,759.9 and 2,283.4 (marked with asterisks) indicate the presence of BoNT B in this sample.

Similar results are obtained from an antibody-extraction of 100 mLD$_{50}$ of BoNT/B spiked into 0.5 mL of milk followed by incubation of the beads with the peptide substrate. However, like with BoNT/A, Fig. 2b is slightly different from Fig. 2a as it is the spectrum of 100 mLD$_{50}$ of BoNT/B spiked into 0.1 mL of stool extract. Nonetheless, the presence of BoNT/B can still be detected in this sample due to the peaks at m/z 1,759.9 and 2,283.4.

2.3   Detection of BONT/E

BoNT/E readily cleaves a peptide of sequence IIGNLRHMALDMGNEIDTQNRQ IDRIMEKADSNKTT with an average molecular weight of 4,042.6 Da, resulting in cleavage products IIGNLRHMALDMGNEIDTQNRQIDR (average MW = 2,924.3) and IMEKADSNKT (average MW = 1,136.3). These peaks are present at

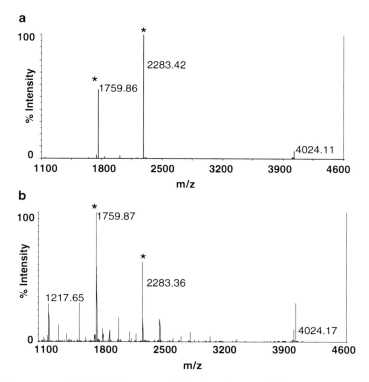

Fig. 3 Mass spectra of BoNT/B from serum (**a**) and stool extract (**b**)

m/z 1,136.6 and 2,922.7 in the mass spectrum in Fig. 4a, which result from the reaction of 10 mLD$_{50}$ of BoNT/E spiked into 0.5 mL of serum, extracted with antibody-coated beads, and analyzed via MALDI-TOF mass spectrometry.

Peaks at m/z 1,136.6 and 2,922.7 (marked with asterisks) indicate that BoNT E is present in this sample. Similar results are obtained from an antibody-extraction of 10 mLD$_{50}$ of BoNT/E spiked into 0.5 mL of milk followed by incubation of the beads with the peptide substrate. Similar to BoNT/A and /B, Fig. 4b demonstrates that extraction from stool extract results in a slightly different mass spectrum. Figure 4b is the spectrum of 100 mLD$_{50}$ of BoNT/E spiked into 0.1 mL of stool extract. Detection of BoNT/E in stool samples is demonstrated through the presence of peaks at m/z 1,136.6 and 2,922.7.

2.4 Detection of BONT/F

The peptide substrate which mimics BoNT/F's natural target is LQQTQAQVDEVVD IMRVNVDKVLER DQKLSELDDRADAL with an average molecular weight of 4,497.0 Da. Cleavage of this peptide by BoNT/F results in the cleavage products

S.R. Kalb et al.

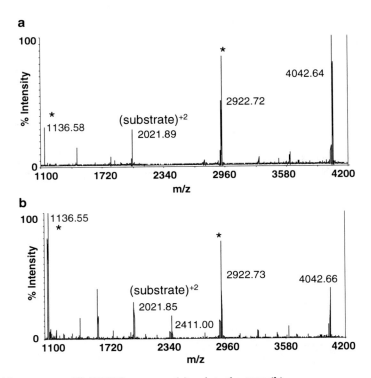

Fig. 4 Mass spectra of BoNT/E from serum (**a**) and stool extract (**b**)

LQQTQAQVDEVVDIMRVNVDKVLERDQ (average MW = 3,169.6 Da) and
KLSELDDRADAL (average MW = 1,345.5 Da). These peaks are present and
marked with asterisks in Fig. 5a, which is the mass spectrum of the reaction of 10
mLD_{50} of BoNT/F spiked into 0.5 mL of serum, extracted with antibody-coated
beads, and analyzed via MALDI-TOF mass spectrometry.

Similar results are obtained from an antibody-extraction of 10 mLD_{50} of
BoNT/F spiked into 0.5 mL of milk followed by incubation of the beads with the
peptide substrate. Figure 5b, the spectrum of 100 mLD_{50} of BoNT/F spiked into
0.1 mL of stool extract, demonstrates that extraction from stool extract results in
a slightly different mass spectrum than extraction from serum or milk. Peaks at *m/z*
3,168.9 and 1,345.9 indicate the presence of BoNT F in a sample.

3 Detection of Multiple Subtypes of BoNT

BoNT serotypes A, B, E, and F can be further differentiated into different subtypes.
It is important to be able to detect multiple subtypes within a serotype as any of the
subtypes could be used in a terrorist event and the specific subtype may be used to
identify the perpetrator and link one event to others.

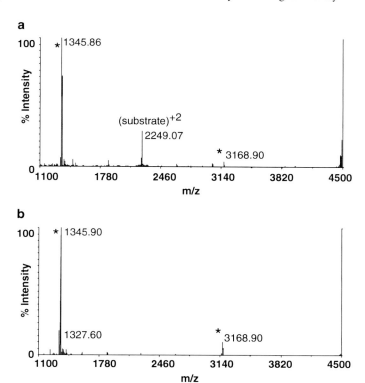

Fig. 5 Mass spectra of BoNT/F from serum (**a**) and stool extract (**b**)

3.1 Detection of Multiple Subtypes of BONT/A

All BoNT/A can currently be classified as BoNT/A1, /A2, /A3, /A4, or /A5. The current definition of a new subtype of BoNT has been established as a variance of 2.5% or more in the amino acid sequence.24 Our laboratory was able to obtain toxins /A1, /A2, /A3, and /A4 and we have tested these toxins to ensure that detection by mass spectrometry is possible. Figure 6 shows the mass spectra acquired from the extraction of BoNT/A1 (6A), /A2 (6B), /A3 (6C), and /A4 (6D) from culture supernatant media in which the bacteria were cultured.

All spectra contain peaks at m/z 1,197.7 and 1,700.0, indicating the presence of BoNT/A in these samples. Furthermore, these spectra indicate that all of these subtypes of BoNT/A demonstrate the same activity against the peptide substrate.

3.2 Detection of Multiple Subtypes of BONT/B

BoNT/B are currently classified as BoNT/B1, /B2, /B3, /B4, or /B5. To date, we have tested BoNT/B1, /B2, /B3, and /B4 and all of these subtypes can be detected by mass

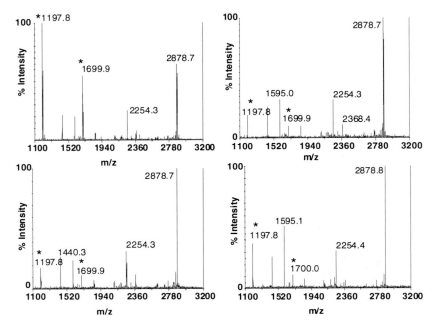

Fig. 6 Mass spectra of BoNT/A subtypes/A1 (**a**), /A2 (**b**), /A3 (**c**), and /A4 (**d**)21

spectrometry. Figure 7 shows the mass spectra acquired from the extraction of BoNT/B1 (7A), /B2 (7B), /B3 (7C), and /B4 (7D) from culture supernatant medium.

Peaks at m/z 1,759.9 and 2,283.4 indicate the presence of BoNT/B in all of the samples. Those peaks also demonstrate identical action of BoNT/B upon the peptide substrate regardless of the subtype.

3.3 Detection of Multiple Subtypes of BoNT/E

Currently, BoNT/E is divided into six subtypes; /E1 through /E6. To date, we have tested BoNT/E1, /E2, /E3, and /E4 and the Endopep-MS method is able to detect those four subtypes. The mass spectra from those reactions are in Fig. 8 and represent BoNT/E1 (8A), /E2 (8B), /E3 (8C), and /E4 (8D) extracted from culture supernatant medium.

Peaks at m/z 1,136.6 and 2,922.7 indicate the presence of BoNT/E in all of the samples. Those peaks also demonstrate that the action of BoNT/E upon the peptide substrate is the same regardless of the subtype.

3.4 Detection of Multiple Subtypes of BoNT/F

BoNT/F are currently classified as BoNT/prot F, /np F, /bv F, or /F baratii. Our laboratory has obtained three of the four toxin subtypes in sufficient quantities, and we

MS DETECTION OF BOTULINUM NEUROTOXIN

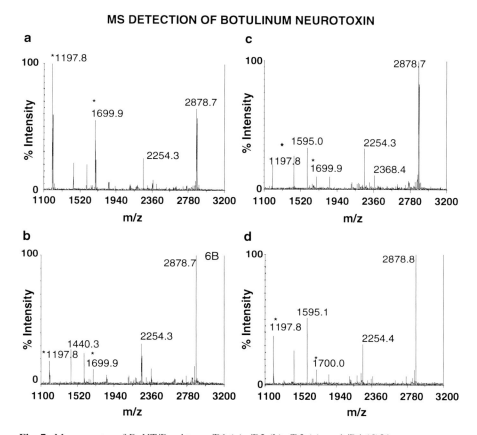

Fig. 7 Mass spectra of BoNT/B subtypes/B1 (**a**), /B2 (**b**), /B3 (**c**), and /B4 (**d**)21

have tested these toxins to ensure that detection by mass spectrometry is possible. Figure 9 shows the mass spectra acquired from the extraction of BoNT/prot F (9A), BoNT /np F (9B), and BoNT /bv F (9C) from culture supernatant medium in which the bacteria were cultured.

All figures contain peaks at m/z 1,345.9 and 3,168.9, indicating the presence of BoNT/F in these samples. Furthermore, these spectra indicate that all of these subtypes of BoNT/F demonstrate the same activity against the peptide substrate.

4   Differentiation of BoNT/A Subtypes

Four of the five currently known subtypes of BoNT/A, have been tested and shown to be detectable with the Endopep-MS assay. However, the assay yields the same data for each of the subtypes; i.e. the Endopep-MS assay cannot be used to differentiate the subtypes of BoNT/A. Differentiation of the subtype can be very useful information for forensic or epidemiologic purposes as it can help determine whether

MS DETECTION OF BOTULINUM NEUROTOXIN

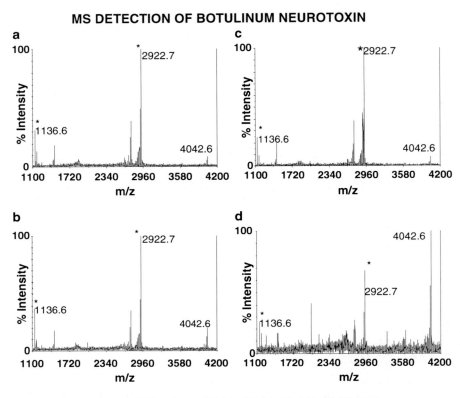

Fig. 8 Mass spectra of BoNT/E subtypes/E1 (**a**), /E2 (**b**), /E3 (**c**), and /E4 (**d**)21

multiple samples originate from a single source and might yield information about the source of the toxin. The mouse bioassay does not provide any information on the subtype of the toxin. Typically, subtype identification is determined through DNA analysis via polymerase chain reaction (PCR),[25–28] or more recently through real-time PCR (RT-PCR).[29] However, these methods can only be used if the bacterium that produces the toxin is present, and BoNT can potentially be present in a sample that does not contain the bacterium. In such a situation, subtype identification would not be possible using traditional, DNA-based methods. Therefore, we have devised a method to identify the subtype using the toxin protein itself rather than bacterial DNA. This method involves tryptic digestion of the toxin and mass spectrometric analysis of the tryptic fragments.

Figure 10 shows the MALDI-TOF mass spectra acquired following tryptic digestion of BoNT/A1 (10A) and /A2 (10B).

It is apparent from these spectra that there are many similarities between these two spectra. BoNT/A1 and /A2 are approximately 90% homologous,[30] so most of the peptides generated from a tryptic digest are homologous. However, there are some peaks which are unique to either BoNT/A1 or /A2, and these can

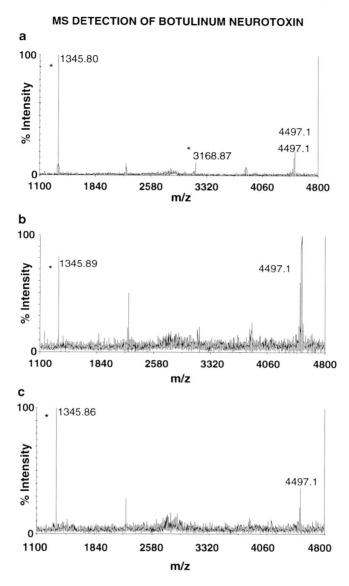

Fig. 9 Mass spectra of BoNT/A subtypes/prot F (**a**), /np F (**b**), and /bv F (**c**)21

be used to identify a subtype as BoNT/A1, /A2, or perhaps another, yet unknown, A subtype. Some of these peaks were subjected to MS/MS for sequence identification, and an example of that is in Fig. 11, which shows the LC-MS/MS spectra of peptides SFGHEVLNLTR from BoNT/A1 and SFGHDVLNLTR from BoNT/A2.

MS DETECTION OF BOTULINUM NEUROTOXIN

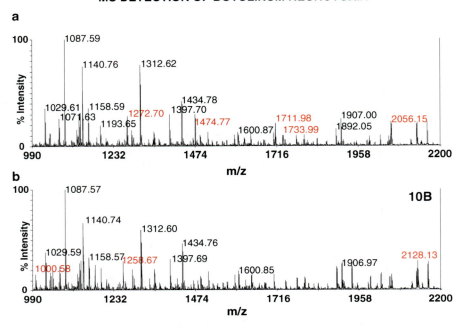

Fig. 10 Mass spectra of tryptic digests of BoNT /A1 (**a**) and /A2 (**b**). Peaks in *red* are unique to that particular subtype

It should be noted that the toxin subtype identification requires much higher levels of toxin than the toxin type identification; however, this level of toxin is comparable to some reports of BoNT/A present in food samples.[31]

It is especially important to note that this technique to differentiate the BoNT subtype is used as an addendum to the Endopep-MS method which identifies the serotype of toxin present in a sample. Once the serotype of toxin is identified by Endopep-MS, the toxin responsible for that activity can be tryptically digested and then analyzed for subtype identification. The subtype identification does not require a separate sample, provided that a sufficient level of toxin is present in the original sample.

5   Conclusions

We have demonstrated that BoNT/A, /B, /E, and /F can be detected via their activity by mass spectrometry. We have shown that this method can be used to detect these toxins in food or clinical samples. This method has three layers of specificity; the first is antibody affinity, the second is enzymatic activity of the toxin upon the peptide substrate, and the third is an accurate mass measurement of

MS DETECTION OF BOTULINUM NEUROTOXIN

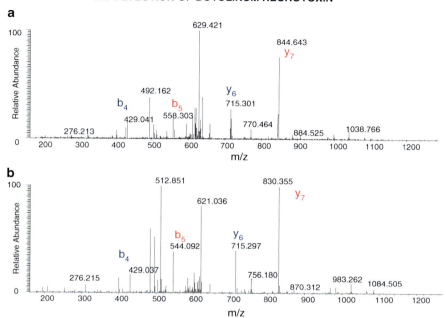

Fig. 11 MS/MS spectra of peptides SFGHEVLNLTR and SFGHDVLNLTR from BoNT/A1 (**a**) and /A2 (**b**) respectively. Fragment ions marked in *blue* indicate common fragment ions; fragment ions marked in *red* indicate fragment ions which can be used to distinguish BoNT/A1 from /A2

the cleavage products of the peptide substrate. Additionally, this method can detect the presence of BoNT/A, /B, /E, or /F in human serum at levels below that of the current standard, the mouse bioassay. Furthermore, this method can be used to detect the presence of the majority of BoNT/A, /B, /E, and /F subtypes. Finally, differentiation of the subtype of toxin is possible through tryptic digestion of the toxin itself and a mass spectral analysis of the tryptic fragments.

References

1. Herrero BA, Ecklung AE, Street CS, Ford DF, King JK (1967) Experimental botulism in monkeys – a clinical pathological study. Exp Mol Pathol 6:84–95
2. Arnon SS, Schechter R, Ingelsby TV, Henderson DA, Bartlett JG, Ascher MS, Eitzen E, Fine AD, Hauer J, Layton M, Lillibridge S, Osterholm MT, O'Toole R, Parker G, Perl TM, Swerdlow PK, Tonat K (2001) Botulinum toxin as a biological weapon: medical and public health management. JAMA 285(8):1059–1070
3. Foran P, Lawrence GW, Shone CC, Foster KA, Dolly JO (1996) Botulinum neurotoxin C1 cleaves both syntaxin and SNAP–25 in intact and permeabilized chromaffin cells: correlation with its blockade of catecholamine release. Biochemistry 35:2630–2636

4. Binz TJ, Blasi S, Yamasaki A, Baumeister E, Link TC, Sudhof R, Jahn R, Niemann H (1994) Proteolysis of SNAP–25 by types E and A botulinal neurotoxins. J Biol Chem 269:1617–1620
5. Blasi J, Chapman ER, Line E, Binz T, Yamasaki S, De Canilli P, Sudhof TC, Niemann H, Jahn R (1993) Botulinum neurotoxin A selectively cleaves the synaptic protein SNAP–25. Nature 160–163
6. Schiavo G, Rossetto O, Catsicas S, Polverino De Laureto P, Dasgupta BR, Benfenati F, Montecucco C (1993) Identification of the nerve terminal targets of botulinum neurotoxin serotypes A, D, and E. J Biol Chem 268:23784–23787
7. Schiavo G, Santucci A, Dasgupta BR, Mehta PP, Jontes J, Benfenati F, Wilson MC, Montecucco C (1993) Botulinum neurotoxins serotypes A and E cleave SNAP–25 at distinct COOH-terminal peptide bonds. FEBS Lett 335:99–103
8. Williamson LC, Halpern JL, Montecucco C, Brown JE, Neale EA (1996) Clostridial neurotoxins and substrate proteolysis in intact neurons: botulinum neurotoxin C acts on synaptosomal-associated protein of 25 kDa. J Biol Chem 271:7694–7699
9. Schiavo G, Benfenati F, Poulain B, Rossetto O, Polverino De Laurento P, Dasgupta BR, Montecucco C (1992) Tetanus and botulinum B neurotoxins block neurotransmitter release by proteolytic cleavage of synaptobrevin. Nature 359:832–835
10. Schiavo G, Malizio C, Trimble WS, Polverino De Laureto P, Milan G, Sugiyama H, Johnson EA, Montecucco C (1994) Botulinum G neurotoxin cleaves VAMP/synaptobrevin at a single Ala-Ala peptide bond. J Biol Chem 269:20213–20216
11. Schiavo G, Shone CC, Rosetto O, Alexander FC, Montecucco C (1993) Botulinum neurotoxin serotype F is a zinc endopeptidase specific for VAMP/synaptobrevin. J Biol Chem 268:11516–11519
12. Yamasaki S, Baumeister A, Binz T, Blasi J, Link E, Cornille F, Roques B, Fykse EM, Sudhof TC, Jahn R, Niemann H (1994) Cleavage of members of the synaptobrevin/VAMP family by types D and F botulinal neurotoxins and tetanus toxin. J Biol Chem 269:12764–12772
13. Yamasaki S, Binz T, Hayashi T, Szabo E, Yamasaki N, Eklund M, Jahn R, Niemann H (1994) Botulinum neurotoxin type G proteolyses the Ala^{81}-Ala^{82} bond of rat synaptobrevin 2. Biochem Biophys Res Commun 200:829–835
14. Blasi J, Chapman ER, Yamasaki S, Binz T, Niemann H, Jahn R (1993) Botulinum neurotoxin C1 blocks neurotransmitter release by means of cleaving HPC–1/syntaxin. EMBO J 12:4821–4828
15. Schiavo G, Shone CC, Bennett MK, Scheller RH, Montecucco C (1995) Botulinum neurotoxin type C cleaves a single Lys-Ala bond within the carboxyl-terminal region of syntaxins. J Biol Chem 270:10566–10570
16. Barr JR, Moura H, Boyer AE, Woolfitt AR, Kalb SR, Pavlopoulos A, McWilliams LG, Schmidt JG, Martinez RA, Ashley DL (2005) Botulinum neurotoxin detection and differentiation by mass spectrometry. Emerg Infect Dis 11(10):1578–1583
17. Boyer AE, Moura H, Woolfitt AR, Kalb SR, Pavlopoulos A, McWilliams LG, Schmidt JG, Barr JR (2005) From the mouse to the mass spectrometer: detection and differentiation of the endoproteinase activities of botulinum neurotoxins A-G by mass spectrometry. Anal Chem 7:3916–3924
18. Kalb SR, Moura H, Boyer AE, McWilliams LG, Pirkle JL, Barr JR (2006) The Use of Endopep-MS for the detection of botulinum neurotoxins A, B, E, and F in serum and stool samples. Anal Biochem 351(1):84–92
19. Kalb SR, Goodnough MC, Malizio CJ, Pirkle JL, Barr JR (2005) Detection of botulinum neurotoxin A in a spiked milk sample with subtype identification through toxin proteomics. Anal Chem 77(19):6140–6146
20. Gaunt PS, Kalb SR, Barr JR (2007) Detection of Botulinum Type E Toxin in channel catfish with visceral toxicosis syndrome using catfish bioassay and endopep mass spectrometry. J Vet Diagn Invest 19(4):349–354
21. Kalb SR, Smith TJ, Moura H, Hill K, Lou J, Garcia-Rodriguez C, Marks JD, Smith LA, Pirkle JL, Barr JR (2008) The Use of Endopep-MS to detect multiple subtypes of botulinum neurotoxins A, B, E, and F. Int J Mass Spec 278:101–108
22. Kalb SR, Lou J, Garcia-Rodriguez C, Geren IN, Smith TJ, Moura H, Marks JD, Smith LA, Pirkle JL, Barr JR (2009) Extraction and inhibition of enzymatic activity of BoNT/A1, /A2, and /A3 by a panel of monoclonal anti-BoNT/A antibodies. PLoS One 4(4):5355

23. Kautter DA, Solomon HM (1977) Collaborative study of a method for the detection of Clostridium botulinum and its toxins in foods. J Assoc Anal Chem 60:541–545

24. Hill KK, Smith TJ, Helma CH, Ticknor LO, Foley BT, Svensson RT, Brown JL, Johnson EA, Smith LA, Okinaka RT, Jackson PJ, Marks JD (2007) Genetic diversity among Botulinum Neurotoxin-producing clostridial strains. J Bacteriol 189:818–832

25. Szabo EA, Pemberton JM, Desmarchelier PM (1992) Specific detection of clostridium botulinum type B by using the polymerase chain reaction. Appl Environ Microbiol 58(1):418–420

26. Szabo EA, Pemberton JM, Desmarchelier PM (1993) Detection of the genes encoding botulinum neurotoxin types A and E by the polymerase chain reaction. Appl Environ Microbiol 59(9):3011–3020

27. Fach P, Hauser D, Guillou JP, Popoff MR (1993) Polymerase chain reaction for the rapid identification of Clostridium botulinum type A strains and detection in food samples. J Appl Bacteriol 75(3):234–239

28. Fach P, Gilbert M, Friffais R, Guillou JP, Popoff MR (1995) PCR and gene probe identification of botulinum neurotoxin A-, B-, E-, F-, and G-producing Clostridium spp. and evaluation in food samples. Appl Environ Microbiol 61(1):389–392

29. Lovenklev M, Holst E, Borch E, Radstrom P (2004) Relative neurotoxin gene expression in clostridium botulinum type B, determined using quantitative reverse transcription-PCR. Appl Environ Microbiol 70(5):2919–2927

30. Willems A, East AK, Lawson PA, Collins MD (1993) Sequence of the gene coding for the neurotoxin of Clostridium botulinum type A associated with infant botulism: comparison with other clostridial neurotoxins. Res Microbiol 144:547–556

31. Kalluri P, Crowe C, Reller M, Gaul L, Hayslett J, Barth S, Eliasberry S, Ferreira J, Holt K, Bengston S, Hendricks K, Sobel J (2003) An outbreak of foodborne botulism associated with food sold at a salvage store in Texas. J Clin Infect Dis 37(11):1490–1495

Functional Assays for Ricin Detection

Eric Ezan, Elodie Duriez, François Fenaille, and François Becher

Abstract In this review, we provide background information on ricin structure, present available functional assays for other toxins that are potential biothreat agents, and finish by describing the functional assay of ricin itself. Using appropriate sample preparation and optimized detection based on *N*-glycosidase activity, we demonstrate that specific detection of whole ricin at a level of around 0.1 ng/mL is possible and applicable to environmental samples.

1 Introduction

Ricin is a toxic protein found in the seeds of the castor oil plant, *Ricinus communis*, at concentrations ranging from 1% to 5% (Fig. 1) and was first isolated by Stillmark in 1888.[1] *Ricinus communis* is widespread in tropical regions, grown as an ornamental plant elsewhere, and is cultivated industrially for the production of castor oil (used in lubricants, hydraulic fluids, paints, inks, soaps, synthetic fibers, and also as a purgative).[2] Ricin constitutes a substantial risk because of its toxicity, widespread availability and ease of extraction. A highly toxic crude extract is easily prepared by treating ground *Ricinus communis* seeds with acetone to eliminate oil and then with acetic acid to extract proteins.[3] These proteins include water-soluble ricin, which is not extracted with the oil, but remains in the slurry, which is detoxified by heat treatment (85°C) and used as fertilizer or as cattle feed.[2] Ricin is a high-molecular-weight protein (~62 kDa) and filters with a molecular weight cutoff of 30 kDa are used to eliminate smaller contaminants.[3]

There are no precise toxicological data on ricin, and its toxicity has been assessed in humans from domestic and industrial accidents, cases of suicide or murder by ingestion, inhalation or injection, and also in animal studies. In humans,

E. Ezan (✉), E. Duriez, F. Fenaille, and F. Becher
CEA, Ibitecs, Service de Pharmacologie et d'Immunoanalyse, 911191, Gif-sur-Yvette, France
e-mail: eric.ezan@cea.fr

J. Banoub (ed.), *Detection of Biological Agents for the Prevention of Bioterrorism*,
NATO Science for Peace and Security Series A: Chemistry and Biology,
DOI 10.1007/978-90-481-9815-3_9, © Springer Science+Business Media B.V. 2011

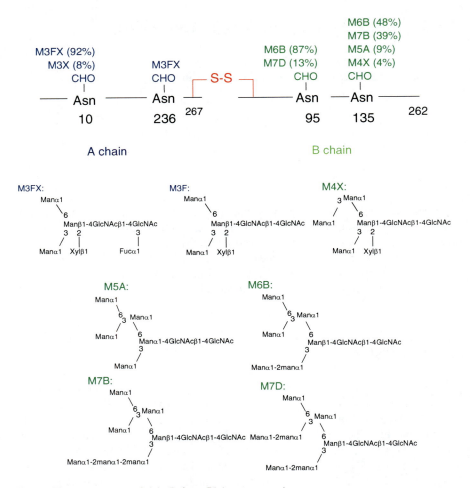

Fig. 1 Glycan structures of ricin D from *Ricinus communis*

the estimated lethal dose is 1 mg by ingestion, 1 μg/kg by injection and 5–10 μg/kg by inhalation, which is a wide range of LD_{50}[4]. As for any agent, the methods of detection of ricin must have a sensitivity cut-off that is consistent with the LD_{50}. To give a concrete example, in a man who drinks 1 L of water a day, a ricin concentration of 1 μg/mL in that water would be needed to reach the LD_{50}. The aim therefore is to develop analytical methods able to detect a quantity of ricin below the LD_{50}, i.e., below 1 μg/mL.

The Centers for Disease Control and Prevention in Atlanta put ricin in category B (Table 1) because there is currently no specific treatment for exposure to it, just treatments of signs and symptoms.[5] Given its toxicity at low doses, ricin is a potential bioterrorism agent and was, for example, used to murder the Bulgarian journalist Georgi Markov in London in 1978, by injection of about 500 μg

Table 1 Detection limits of assays of enzymatic activity of ricin by measurement of released adenine

Method of detection	Detec-tion limit (ng/mL)	Nature of Subs.	Type of substrate	Enzym-atic reaction time (h)	Ref.
Colorimetry	1,000	RNA DNA	Natural	2	Heisler et al. 2002[60]
LC-fluorescence	5,000	RNA DNA	Natural Natural	2	Barbieri et al. 1997[61]
LC-radioactivity	1,000	DNA	Natural	2	Brigotti et al. 1998[62]
LC-MS/MS	100	RNA	Synthetic	5	Hines et al. 2004[59]
Magnetic beads and measurement of electro-chemoluminescence	0.1	RNA	Synthetic	3	Keener et al. 2006[63]

(i.e., equivalent of five seeds).[6] Among potential pathways to exposure, aerosol release of ricin is likely to be favored, and would constitute the greatest risk.[2] More simply, ricin could be used to adulterate food or beverages. So the diversity of sample types to analyze leads to methodological complications in developing specific assays.

Numerous proteins structurally and functionally related to ricin have been characterized from a great variety of plants. Like ricin, these proteins irreversibly inactivate eukaryotic ribosomes, and so collectively are known as ribosome-inactivating proteins(RIPs).[7] Abrin of *Abrus precatorius* and volkensin of *Adenia volkensii* are examples.[8] Ricin is a glycoprotein of molecular weight ~62 kDa comprising two polypeptide chains, A and B, respectively of 267 and 262 amino acids, linked by a disulfide bridge.[9,10] Ricin binds to eukaryotic cells by its B chain, a lectin which binds to terminal β-D-galactopyranose residues of cell surface glycolipids and glycoproteins.[7,11] After binding, ricin is internalized by clathrin-dependent endocytosis and is transported to an intracellular compartment of the Golgi apparatus.[12] To reach the ribosomes in the cytosol, ricin must first cross the endomembrane system to be internalized in the endoplasmic reticulum, where the A and B chains are cleaved at the interchain disulfide bridge.[13] Chain A is then recognized by a transmembrane complex of the endoplasmic reticulum, Sec61p, which facilitates its translocation to the cell cytosol.[14] Once in the cytosol, the A chain, which has toxic activity, catalyzes the cleavage of the N-glycoside bond of adenosine 4324, specific to the GAGA sequence of the 28S ribosomal RNA of the large eukaryotic 60S subunit. This depurination irreversibly inactivates the ribosome by altering the binding sites of the elongation factor EF-2.[15] As rRNA is modified, protein synthesis is stopped, and this leads to apoptosis of the cell. Note that a single ricin toxin A chain in the cytosol can inactivate a 1,000 ribosomes a minute.[7] In addition, animal experiments (e.g., in mice) have shown that the toxicity of the ricin A chain is significantly lower (~fivefold) than that of whole ricin because of problems of internalization.[16] As for any potential bioterrorist agent, the formal establishment of poisoning by ricin is based on identification of the toxin in exposed individuals or at the presumed scene of exposure, meaning that there is a wide range of types of samples to analyze (e.g., solid or liquid foodstuffs, or both, soil, plasma). The use of two or more different technologies is required for unambiguous identification of ricin.[17] Furthermore, ricin toxicity depends on maintenance of its native structure, particularly the disulfide bridge between the A and B chains.[13] For this reason, it is important to study not only its biological activity, but also its structural integrity. Various analytical methods have been developed to detect and quantify ricin, based on two types of approaches.

Sensitive methods for detection of ricin usually rely on immunoassay. ELISA generally shows high sensitivity with reported detection limits between 0.1 and 80 ng/mL.[18,19] Recently, a more sensitive immuno-polymerase chain reaction assay was reported with a detection limit of 10 fg/mL.[20] However, the major limitation of these assays is their detection of both functional and nonfunctional ricin. Measuring the functional toxin specifically is important because only active toxin poses a threat to human health and life.

Other alternatives, such as mass spectrometric methods, can generally achieve rapid, sensitive, and highly specific characterization of proteins. A recently published method uses both MALDI-TOF- and LC-ESI-MS/MS.[21] MALDI-TOF-MS was used to screen for ricin peptides after trypsinization, and LC-ESI-MS/MS detected the peptides in the multiple reaction monitoring (MRM) mode. The sensitivity was 50, 100 and 1,000 ng/mL, as described for analysis of other proteins or toxins with similar technology.[22,23] These methods appear very specific, but less sensitive, and like immunoassays they do not specifically measure functional ricin.

Clearly, there is a need for specific detection of active ricin reaching the sensitivity of the best ELISA, i.e., at least 0.1 ng/mL. Immunorecognition of antibodies combined with mass spectrometric detection is considered one of the most specific analytical approaches, and has been applied to the detection of peptides/proteins in biological samples such as botulinum toxins.[24]

In this review, we provide background information on ricin structure, present available functional assays for other toxins that are potential biothreat agents, and finish by describing the functional assay of ricin itself. Using appropriate sample preparation and optimized detection based on N-glycosidase activity, we demonstrate that specific detection of whole ricin at a level of around 0.1 ng/mL is possible and applicable to environmental samples.

2   Ricin Structure

Ricin is a glycoprotein of molecular weight ~62 kDa comprising two polypeptide chains, A and B, respectively of 267 and 262 amino acids,[10] which are linked by a disulfide bridge.[9]

There are two subtypes of ricin – D and E – of similar overall toxicity, but activities that differ slightly because of a difference in affinity for certain cells.[25] This seems to be the result of structural variants of the B chain,[21,26] since ricins D and E differ only by the amino acid composition of the B chain (93% homology). The B chain of ricin E is a hybrid of the B chains of ricin D and of *Ricinus communis* agglutinin (RCA).[10] RCA is a protein of sequence analogous to that of ricin that is also found in ricin seeds, but is not toxic. The A chains of RCA and ricin share 94% homology, and the B chain of RCA shares respectively 85% and 89% homology with the B chains of ricins D and E.[21] Ricin D and RCA seem to be present in all ricin seeds, whereas ricin E appears to be present solely in small seeds.[10] Also, ricins D and E as well as RCA would appear to be present in analogous proportions.[27]

Ricin is a highly heterogeneous protein in terms of posttranslational modifications,[26] and four consensus sequences of N-glycosylation Asn-X-Ser/Thr are present in its primary sequence. The corresponding sites of N-glycosylation are Asn-10 and Asn-236 of the A chain and Asn-95 and Asn-135 of the B chain.[28,29] Figure 1 shows the glycan composition and structures of ricin D from *R. communis*.[28,29]

It is suspected that ricins of different origins have distinct glycan motifs and primary structures. In a study of seeds from five varieties of *R. communis* – *R. zanzibariensis*, *R. carmencita*, *R. impala*, *R. sanguineus* and *R. gibsonii* – Despeyroux et al. combined mass spectrometry (ESI-MS), capillary-zone electrophoresis (CZE), capillary isoelectric focusing (CIEF) and multianalyte resonant mirror to distinguish *R. zanzibariensis* clearly from the four other varieties[26]. These differences cannot be attributed solely to variants of glycoside motifs, but to primary sequence variations also. Databases give different amino acid sequences, probably because these sequences were determined using different seed varieties. Ricin from *Ricinus sanguineus* seeds differs from ricin of *Ricinus communis* seeds by two amino acids in the A chain: asparagine is replaced by serine in position 136,[30] and isoleucine by valine in position 173. Despeyroux et al. also found functional differences between ricins.[26]

We recently developed a method of detecting ricin through off-line coupling of immunocapture and MALDI-TOF-MS analysis after tryptic digestion[31]. In applying this method to 17 varieties of ricin seed from various geographical areas (Spain, Tanzania, Pakistan, India, China), the antibodies used recognized with similar affinities the B chains of the different ricins, and the three diagnostic peptides were also found in the different species. An in-depth study of the mass spectra of the MALDI-TOF-MS analysis of ricin extracts from different varieties of *Ricinus* did not reveal discriminant peptides, potentially corresponding to structural variants.

Figure 2 illustrates the great resemblance of the two mass spectra of seed extracts (after immunocapture) of *R. communis* and *R. zanzibariensis*. The only intensity variations are attributable to the enzymatic digestion and to the mass spectrometry analysis itself.

With or without prior immunocapture, the MALDI-TOF-MS analysis of purified ricin samples from *R. communis*, *R. impala* and *R zanzibariensis* did not provide evidence of important structural differences. Approaches involving nanoLC-ESI-MS/MS are under development to confirm or refute the existence of structural variants.

3 Functional Assays for Toxins

Numerous toxins consist of a component that binds to cell surfaces and a catalytic unit that modifies cell homeostasis, resulting in deleterious effects on cell targets. The exquisite specificity of their enzymatic components and high turnover provide the basis for the development of assays for pharmacological study of these toxins. Since these toxins or their expression vectors are also considered as potential terrorist weapons, there is considerable interest in using these activities as a means to monitor infection in human plasma as well as the biological threat in various environmental media.[32] For instance, this was applied to the development of bioassays for botulism toxins and anthrax toxins with the lowest limit of detection in the pico- and femtomolar range.

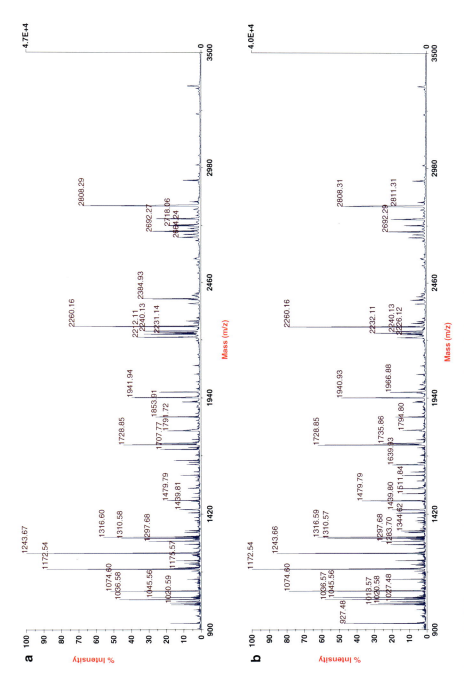

Fig. 2 MALDI-TOF mass spectra of seed extracts from (**a**) *R. communis* and (**b**) *R. zanzibariensis*, after immunocapture

Botulinum neurotoxins (BoNTs/A-G) produced by the genus *Clostridium* are responsible for the paralytic illness botulism, and are one of the most toxic bioweapons. Because of this extreme toxicity, their availability and ease of preparation, BoNTs have been identified as a major biothreat agent. Therefore, in suspected intentional contamination, analytical techniques for prompt detection and diagnosis are essential, and must identify the causal agent rapidly to expedite suitable confinement and treatment. Currently, the widely accepted test for the identification of BoNTs in both clinical specimens and food is the mouse bioassay. Its main advantage is sensitivity, with detection of 10–20 pg/mL of toxin, which is estimated to correspond to one mouse LD50.[33,34] However, although mice often exhibit signs of botulism within a few hours after a BoNT sample injection, 4 days are required to confirm a negative result.

There is clearly a place for alternative in vitro assays. One strategy has been to make use of the catalytic activity carried by the light chain of the toxin. BoNTs are composed of two chains, a heavy chain ($\approx$100 kDa) and a light chain ($\approx$50 kDa) held together by a disulfide bond.[35] The heavy chain is responsible for binding to the presynaptic membrane and translocation of the light chain into the cytosol of neuronal cells. The light chain cleaves specific proteins (SNAP-25, VAMP, syntaxin) involved in forming the soluble *N*-ethylmaleimide–sensitive fusion attachment protein receptor (SNARE) complex.[36] This complex is required for fusion of the synaptic vesicle with the presynaptic plasma membrane and communication between neurons.

Protein cleavage by BoNTs has been exploited for the development of sensitive in vitro assays, using peptide substrates mimicking the natural target of BoNTs. Peptide cleavage has been monitored either by fluorimetry[37] or ELISA.[38–41] Recently, it has been demonstrated that mass spectrometry is an alternative that can differentiate toxin types specifically (Endopep-MS assay). Each BoNT type cleaves a unique site on substrate peptides, generating products with different masses. Based on MALDI-TOF-MS or LC-ESI-MS/MS analysis of the product fragments, the Endopep-MS assay differentiates all seven BoNTs (A, B, C, D, E, F and G types). Sensitivity of between 0.039 and 0.625 LD_{50}/mL in buffer samples, so better than that of the mouse assay, was demonstrated for types A, B, E and F.[42] This Endopep-MS assay has been applied to environmental and clinical samples, using an antibody capture step to purify and concentrate the toxins.[24,43] This step, using antibody-coated magnetic beads, was necessary to remove most proteases from the sample and to concentrate the toxin from higher volumes such as 0.5 mL of serum. Sensitivity similar to that of the mouse bioassay was obtained in spiked serum, with a 4-h detection time. These results should be compared with those of the reference mouse assay, where 4 days are required to obtain definitive results in the case of negative findings.

Bacillus anthracis, which causes anthrax, is another major agent considered in CBRN (chemical, biological, radiological, and nuclear) risk management, notably because of its ease of diffusion as spores, the high mortality (close to 100%)

of inhalation anthrax, and the nonspecific symptoms of gastrointestinal anthrax and inhalation anthrax, which delay diagnosis. Anthrax toxin comprises three proteins that together account for its virulence: protective antigen (PA), a common component in receptor binding and internalization of two enzymatically active moieties – a metalloprotease (lethal factor, LF) and an adenylate cyclase (edema factor, EF).

Diagnostic techniques designed for use when *Bacillus anthracis* infection is suspected include detection of the bacteria by phenotypic (API strips, morphological characteristics of cultured colonies) or genomic (real-time PCR) methods, or toxin detection. Toxins in serum can be quantified by Western blot[44] or more sensitively by ELISA.[45,46] Antibodies to PA or EF can be detected by an indirect microhemagglutination test,[47] antibodies to PA and LF by electrophoretic immunotransblot,[48] antibodies to PA,[49] EF, and LF[50] and poly-D-glutamic acid capsule by ELISA,[50] and antibodies to PA by fluorescent covalent microsphere immunoassay.[51] However, DNA-based and antibody-based techniques do not indicate whether the pathogens are still viable or whether the anthrax toxins are still functional.

Indirect detection of *B. anthracis* in serum has been developed, based on the activity of EF and LF. LF is a zinc endoprotease, which acts on the N-terminal part of six mitogen-activated protein kinases.[52]

Its action blocks the signaling pathway and deregulates production of cytokines. EF is an adenylate cyclase, which catalyzes the transformation of adenosine triphosphate (ATP) to cyclic adenosine monophosphate (cAMP), in the presence of calmodulin and calcium. This excess of cAMP results in edema. These two enzymes have high catalytic activities which explains the potential gain in sensitivity when using an activity-based assay.[53,54]

Detection based on the enzymatic approach has a sensitivity of between 1 and 10 pg/mL for EF depending on the matrix, which represents a 50- to 1,000-fold gain in sensitivity over that of classic immunological tests.[55] The cAMP produced by the toxin from ATP is quantified using an immunoassay (ELISA). The enzymatic reaction can be done directly in serum, without prior extraction of the toxins. The cAMP formed, greatly exceeds endogenous cAMP. Only EF catalyzes the ATP transformation reaction this efficiently, which gives the test its specificity. The incubation of the serum with the substrates is brief (10 min for LF and 30 min for EF) and the time of analysis is short. Results can be obtained in less than 1 h.

The Centers for Disease Control and Prevention in Atlanta have developed MALDI-TOF mass spectrometry to detect products of LF,[56] whose particular cleavage of peptide substrate distinguishes it from other plasma proteases or toxins. LF is first isolated by immunocapture using antibody-coated magnetic beads to eliminate endogenous antiprotease, which could degrade the substrate. The enzymatic reaction is then performed with a 45-amino-acid peptide mimicking the natural protein substrate. The entire analytical method time requires 4 h and detects LF levels as low as 0.05 ng/mL in serum from infected monkeys.

4   Functional Assay for Ricin

Like the other toxins mentioned above, ricin can be detected in its bioactive form,[13] by using its catalytic activity for indirect detection. However, RIPs have the same catalytic activity, so this type of method requires specificity.

To detect ricin using its functional activity, two types of tests are described in the literature: biological tests, i.e., in vivo lethal dose tests,[16] (reference test in mice) and in vitro cytotoxicity tests.[11,57,58] The LD_{50} is 75 ng and 500 pg/mL, respectively for these two tests.[11,16] The two main drawbacks of these tests are that they are time-consuming (3–10 days) and problems of standardization and reproducibility, linked to variability between animals and cultured cells.

In a second type of test, based on biochemical methods using the N-glycosidase activity of ricin,[59–63] bioactive ricin is detected by measuring the product of its enzymatic activity, adenine, or the degraded substrate, the depurinated oligonucleotide.

Kalb's team has developed the detection of functional ricin by MALDI-TOF-MS assay of the depurinated oligonucleotide,[43] which differs in mass by 118 Da from the initial substrate, owing to the release of adenine. Note that immunocapture was performed upstream to increase specificity. The limit of detection of ricin is 1.2 ng/mL. The feasibility of the method developed by Kalb et al. has been evaluated on clinical samples (e.g., serum and saliva) and on food samples spiked with 2 μg/mL ricin.[43,64] However, no information on the various detection limits in these complex media is given.

Various teams have used the strategy of quantifying the adenine released by ricin from synthetic DNA or RNA substrate.[65] This work has led to quantification of adenine by colorimetry,[60] by chromatography coupled to fluorescence,[61] to radioactivity[62] or to mass spectrometry[59] and more recently by electrochemiluminescence.[63] Table 1 summarizes the main characteristics of these methods.

Of current methods of detection of bioactive ricin, only those of Hines et al.[59] by LC-ESI-MS/MS quantification of adenine, and Keener et al.[63] by measurement of electroluminescence, are sufficiently sensitive, i.e., of the order of ng/mL ricin. The advantage of the analysis by mass spectrometry is the inherent specificity of the detector itself. Nevertheless, these methods have important limitations. They lack specificity for ricin because there are other proteins able to depurinate an RNA or DNA substrate.[63] The analytical signal recorded therefore does not confirm the presence of ricin, but shows that RIPs are present, particularly in complex media. These methods have been developed using buffered solutions of ricin, and have not been applied to biological or environmental samples, and so cannot be used to extract ricin from complex media, in which certain compounds, notably endogenous adenine (e.g., milk) and RNA degradation enzymes (RNAses), may interfere with assays.

The sensitivity of detection of bioactive ricin depends on the mode of detection and also on the substrate of the enzymatic reaction itself. Key among factors affecting the efficiency of the depurination reaction is the nature of the substrate. In the methods described above, the specific substrate used is natural or synthetic DNA or RNA. Table 1 shows that methods using synthetic RNA as substrate seem to be the most sensitive[59,63] because ricin is catalytically more efficient when the substrate more closely mimics its endogenous substrate. However, experimental manipulation of RNA calls for special precautions owing to its lability, notably its degradation by RNAses.

Our objective has been to develop a biochemical method of detection specific to functional ricin, i.e., whole ricin with N-glycosidase activity. In comparison with existing methods, additional specificity has been provided by the combined use of mass spectrometry and immunocapture using antibodies directed against the B chain of ricin. Immunocapture also extracts and concentrates ricin from complex samples. In the event of a biothreat, the response time is vital, and special attention has been paid to minimizing the overall analysis time.

The first step of the assay consists of specific capture of ricin by its B chain (Fig. 3), on magnetic beads coated with antibodies. The captured ricin is then incubated with a 14 mer RNA substrate containing the GAGA sequence and mimicking the natural RNA ribosomal substrate of the toxin. Finally, detection is based upon the LC-ESI-MS/MS analysis of adenine released by the ricin A chain. During development of the method of detecting bioactive ricin, attempts at optimization have focused on three main points: chromatographic analysis, enzymatic reaction conditions, and immunocapture.

Utilization of LC-ESI-MS/MS (MRM mode) confers great sensitivity on the assay of adenine. This assay was developed by Hines et al., 2004, but lack of reproducibility (i.e., poor chemical stability of the column regarding pH) and of sensitivity (i.e., limit of detection 2.4 ng/mL adenine) in their conditions prompted us to optimize the chromatographic analysis of the released adenine.[59] Being a small, polar, positively charged molecule, adenine is not easily retained chromatographically

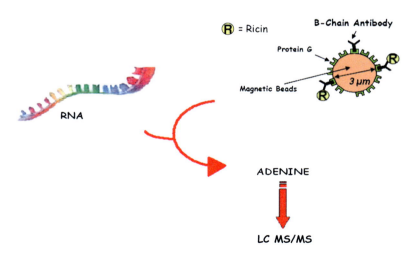

Fig. 3 Detection of functional ricin by combination of immunocapture and LC-ESI-MS/MS

and it was necessary to test the efficiency of different columns and elution conditions. Chromatographic analysis at the pH of adenine neutralization on an Atlantis® C18 column was chosen for further development and enabled us to achieve a limit of detection of 1 ng/mL (500 pg) ricin by direct LC-MS/MS analysis and of 0.1 ng/mL after immunocapture. So, our method of detection of ricin is approximately ten times more sensitive than that analyzing the depurinated oligonucleotide by MALDI-TOF-MS after immunocapture.[64]

Another determinant step is depurination of the ricin, and such methods use the enzymatic activity of ricin to amplify the signal to be detected. Optimization of the enzymatic reaction conditions increases sensitivity. The natural substrate of ricin is eukaryotic 28S rRNA, which contains the nucleotide sequence *GAGA* in the sarcin-ricin loop. Ricin can also depurinate a synthetic oligonucleotide of RNA or DNA that bears this *GAGA* sequence. We chose RNA as substrate as it gave the best limits of detection. The mechanism and specificity of the A chain of ricin were studied by Chen et al. using ten to 18-nucleotide RNA strands containing the specific motif, the *GAGA* loop.[65] A pH of 4 is optimal for the action of ricin on strands of synthetic RNA. Although this pH is far from physiological, ricin is enzymatically active inside the cell, or certain cell compartments (e.g., lysosomes) may be at a pH well below physiological.[66] At the optimal pH for depurination, the preferred substrate is an RNA of 14 bases, which is hydrolyzed at a rate of 219 moles per min, with a k_{cat}/K_m ratio of $4.5.10^5$ $M^{-1}s^{-1}$ (i.e., $K_m = 8.1$ μM) in stable conditions of catalysis. Smaller or longer substrates have lower K_m values, but all are close to ~5 μM. Substrates of ten and 18 bases have k_{cat}/K_m ratios close to 10^4 $M^{-1}s^{-1}$.[65] In our method, we use a synthetic RNA substrate of 14 bases, which enables the best catalytic efficiency. Although Chen[65] established the optimal conditions of depurination, adjustments have made the in vitro enzymatic reaction more efficient, notably by optimizing substrate concentration. Increased release of adenine up to 3.55 nmol of RNA has been observed, a quantity for which ricin seems to be at peak efficiency. So, we have fixed the quantity of RNA used for the depurination reactions at 3.55 nmol (i.e., 89 μM), which corresponds to ~10 K_m according to Chen[65] for an identical 14-base RNA substrate. Complementary results have shown that the utilization of a synthetic RNA substrate is beneficial (gain by a factor of ~5) by comparison with a synthetic DNA substrate. These results are consistent with those of Chen.[65] The synthetic RNA substrate mimics more closely the endogenous substrate of ricin. As a function of the quantity of adenine released, the native form (i.e., whole ricin) has greater enzymatic efficiency than the A chain, at equivalent molar concentrations. Barbieri group has observed the same tendency (enzymatic activity of ricin three times that of the ricin A chain).[61] Although these analyses were done in vitro, the structural integrity of ricin seems to play a part in its activity.

The optimum incubation time for maximum sensitivity was found to be 24 h. This delays the results of the assay, which may be prejudicial in the case of bioterrorist attack. However, a preliminary result at threefold lower sensitivity can be given after 6 h, i.e., 4 h of incubation and two additional hours for sample processing, and confirmed 20 h later if more sensitivity is needed.

We found that three other samples of ricin purified from *Ricinus communis* (no. 2), *Ricinus impala* and *Ricinus zanzibariensis* had different activities at equivalent quantities of ricin. These results could be explained by sample instability or non-specific binding, or because of variations in the activity of ricins dependent on naturally occurring sequence variants.[21,26] For example, there are ricin isoforms depending on the variety of seeds from *R. sanguineus*[30] and the Despeyroux team have found numerous glycoforms.[26]

The ultimate aim of this method is to detect ricin in complex biological or environmental samples, which often contain an abundance of proteins and enzymes that degrade nucleic acids (i.e., RNAse, DNAse) and may interfere with detection. So, sample treatment is needed to extract the ricin from complex samples. The advantage of immunoaffinity extraction is that it increases specificity for RIPs, compared with the abovementioned methods of detection.[59-63] Immunoaffinity extraction of ricin on conventional supports like agarose gel or protein A/G magnetic beads allows direct depurination of the captured ricin. Nonetheless, the second mode favors the incubation time and ease of use by using magnetic beads functionalized with antibodies, directly suspended in the biological sample. The advantage of this type of support is that it presents a greater surface area for capture and does not need chemical reactions to graft the antibodies. Before this study, this format of extraction by immunoaffinity proved effective in isolating and concentrating other protein toxins.[24] Immunocapture of the B chain was therefore set up as illustrated on Fig. 3.[67]

The feasibility of ricin detection by LC-ESI-MS/MS analysis coupled with immunocapture by anti-B chain antibodies has been demonstrated using food samples, such as milk, tap water and mineral water. On analysis of milk samples spiked with ricin, no inhibition by endogenous lactose of antigen-antibody recognition during immunocapture was noted, suggesting that this method could be used to detect ricin in other matrices with the same efficiency. Validation of the method with a wider range of samples should be envisaged, together with tests of ricin stability. However, the results obtained with three other samples of ricin purified from different seed varieties suggest that there may be between-variety differences in enzymatic activity. In view of these results, completed by analysis of other samples of ricin purified from different seed varieties, it may be interesting to evaluate and confirm these differences in enzymatic activity and to determine their origin for a better understanding of the biological process. These differences point to valuable studies, notably in vivo, to determine the LD_{50} of ricin as a function of seed variety.

5   Conclusion

The presence of ricin in the environment in the wake of a suspected bioterrorist incident should be confirmed using several analytical methods. Immunoassays are highly sensitive, and mass spectrometric methods very specific, but less sensitive.

However, none can distinguish between functional and nonfunctional ricin. Our approach based on immunocapture by anti-B chain antibodies coupled to mass spectrometry determination of the release of adenine by the A chain allows sensitive and specific determination of functional ricin. This is the first method capable of specifically detecting functional ricin with sensitivity similar to that of enzyme immunoassay and easily applicable to environmental samples. The assay requires 26 h, which may appear long in the event of a bioterrorism incident. However, a preliminary response can be given after 6 h with threefold lower sensitivity, but still below the ng/mL scale. Means for shortening assay time or enhancing sensitivity or both must be considered. The first possibility is to work with a larger sample volume. The catalytic efficiency of the depurination reaction by ricin could then be increased by developing a modified RNA substrate. For instance, introduction of a 2′-deoxyribonucleoside at the second position of the GAGA sequence, i.e., GdAGA, increases the catalytic constant.[68] A last approach would be to optimize the sensitivity of adenine detection using improved analytical methods, such as nanoLC/MS.

We have demonstrated the feasibility of using immunocapture and mass spectrometry. The method should now be validated in a separate study to establish more precisely the sensitivity and precision and also to confirm the specificity when applied to environmental matrices or even clinical samples.

References

1. Lord JM, Roberts LM, Robertus JD (1994) Ricin: structure, mode of action, and some current applications. FASEB J 8:201–208
2. Audi J, Belson M, Patel M, Schier J, Osterloh J (2005) Ricin poisoning: a comprehensive review. JAMA 294:2342–2351
3. Ovenden SP et al (2009) De novo sequencing of RCB-1 to -3: peptide biomarkers from the castor bean plant Ricinus communis. Anal Chem 81:3986–3996
4. Bradberry SM, Dickers KJ, Rice P, Griffiths GD, Vale JA (2003) Ricin poisoning. Toxicol Rev 22:65–70
5. Burnett JC, Henchal EA, Schmaljohn AL, Bavari S (2005) The evolving field of biodefence:therapeutic developments and diagnostics. Nat Rev Drug Discov 4:281–297
6. Crompton R, Gall D (1980) Georgi Markov–death in a pellet. Med Leg J 48:51–62
7. Olsnes S, Kozlov JV (2001) Ricin. Toxicon 39:1723–1728
8. Barbieri L et al (2004) Enzymatic activity of toxic and non-toxic type 2 ribosome-inactivating proteins. FEBS Lett 563:219–222
9. Lappi DA, Kapmeyer W, Beglau JM, Kaplan NO (1978) The disulfide bond connecting the chains of ricin. Proc Natl Acad Sci USA 75:1096–1100
10. Araki T, Funatsu G (1987) The complete amino acid sequence of the B-chain of ricin E isolated from small-grain castor bean seeds. Ricin E is a gene recombination product of ricin D and Ricinus communis agglutinin. Biochim Biophys Acta 911:191–200
11. Frankel A et al (1996) Characterization of single site ricin toxin B chain mutants. Bioconjug Chem 7:30–37
12. Sandvig K, van Deurs B (1999) Endocytosis and intracellular transport of ricin: recent discoveries. FEBS Lett 452:67–70
13. Lord MJ et al (2003) Ricin. Mechanisms of cytotoxicity. Toxicol Rev 22:53–64

14. Roberts LM, Smith DC (2004) Ricin: the endoplasmic reticulum connection. Toxicon 44: 469–472
15. Moazed D, Robertson JM, Noller HF (1988) Interaction of elongation factors EF-G and EF-Tu with a conserved loop in 23S RNA. Nature 334:362–364
16. Fu T et al (1996) Ricin toxin contains three lectin sites which contribute to its in vivo toxicity. Int J Immunopharmacol 18:685–692
17. Brinkworth CS, Pigott EJ, Bourne DJ (2009) Detection of intact ricin in crude and purified extracts from castor beans using matrix-assisted laser desorption ionization mass spectrometry. Anal Chem 81:1529–1535
18. Poli MA, Rivera VR, Hewetson JF, Merrill GA (1994) Detection of ricin by colorimetric and chemiluminescence ELISA. Toxicon 32:1371–1377
19. Yan-Kenigsberg J, Bertocchi A, Garber EA (2008) Rapid detection of ricin in cosmetics and elimination of artifacts associated with wheat lectin. J Immunol Methods 336:251–254
20. Lubelli C et al (2006) Detection of ricin and other ribosome-inactivating proteins by an immuno-polymerase chain reaction assay. Anal Biochem 355:102–109
21. Fredriksson SA et al (2005) Forensic identification of neat ricin and of ricin from crude castor bean extracts by mass spectrometry. Anal Chem 77:1545–1555
22. Van Baar BL, Hulst AG, Wils ER (1999) Characterisation of cholera toxin by liquid chromatography – electrospray mass spectrometry. Toxicon 37:85–108
23. Van Baar BL, Hulst AG, de Jong AL, Wils ER (2002) Characterisation of botulinum toxins type A and B, by matrix-assisted laser desorption ionisation and electrospray mass spectrometry. J Chromatogr A 970:95–115
24. Kalb SR, Goodnough MC, Malizio CJ, Pirkle JL, Barr JR (2005) Detection of botulinum neurotoxin A in a spiked milk sample with subtype identification through toxin proteomics. Anal Chem 77:6140–6146
25. Oda T, Komatsu N, Muramatsu T (1997) Cell lysis induced by ricin D and ricin E in various cell lines. Biosci Biotechnol Biochem 61:291–297
26. Despeyroux D et al (2000) Characterization of ricin heterogeneity by electrospray mass spectrometry, capillary electrophoresis, and resonant mirror. Anal Biochem 279:23–36
27. Lin TT, Li SL (1980) Purification and physicochemical properties of ricins and agglutinins from Ricinus communis. Eur J Biochem 105:453–459
28. Kimura Y et al (1988) Structures of sugar chains of ricin D. J Biochem 103:944–949
29. Kimura Y, Kusuoku H, Tada M, Takagi S, Funatsu G (1990) Structural analyses of sugar chains from ricin A-chain variant. Agric Biol Chem 54:157–162
30. El-Nikhely N, Helmy M, Saeed HM, bou Shama LA, bd El-Rahman Z (2007) Ricin A chain from Ricinus sanguineus: DNA sequence, structure and toxicity. Protein J 26:481–489
31. Duriez E, Fenaille F, Tabet JC, Lamourette P, Hilaire D, Becher F, Ezan E (2008) Detection of ricin in complex samples by immunocapture and matrix-assisted laser desorption/ionization time-of-flight mass spectrometry. J Proteome Res 7:4154–4163
32. Demirev PA, Fenselau C (2008) Mass spectrometry in biodefense. J Mass Spectrom 43:1441–1457
33. Sharma SK, Whiting RC (2005) Methods for detection of Clostridium botulinum toxin in foods. J Food Prot 68:1256–1263
34. Notermans S, Nagel J (1989) Botulinum neurotoxin and tetanus toxin. In: Simpson LL (ed) Assays for botulinum and tetanus toxins. Academic, San Diego, CA, pp 319–331
35. Montecucco C, Schiavo G (1995) Structure and function of tetanus and botulinum neurotoxins. Q Rev Biophys 28:423–472
36. Eswaramoorthy S, Kumaran D, Keller J, Swaminathan S (2004) Role of metals in the biological activity of Clostridium botulinum neurotoxins. Biochemistry 43:2209–2216
37. Schmidt JJ, Stafford RG (2003) Fluorigenic substrates for the protease activities of botulinum neurotoxins, serotypes A, B, and F. Appl Environ Microbiol 69:297–303
38. Ekong TA, Feavers IM, Sesardic D (1997) Recombinant SNAP-25 is an effective substrate for Clostridium botulinum type A toxin endopeptidase activity in vitro. Microbiology 143:3337–3347

39. Hallis B, James BA, Shone CC (1996) Development of novel assays for botulinum type A and B neurotoxins based on their endopeptidase activities. J Clin Microbiol 34:1934–1938

40. Sharma SK, Ferreira JL, Eblen BS, Whiting RC (2006) Detection of type A, B, E, and F Clostridium botulinum neurotoxins in foods by using an amplified enzyme-linked immunosorbent assay with digoxigenin-labeled antibodies. Appl Environ Microbiol 72:1231–1238

41. Wictome M, Newton K, Jameson K, Hallis B, Dunnigan P, Mackay E, Clarke S, Taylor R, Gaze J, Foster K, Shone C (1999) Development of an in vitro bioassay for Clostridium botulinum type B neurotoxin in foods that is more sensitive than the mouse bioassay. Appl Environ Microbiol 65:3787–3792

42. Boyer AE, Quinn CP, Woolfitt AR, Pirkle JL, McWilliams LG, Stamey KL, Bagarozzi DA, Hart JC Jr, Barr JR (2007) Detection and quantification of anthrax lethal factor in serum by mass spectrometry. Anal Chem 79:8463–8470

43. Kalb SR et al (2006) The use of Endopep-MS for the detection of botulinum toxins A, B, E, and F in serum and stool samples. Anal Biochem 351:84–92, Kalb, S. R., Woolfitt, A. R. & Barr, J. R

44. Molin FD, Fasanella A, Simonato M, Garofolo G, Montecucco C, Tonello F (2008) Ratio of lethal and edema factors in rabbit systemic anthrax. Toxicon 52:824–828

45. Sastry KS, Tuteja U, Santhosh PK, Lalitha MK, Batra HV (2003) Identification of Bacillus anthracis by a simple protective antigen-specific mAb dot-ELISA. J Med Microbiol 52:47–49

46. Mabry R, Brasky K, Geiger R, Carrion R Jr, Hubbard GB, Leppla S, Patterson JL, Georgiou G, Iverson BL (2006) Detection of anthrax toxin in the serum of animals infected with Bacillus anthracis by using engineered immunoassays. Clin Vaccine Immunol 13:671–677

47. Buchanan TM, Feeley JC, Hayes PS, Brachman PS (1971) Anthrax indirect microhemagglutination test. J Immunol 107:1631–1636

48. Harrison LH, Ezzell JW, Abshire TG, Kidd S, Kaufmann AF (1989) Evaluation of serologic tests for diagnosis of anthrax after an outbreak of cutaneous anthrax in Paraguay. J Infect Dis 160:706–710

49. Quinn CP, Semenova VA, Elie CM, Romero-Steiner S, Greene C, Li H, Stamey K, Steward-Clark E, Schmidt DS, Mothershed E, Pruckler J, Schwartz S, Benson RF, Helsel LO, Holder PF, Johnson SE, Kellum M, Messmer T, Thacker WL, Besser L, Plikaytis BD, Taylor TH Jr, Freeman AE, Wallace KJ, Dull P, Sejvar J, Bruce E, Moreno R, Schuchat A, Lingappa JR, Martin SK, Walls J, Bronsdon M, Carlone GM, Bajani-Ari M, Ashford DA, Stephens DS, Perkins BA (2002) Specific, sensitive, and quantitative enzyme-linked immunosorbent assay for human immunoglobulin G antibodies to anthrax toxin protective antigen. Emerg Infect Dis 8:1103–1110

50. Sirisanthana T, Nelson KE, Ezzell JW, Abshire TG (1988) Serological studies of patients with cutaneous and oral-oropharyngeal anthrax from northern Thailand. Am J Trop Med Hyg 39:575–581

51. Biagini RE, Sammons DL, Smith JP, Page EH, Snawder JE, Striley CA, MacKenzie BA (2004) Determination of serum IgG antibodies to Bacillus anthracis protective antigen in environmental sampling workers using a fluorescent covalent microsphere immunoassay. Occup Environ Med 61:703–708

52. Vitale G, Bernardi L, Napolitani G, Mock M, Montecucco C (2000) Susceptibility of mitogen-activated protein kinase kinase family members to proteolysis by anthrax lethal factor. Biochem J 352(Pt 3):739–745

53. Turk BE, Wong TY, Schwarzenbacher R, Jarrell ET, Leppla SH, Collier RJ, Liddington RC, Cantley LC (2004) The structural basis for substrate and inhibitor selectivity of the anthrax lethal factor. Nat Struct Mol Biol 11:60–66

54. Drum CL, Yan SZ, Sarac R, Mabuchi Y, Beckingham K, Bohm A, Grabarek Z, Tang WJ (2000) An extended conformation of calmodulin induces interactions between the structural domains of adenylyl cyclase from Bacillus anthracis to promote catalysis. J Biol Chem 275:36334–36340

55. Duriez E, Goossens PL, Becher F, Ezan E (2009) Femtomolar detection of the anthrax edema factor in human and animal plasma. Anal Chem 81:5935–5941
56. Boyer AE, Moura H, Woolfitt AR, Kalb SR, McWilliams LG, Pavlopoulos A, Schmidt JG, Ashley DL, Barr JR (2005) From the mouse to the mass spectrometer: detection and differentiation of the endoproteinase activities of botulinum neurotoxins A-G by mass spectrometry. Anal Chem 77:3916–3924
57. McGuinness CR, Mantis NJ (2006) Characterization of a novel high-affinity monoclonal immunoglobulin G antibody against the ricin B subunit. Infect Immun 74:3463–3470
58. Mantis NJ, McGuinness CR, Sonuyi O, Edwards G, Farrant SA (2006) Immunoglobulin A antibodies against ricin A and B subunits protect epithelial cells from ricin intoxication. Infect Immun 74:3455–3462
59. Hines HB, Brueggemann EE, Hale ML (2004) High-performance liquid chromatography-mass selective detection assay for adenine released from a synthetic RNA substrate by ricin A chain. Anal Biochem 330:119–122
60. Heisler I, Keller J, Tauber R, Sutherland M, Fuchs H (2002) A colorimetric assay for the quantitation of free adenine applied to determine the enzymatic activity of ribosome-inactivating proteins. Anal Biochem 302:114–122
61. Barbieri L et al (1997) Polynucleotide:adenosine glycosidase activity of ribosome-inactivating proteins: effect on DNA, RNA and poly(A). Nucleic Acids Res 25:518–522
62. Brigotti M et al (1998) A rapid and sensitive method to measure the enzymatic activity of ribosome-inactivating proteins. Nucleic Acids Res 26:4306–4307
63. Keener WK, Rivera VR, Young CC, Poli MA (2006) An activity-dependent assay for ricin and related RNA N-glycosidases based on electrochemiluminescence. Anal Biochem 357:200–207
64. Kalb SR, Barr JR (2009) Mass spectrometric detection of ricin and its activity in food and clinical samples. Anal Chem 81:2037–2042
65. Chen XY, Link TM, Schramm VL (1998) Ricin A-chain: kinetics, mechanism, and RNA stem-loop inhibitors. Biochemistry 37:11605–11613
66. De DC (2005) The lysosome turns fifty. Nat Cell Biol 7:847–849
67. Becher F, Duriez E, Volland H, Tabet JC, Ezan E (2007) Detection of functional ricin by immunoaffinity and liquid chromatography-tandem mass spectrometry. Anal Chem 79:659–665
68. Amukele TK, Schramm VL (2004) Ricin A-chain substrate specificity in RNA, DNA, and hybrid stem-loop structures. Biochemistry 43(17):4913–4922

Challenges of Detecting Bioterrorism Agents in Complex Matrices

Erica M. Hartmann and Rolf U. Halden

Abstract This chapter offers an overview of the shift from the use of mass spectrometry for studying purified bioterrorism agents to the development of methods for rapid detection thereof in environmental and clinical samples. We discuss the difficulties of working with such complex matrices and present methods for quickly and effectively reducing complexity through sample preparation. Finally, we examine a success story wherein the common pathogen and potential bioterrorism agent norovirus is detected at clinically relevant levels in human stool.

Keywords Environmental samples • Clinical samples • Norovirus • Stool

1 Introduction

1.1 The Target Signal: Non-Target Signal Problem

Historically, we have used mass spectrometry (MS) to analyze bioterrorism agents as pure samples, but we are now in the position to examine them in more realistic settings, such as environmental or clinical samples. One major obstacle that has traditionally stood in the way of such analyses is the target signal-to-non-target signal (S:N) problem. Probing for peptide biomarkers in environmental samples equates to searching for the proverbial needle in a haystack, even when starting with relatively concentrated samples. These biomarkers must be reproducible and unique to the bioterrorism agent in question. Screening for suitable characteristic biomarkers can

E.M. Hartmann
The Biodesign Institute at Arizona State University, 1001 S. McAllister Ave, Tempe, AZ, USA

R.U. Halden (✉)
The Biodesign Institute at Arizona State University, School of Sustainable Engineering and the Built Environment, Arizona State University, 1001 S. McAllister Ave, Tempe, AZ, USA
e-mail: rhalden@jhsph.edu

involve the investigation of hundreds of candidate proteins and peptides before finding success. Matrix effects, i.e. the presence of legitimate, albeit undesirable signals from non-target substances in the sample, make it vastly more difficult to find these bio-markers in more complex samples. As the number of non-target peaks increases, signal from the target is more difficult to recognize and observe (see Fig. 1).

To avoid ambiguity, it is appropriate to pause here for a few definitions. In this chapter, we use the term "non-target signal" to encompass both true background "noise" and undesirable, interfering signals from sources other than the biomarker(s) of the bioterrorism agent of interest. Detector response from the required after biomarker(s) is referred to as the "target signal." The "sample matrix" is comprised of everything present in the sample aside from the targets themselves.

Throughout this chapter, we will refer to the *spectrum of sample matrix complexity*. The spectrum ranges from the simplest sample possible, a purified protein, to the ulti-mate challenge of environmental or clinical samples, bearing in mind that not all environmental or clinical matrices are equally complex.

While there are many methods of overcoming the S:N problem, many of them involve labor- and time-intensive sample preparation. Since MS-based techniques should be rapid and designed for high-throughput, these additional measures taken

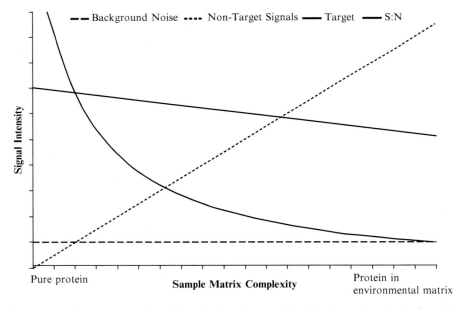

Fig. 1 A graphical representation of analytical challenges arising from increasing levels of sample matrix complexity. As the sample becomes more complex, phenomena such as ion suppression may reduce the intensity of the signal from a given target. At the same time, signals from non-target masses increase in number and intensity. The ratio of target signal to other signal sources (back-ground chemical noise plus all non-target, signal producing sample constituents) decreases corre-spondingly, thereby making it progressively more challenging to identify and quantify the target in samples of increasing complexity

for sample purification may diminish significantly the intrinsic advantages of MS. In the quest for the optimal MS assay, it is therefore desirable to evaluate sample preparation based not only on its effectiveness but also on the time it requires.

1.2 Potential Matrices

A general list of sample matrices of interest for the analysis of bioterrorism agents includes air/aerosols, water, soil, sediment, food,[1,2] as well as biological specimens of exhaled breath condensate, urine, saliva, blood and stool. Each of these matrices presents its own unique challenges and considerable differences exist with respect to non-target sample content that may interfere with successful analysis of the target. Matrix signals result from cells and tissues of non-target organisms, proteins, peptides, as well as from other sample constituents including surfactants and/or salts.[3]

2 Potential Solutions

As with many obstacles, there are multiple avenues for addressing the S:N problem. The following discussion of potential solutions is not meant to be exhaustive. Rather, it highlights a few areas where progress has already been made and then focuses on sample preparation techniques because these represent an accessible and practical avenue in many situations.

2.1 Instrumentation

Since the invention of soft ionization, the field of biopolymer MS has already seen progressive technical amelioration over many generations of instrumentation. In general, mass spectrometers have become less expensive and smaller.[4] At the same time, resolution, mass accuracy and sensitivity have all improved.[3,4]

The main soft ionization techniques currently used are electrospray ionization (ESI) and matrix-assisted laser desorption/ionization (MALDI). A variation on ESI known as nanospray has also been developed. Nanospray grants favorable ionization efficiencies, low limits of detection and enhanced signal intensity.[5] MALDI has seen a number of technical improvements as well. In addition to the hardware, the selection of MALDI matrix–not to be confused with sample matrix–also dramatically impacts the mass range and ionization efficiencies.[6]

These ionization techniques can then be coupled to a suite of MS detectors. MALDI is most commonly paired with a time-of-flight (TOF) detector. While the flight tubes for these TOFs can be quite long, reflectors and, for tandem MS, orthogonal systems have helped to condense the dimensions of these systems. Additional mass analyzers include linear ion traps (LTQ), triple quadrupoles, quadrupole ion traps, orbitraps, Fourier-transform ion cyclotron resonance (FTICR) systems, and others.[4,7]

The advent of tandem MS (MS/MS) has also greatly expanded the realm of protein investigation, making it possible to study all levels of protein structure, from primary to quaternary.[7] It is now possible to determine many post-translational modifications and partial or full amino acid sequences by MS/MS.[3] This latter development has been particularly beneficial for protein identification. The generation of fragment ions for analysis in the second MS can be as simple as post-source decay in MALDI. Here, a small population of ions spontaneously dissociate, when induced with high energy during passage through the first flight tube. For more complete fragmentation and the formation of predictable fragment ion series, a host of dissociation techniques, including collision induced dissociation, surface-induced dissociation, black-body infrared radioactive dissociation and electron-capture-induced dissociation[4] are available. For improved performance and other advantages, MS detectors can be paired in hybrid MS/MS instruments, resulting in mass accuracy below one part per million.[7]

Whereas MS/MS historically was reserved for specialists, this technique is now increasingly accessible.[4,7] Software packages that take advantage of constantly increasing computing power make it relatively easy to analyze the thousands of spectra that are generated in an MS/MS run. Especially with on-line systems, where a pre-separation step such as liquid chromatography (LC) is incorporated into the analysis, much of the process is or can be automated.[3] This degree of automation goes hand in hand with the high-throughput capabilities of MS and MS/MS analyses of proteins.

The increased capacity for proteomics studies is also spurred on by the ever expanding genomic databases and developments in bioinformatics.[3,4] Without the vast genomic dataset, many of the techniques and studies discussed in this chapter would not be possible.

It is also worth noting that several methods for peptide quantitation have been crafted, including isotope-coded affinity tags[8] and stable isotope labelling.[9,10] Bioinformatics-based quantitation, such as *Exponentially Modified Protein Abundance Index* (emPAI)[11,12] and spectral counting,[1,7] are also being employed. The ability to both detect and quantify targets greatly adds to the appeal of using MS for studying protein and peptide biomarkers.

2.2   Target Selection

One approach to detecting bioterrorism agents using MS is fingerprinting whole cells to obtain mass spectral "barcodes".[13,14] This technique can be very informative, providing identification at the sub-strain level for pure samples[15] and at the species level for mixtures.[14] For bacterial source tracking the reproducibility of MS-based techniques can surpass that of DNA fingerprinting.[15] As an added benefit, MS of whole cells can also be performed with minimal sample preparation requirements.

However, whole cells cannot be identified without extremely pure samples and a library of standard spectra, a bioterrorism monitoring scenario representing the exception rather than the norm. The need for purity, a minimum amount of biomass, and dependence of the method's outcome on the growth condition (vegetative state) of the biomass assayed often necessitate cultivation of the sample. Because of the wide mass range required for whole cell fingerprinting, the resolution of the peaks is relatively low, and the number of reproducibly detectable peaks typically is limited.[16] This results in poor statistical power of the identification. Ultimately, this technique is limited by the size of the fingerprint (microbial barcode) library.

MS also has been combined with bioinformatics-driven proteomics. In the top-down approach, intact proteins are introduced into the first MS, fragmented in a collision cell, and the fragment ions are analyzed in the subsequent MS. This approach yields better peak resolution than whole cells because the mass range queried is smaller. The greater resolution allows for the determination of post-translational modifications and amino acid sequence information.[17] This technique also includes as targets genetically engineered novel pathogens if the latter express a known toxin or virulence factor, as proteins can potentially be identified from transformed organisms.[18]

Top-down approaches are potentially powerful but require expensive high-resolution instrumentation, such as an FTICR-MS, and the analysis of fragmentation spectra can be challenging and time-consuming because fragmentation patterns are complicated and few databases exist.[17] Several improvements are required for this approach, including better fragmentation and faster and more reproducible methods for introducing the sample into the MS.[7] Furthermore, this technique can require more sample preparation because cells typically have to be lysed, and the protein content of the cell separated from the lipids and nucleic acids.

In contrast, the bottom-up proteomics approach currently is more common, easier to perform, and more tools and inexpensive techniques are available for performing such analyses.[3] In this technique, proteins are digested to render multiple peptides. These peptides provide multiple reference ions, which boost the statistical power of target identifications. Identifications can be obtained using peptide mass fingerprinting (PMF), in which observed peptide masses are compared with theoretically generated mass values, or using tandem mass spectrometry, in which characteristic peptides are further fragmented.

Although the bottom-up approach confers many advantages, it can also involve the longest sample preparation protocols, primarily due to time required for protein digestion. Similar to the top-down approach, this technique also is limited by the available genomic dataset.

All of these methods are valid approaches to identifying bioterrorism agents. Depending on the situation, individual methods may be more or less suited to the task at hand than others. As previously mentioned, time is a very important factor in determining which methods will be used for monitoring of bioterrorism agents in environmental media.

2.3   Sample Preparation

Regardless of the target selected, some sample preparation is necessary. Some steps may simply be required for proper functioning of the mass spectrometer. For example, desalting may be used to prevent the formation of adducts that would shift the observed peaks. Other steps are more geared towards boosting the signal of the target relative to non-target sample constituents. These latter steps often are essential and thus shall be discussed further.

Many different avenues of purification and signal amplification have been investigated, and a few will be discussed further in Section 3. Commonly employed purification options include size fractionation by sorting, physical screening, chemical treatment (e.g., precipitation), concentration via affinity and chromatographic separation. Additional options exist but, in the interest of brevity, are not discussed here. Culturing may also be used to increase the biomass available for analysis, thereby amplifying the target mass and improving the target to non-target ratio within the sample matrix. However, this approach may not always be practical and may be time-prohibitive.

As with target selection, the choice of sample preparation methods greatly impacts the time it takes to get from sample collection to identification. Some methods may also make the protocol prohibitively costly or difficult. It is therefore crucial to consider not just the efficacy of the preparation but also whether or not it is appropriate for the situation, i.e. routine high-throughput monitoring for bioterrorism agents.

3   Overview of Non-Bioterrorism Agent Studies

While not concentrating explicitly on the issue of bioterrorism monitoring, the following studies can serve to highlight the challenges of the task at hand. All studies discussed in this section concentrate on the identification of proteins and bacteria in complex sample matrices using MS. The methods used in these studies also are directly applicable to the monitoring for bioterrorism agents. For a brief comparison of these methods see Table 1.

3.1   Detection of Toluene Dioxygenase from
Pseudomonas putida F1

This study aimed to identify the toluene dioxygenase (gi|148548093) expressed by *Pseudomonas putida* F1 as a marker for growth on toluene.[19]

Previous work indicated that it is possible to identify, with almost no sample preparation, catabolic biomarkers for biodegradation from a pure culture of aerobic,

Table 1 Comparison of case study methods. Mass spectrometry (MS), sodium dodecyl sulfate polyacrylamide gel electrophoresis (SDS PAGE), strong cation exchange (SCX), solid phase extraction (SPE), reverse-phase high performance (RP-HP) liquid chromatography (LC), matrix-assisted laser desorption/ionization (MALDI), time-of-flight (TOF), quadrupole (Q), electrospray (ESI), linear ion trap (LTQ), peptide mass fingerprinting (PMF), tandem mass spectrometry (MS/MS), National Center for Biotechnology Information non-redundant protein sequence database (NCBInr), environmental protein coding sequences (eCDs).

Section	3.1	3.2	3.3	4
Matrix	Bacterial cell	Bovine milk	Sargasso seawater	Human stool
Target	Toluene dioxygenase	*Escherichia coli* proteins	SAR11 proteins	Norovirus capsid proteins
Chemical treatment				Vertrel XF
Separation from matrix	Centrifugation	Sucrose gradient centrifugation	Filtration	Size fractionation using 2-step filtration
Protein separation	SDS-PAGE			
Peptide separation #1		SCX	SPE or SCX	
Peptide separation #2		RP-HPLC	LC	
MS	MALDI-TOF	Nanospray-Q-TOF	ESI-LTQ	Nanospray-Q-TOF
Analysis software	MASCOT PMF	MASCOT MS/MS	SEQUEST	MASCOT MS/MS
Database	NCBInr	SWISSPROT [*E. coli*]	In-house eCDs	NCBInr
Time estimate	9 h	20 h	12 h	8 h[a]

[a] Estimated time required when employing rapid digestion strategies.

biodegradative bacteria grown on minimal media supplemented with the substrate of interest.[20] The procedure involved the extraction and digestion of the whole soluble proteome of *Sphingomonas wittichii* RW1, a dioxin mineralizing bacterium that, for convenience, was grown on inexpensive and readily available dibenzofuran for expression of the dioxin degradation pathway. This digest was then analyzed using MALDI-TOF MS. The dioxin dioxygenase could then be identified from these spectra using PMF.[20]

Successful detection of a defined target protein in bacterial whole cell extracts by PMF is extraordinary because bacteria express hundreds, if not thousands, of proteins at any given point in time. That a single protein can be identified from the whole soluble proteome without any sort of separation implies that this protein is very highly expressed, yields tryptic peptides in the mass range examined, and ionizes favorably using the chosen ionization technique, in this case MALDI. Indeed, when the protocol was applied to other bacteria grown on other toxic substrates for expression of distinct catabolic enzymes other than the dioxin dioxygenase, their signals, while present, were drowned out by non-target proteins.[19]

The problem of competing signals from other proteins within the organism is the first increment in the spectrum of sample matrix complexity from purified protein to environmental samples. To reduce the sample complexity, the whole cell lysates were separated using SDS PAGE, and individual bands were excised for MALDI-TOF MS analysis.[19] This simple step was enough to allow for the identification of several proteins–including the target enzyme for *P. putida* F1.[19] However, the additional time required to run the gel, stain it, excise the bands, destain them, and extract the sample from the gel amounted to about 4 h. The gel-based step also theoretically raises the limit of detection.[21] although quantitative work was not done.

Although these disadvantages reduce the attractiveness of this technique for the detection of bioterrorism agents, the results demonstrate that it is possible to obtain meaningful identifications of specific proteins in complex digests of pure cultures when employing only a single separation step, i.e., one-dimensional SDS PAGE. More rapid separation steps that result in less sample loss would be more amenable to counterterrorism work.

3.2   *Characterization of Mastitic Escherichia coli in Bovine Milk*

The goal of this study was to determine if growth on milk, as opposed to Luria-Bertani (LB) broth, influenced the expression of proteins linked with pathogenicity in *Escherichia coli*.[12] To that end, the researchers cultured *E. coli* on both bovine milk and LB broth. They collected two fractions of the proteome, cytosolic and membrane-associated.[12]

To separate the targets from interfering proteins from the media, especially caseins, the researchers washed the harvested cells twice with cold Dulbecco's Phosphate Buffered Saline, centrifuged twice in a sucrose gradient, and then washed six more times. Although Western blots indicated that the caseins were removed from the *E. coli* cultures, proteins of bovine origin were still identified.[12]

Prior to MS analysis, samples were further fractionated using strong cation exchange (SCX) followed by reverse-phase high performance (RP-HP) liquid chromatography (LC). These steps are useful to separate proteins from within the target organism, as was seen in the previous example. The researchers successfully identified 633 proteins from *E. coli*, several of which had biologically relevant functions for growth in milk and some may be involved in pathogenesis. However, they also identified 25 bovine proteins. Using emPAI, they determined that there were over 100 bacterial proteins that were more abundant than the most abundant bovine protein identified in the cytosolic fraction. However, the most abundant bovine protein in the membrane-associated fraction was the 24th most abundant protein identified in that fraction.[12]

Returning to the spectrum of sample matrix complexity, we have added interfering signals from the matrix, i.e., bovine proteins, as well as signals from endogenous proteins. The procedure to counter these non-target sample constituents takes an estimated 20 h from sample collection to protein identification, excluding the substantial time required for microbial cultivation. It would therefore take roughly 3 days for an analyst to perform this protocol. Much of the time involved in this protocol is devoted to the column-based SCX and RP-HPLC separations. The most interesting step, in terms of overcoming challenges presented by a complex matrix, is the additional centrifugation performed to separate in a sucrose gradient the milk proteins from the target proteins, a process requiring approximately 160 min. These results demonstrate that, for relatively simple liquid sample matrices, centrifugation may be sufficient to remove a sufficient amount of non-target interferences to enable successful identification of the biological agent of interest.

3.3 Metaproteomics of Bacteria in the Sargasso Sea

The purpose of this study was to observe proteins expressed by SAR11, a clade (phylogenetically related group) of abundant marine bacteria.[1] Samples were taken from the Sargasso Sea, where bacterial growth is often nutrient-limited. These samples contain interferences from the medium, i.e., seawater, and from other bacteria, especially of the *Synechococcus* and *Prochlorococcus* genera. The presence of multiple SAR11 proteins also confounds the analysis because the multiple targets must be unequivocally identified, and some peptides are not unique to a single protein.[1] As such, this study used a true environmental sample, one towards the far extreme of the spectrum of sample matrix complexity and target-to-non-target ratios.

To separate and concentrate the bacterial biomass, samples were passed through tandem Millipore Pellicon systems with 30 kDa regenerated cellulose filters.[1] No attempt was made to separate bacterial species prior to MS analysis. Cells were lysed and their contents digested. Resultant peptides were processed using either solid phase extraction (SPE) or SCX and then separated using LC coupled to ESI-MS/MS.[1] This procedure takes approximately 12 h, excluding sample collection.

To identify proteins from the SAR11 clade, the investigators constructed a database of environmental protein coding sequences (eCDSs). One problem that they encountered was that not all of these eCDSs were unique to SAR11. To test the specificity, they queried the observed SAR11 peptides against similarly constructed databases for *Synechococcus* and *Prochlorococcus* as well as a database compiled from the rest of the metagenomic data from the Sargasso Sea. Of the total of 2,215 peptides they used to identify SAR11 proteins, 24 overlapped with the *Synechococcus* eCDSs, 20 with the *Prochlorococcus* eCDSs, and 1,226 with the remaining Sargasso Sea eCDSs. Despite this overlap, the investigators were able to confidently identify 236 proteins of SAR11 origin.[1]

Again, we see the use of column-based techniques to separate out both peptides from within the target organism and those from without. In contrast with the previous study, additional signals from the media were removed by filtration, as opposed to centrifugation. To deal with the additional species, the investigators added a bioinformatics component to the analysis. The results of this study demonstrate how bioinformatics, when used in complement with sample preparation, can be used to compensate for sample matrix complexity.

4   Case Study: Norovirus Detection in Stool

This is a landmark study because it tackled both an extremely difficult target and an extremely difficult sample matrix. Notable here is the simplicity of the method, which also is very rapid and enabled successful detection at clinically relevant copy numbers of the bioterrorism agent. Because of the relevance of the target and the speed of the method, this study highlights the possibilities and current limitations of MS-based monitoring of bioterrorism agents in complex samples.

Norovirus refers to the *Norovirus* genus of viruses that cause acute gastroenteritis.[22] The CDC has classified norovirus as a Class B bioterrorism agent, meaning that it is "moderately easy to disseminate" with "moderate morbidity rates" and necessitates "specific enhancements of CDC's diagnostic capacity and enhanced disease surveillance."[23] It is estimated that as few as ten virus particles can cause an infection, although clinical virus titers in stool range from 100 to 1,000 fmol/ml.[22]

One of the difficulties of working with human norovirus is that it cannot be cultured outside of human hosts. This limitation rules out culturing of samples to increase the amount of target copies. On the upside, norovirus capsids are assembled from a single protein that occurs at a high copy number. To protect analysts from infection, this study made use of virus-like particles (VLPs), each consisting of 180 identical capsid proteins which are identical to the bioterrorism agent except for the fact that they are devoid of any viral, infectious RNA. Capsid proteins comprising virus capsids are attractive targets for protein-based detection methods because they frequently occur in multiple copies per virus particle. In the case of the human norovirus, the sequence of the protein (gi|34223984) already had been determined and entered into online genomic databases. An in silico tryptic digest showed 15 possible fragments, equating to a potential sequence coverage of up to 58.8% in the 500 to 5,000 m/z range.[24]

Along the continuum of sample matrix composition, stool is localized at the far side of extreme complexity.[25] In addition to the target, stool samples may contain host (human) cells, animal and plant (food) cells, microbial cells from the gut community,[25] and non-proteinaceous organic and inorganic chemicals.[26]

Before attempting to identify the VLPs in samples of increasing complexity, the investigators examined the pure, intact protein using 1D MS.[24] With this method, monomers and dimers of the capsid protein were detectable at 60 pmoles. This relatively high detection limit was due to the aggregation of the capsid proteins. Nevertheless, the resultant spectra demonstrated the purity of the synthesized VLPs.[24]

Using the PMF approach with trypsin digestion, the investigators next assayed dilutions of the VLP standard and confidently detected the capsid protein down to levels of 50 fmoles. Detection at 100 fmoles were highly reproducible.[24] These detection limits are comparable to the clinically relevant range in which norovirus may occur in watery stool of acutely ill individuals.

The investigators then applied PMF to VLPs spiked into processed human stool extract.[24] As could be expected, the S:N ratio was unfavorable and effectively prevented successful identification of the target. However, a putative target peak of m/z 1,495.8 was consistently observable in norovirus-fortified samples. To capitalize on this finding, sample splits were introduced into a nanospray-ESI-MS/MS instrument to force fragmentation and identification of the putative 1,495.8 Da target peptide. In the nanospray single MS spectrum, the peptide was observable at m/z 748.4 as the double charged ion. The investigators therefore selected this peak for collision induced dissociation using MS/MS. Analysis of the VLP-fortified stool extract with this approach produced a nano-ESI-MS/MS spectrum showing 16 fragments corresponding to eight of the 12 y ion series fragments of the capsid protein peptide of sequence TLPDTIEVPLEDVR.[24]

This MS/MS-based method yielded a detection limit of $3 \cdot 10^8$ viruses per sample. This detection limit translates to a sample volume requirement of approximate 125 μl of stool to enable successful analysis using this approach.[24]

The entire method would take about 8 h to complete when leveraging rapid digestion techniques. As performed, the sample processing scheme involved dilution of stool samples in ammonium bicarbonate buffer, extraction with Vertrel XF to remove lipids, passage through a 0.22 μm filter followed by a 100 kDa MW cutoff filter, and concentration down to 100 μl in final sample volume.[24] For the purposes of this study, samples were digested overnight with trypsin.[24] More rapid digestion procedures, such as immobilized trypsin columns, could easily be substituted for the overnight digestion. This would allow for a significant reduction prep time, thereby facilitating execution of the analysis in a single work shift, taking no longer than 8 h.

The approach taken here for increasing S:N ratios differs from the aforementioned studies. The investigators did not rely on gels or LC-separation to concentrate their target and remove undesirable sample constituents. Instead, they reduced chemical interferences by treating with Vertrel XL and they concentrated their target by employing a physical screening approach involving size fractionation by sequential filtration. In the two-step filtration process, the first step serves to remove unwanted, large cells from the sample. The second step served to separate the VLPs present in the initial filtrate by passing it through a molecular cutoff filter that retained the VLPs but not the smaller dissolved molecules stemming from stool and cell debris. Finally, MS/MS and bioinformatics tools were used to cope with other endogenous and exogenous proteins present.

As an extension of this method, the investigators also developed a quantitative method using stable isotope-labeled standards and single reaction monitoring. They were able to detect their target at 500 attomoles.[10] However, this method was not tested in environmental samples and thus remains to be proven as a viable method for the detection of norovirus in stool.

5   Conclusions

During monitoring of environmental samples for bioterrorism agents, the concentration of the target and the complexity of the sample vary widely. Best-case scenarios of samples to be analyzed involve pure or semi-pure powders of dry microbial spores or vegetative cells. In these instances, traditional methods including mass spectral "barcoding" of samples may be sufficient to enable successful analysis. More likely, however, is that the analyst is challenged with the detection of minute target quantities in relatively dilute samples of great complexity. In this situation, sophisticated MS equipment and sample preparation techniques may still allow to determine a given biological agent reproducibly and with confidence. The successful detection of norovirus particles in stool at clinically relevant concentrations served to illustrate the applicability of MS approaches even in these extremely unfavorable conditions.

However, the analysis of dilute environmental samples for bioterrorism agents is still in its infancy. More work will be required in future years to abridge existing sample preparation protocols and to introduce new ones to advance the field. Insights can be gained from reviewing medical and environmental studies that do not fall into the domain of monitoring for bioterrorism agents. Alongside with the development of streamlined sample processing techniques, advances in instrument development, bioinformatics and computing power will be critical to propel the research field of bioterrorism monitoring forward.

Acknowledgements The authors gratefully acknowledge Drs. Tzu-Chiao Chao and Nicole Hansmeier for their help in editing this manuscript.

References

1. Sowell SM, Wilhelm LJ, Norbeck AD, Lipton MS, Nicora CD, Barofsky DF, Carlson CA, Smith RD, Giovanonni SJ (2009) Transport functions dominate the SAR11 metaproteome at low-nutrient extremes in the Sargasso Sea. ISME J 3(1):93–105
2. Pellerin C (2000) The next target of bioterrorism: your food. Environ Health Perspect 108(3):A126–A129
3. Apweiler R, Aslanidis C, Deufel T, Gerstner A, Hansen J, Hochstrasser D, Kellner R, Kubicek M, Lottspeich F, Maser E, Mewes HW, Meyer HE, Muellner S, Mutter W, Neumaier M, Nollau P, Nothwang HG, Ponten F, Radbruch A, Reinert K, Rothe G, Stockinger H, Tarnok A, Taussig MJ, Thiel A, Thiery J, Ueffing M, Valet G, Vandekerckhove J, Verhuven W, Wagener C, Wagner O, Schmitz G (2009) Approaching clinical proteomics: current state and future fields of application in fluid proteomics. Clin Chem Lab Med 47(6):724–744
4. Griffiths WJ, Jonsson AP, Liu S, Rai DK, Wang Y (2001) Electrospray and tandem mass spectrometry in biochemistry. Biochem J 355(3):545–561
5. Stutz H (2005) Advances in the analysis of proteins and peptides by capillary electrophoresis with matrix-assisted laser desorption/ionization and electrospray-mass spectrometry detection. Electrophoresis 26(7–8):1254–1290
6. Renato Z, Richard K (1998) Ion formation in MALDI mass spectrometry. Mass Spectrom Rev 17(5):337–366
7. Cravatt BF, Simon GM, Yates JR (2007) The biological impact of mass-spectrometry-based proteomics. Nature 450(7172):991–1000
8. Shiio Y, Aebersold R (2006) Quantitative proteome analysis using isotope-coded affinity tags and mass spectrometry. Nat Protocols 1(1):139–145
9. Ong S-E, Blagoev B, Kratchmarova I, Kristensen DB, Steen H, Pandey A, Mann M (2002) Stable Isotope Labeling by Amino Acids in Cell Culture, SILAC, as a simple and accurate approach to expression proteomics. Mol Cell Proteomics 1(5):376–386
10. Colquhoun DR (2007) Public health applications of quantitative biomarkers. The Johns Hopkins University, Baltimore, MD
11. Ishihama Y, Oda Y, Tabata T, Sato T, Nagasu T, Rappsilber J, Mann M (2005) Exponentially modified protein abundance index (emPAI) for estimation of absolute protein amount in proteomics by the number of sequenced peptides per protein. Mol Cell Proteomics 4(9):1265–1272
12. Lippolis JD, Bayles DO, Reinhardt TA (2009) Proteomic changes in Escherichia coli when grown in fresh milk versus laboratory media. J Proteome Res 8(1):149–158
13. Von Seggern CE, Halden RU (2009) Detection of bioterrorism agents and related public health threats utilising matrix-assisted laser desorption/ionisation mass spectrometry. Int J Health Sci 2(2):197–203
14. Wahl KL, Wunschel SC, Jarman KH, Valentine NB, Petersen CE, Kingsley MT, Zartolas KA, Saenz AJ (2002) Analysis of microbial mixtures by matrix-assisted laser desorption/ionization time-of-flight mass spectrometry. Anal Chem 74(24):6191–6199
15. Siegrist TJ, Sandrin TR (2007) Discrimination and characterization of environmental strains of E. coli by MALDI-TOF MS. J Microbiol Methods 68(3):554–562
16. Wunschel SC, Jarman KH, Petersen CE, Valentine NB, Wahl KL, Schauki D, Jackman J, Nelson CP, White VE (2005) Bacterial analysis by MALDI-TOF mass spectrometry: an inter-laboratory comparison. J Am Soc Mass Spectrom 16(4):456–462
17. Zabrouskov V, Senko MW, Du Y, Leduc RD, Kelleher NL (2005) New and automated MSn approaches for top-down identification of modified proteins. J Am Soc Mass Spectrom 16(12):2027–2038
18. Shiaw-Lin W, Ian J, William SH, Barry LK (2004) A new and sensitive on-line liquid chromatography/mass spectrometric approach for top-down protein analysis: the comprehensive analysis of human growth hormone in an E. coli lysate using a hybrid linear ion trap/Fourier transform ion cyclotron resonance mass spectrometer. Rapid Commun Mass Spectrom 18(19):2201–2207

19. Hartmann EM, Colquhoun DR, Halden RU (2009) Identification of putative biomarkers for toluene degrading burkholderia and pseudomonads using matrix-assisted laser desorption/ ionization time-of-flight mass spectrometry and peptide mass fingerprinting (submitted)
20. Halden RU, Colquhoun DR, Wisniewski ES (2005) Identification and phenotypic characterization of Sphingomonas wittichii strain RW1 by peptide mass fingerprinting using matrix-assisted laser desorption ionization-time of flight mass spectrometry. Appl Environ Microbiol 71(5):2442–2451
21. Link AJ, Eng J, Schieltz DM, Carmack E, Mize GJ, Morris DR, Garvik BM, Yates JR (1999) Direct analysis of protein complexes using mass spectrometry. Nat Biotechnol 17(7): 676–682
22. Noroviruses (2009). http://www.cdc.gov/ncidod/dvrd/revb/gastro/norovirus-factsheet.htm. Accessed 8 Aug 2009
23. Bioterrorism agents/diseases (2009). http://www.bt.cdc.gov/agent/agentlist-category.asp. Accessed 8 Aug 2009
24. Colquhoun DR, Schwab KJ, Cole RN, Halden RU (2006) Detection of norovirus capsid protein in authentic standards and in stool extract by Matrix-Assisted Laser Desorption Ionization (MALDI) and nanospray mass spectrometry. Appl Environ Microbiol 72(5):2442–24451
25. Oleksiewicz MB, Kjeldal HØ, Klenø TG (2005) Identification of stool proteins in C57BL/6J mice by two-dimensional gel electrophoresis and MALDI-TOF mass spectrometry. Biomarkers 10(1):29–40
26. Rang HP, Dale MM (1991) Absorption, fate and distribution of drugs. In: Rang HP, Dale MM, Ritter JM, Moore PK (eds) Pharmacology, 2nd edn. Churchill Livingstone, Edinburgh, p 955

DESI-MS/MS of Chemical Warfare Agents and Related Compounds

Paul A. D'Agostino

Abstract Solid phase microextraction (SPME) fibers were used to headspace sample chemical warfare agents and their hydrolysis products from glass vials and glass vials containing spiked media, including Dacron swabs, office carpet, paper and fabric. The interface of the Z-spray source was modified to permit safe introduction of the SPME fibers for desorption electrospray ionization mass spectrometric (DESI-MS) analysis. A "dip and shoot" method was also developed for the rapid sampling and DESI-MS analysis of chemical warfare agents and their hydrolysis products in liquid samples. Sampling was performed by simply dipping fused silica, stainless steel or SPME tips into the organic or aqueous samples. Replicate analyses were completed within several minutes under ambient conditions with no sample pre-treatment, resulting in a significant increase in sample throughput. The developed sample handling and analysis method was applied to the determination of chemical warfare agent content in samples containing unknown chemical and/or biological warfare agents. Ottawa sand was spiked with sulfur mustard, extracted with water and autoclaved to ensure sterility. Sulfur mustard was completely hydrolysed during the extraction/autoclave step and thiodiglycol was identified by DESI-MS, with analyses generally being completed within 1 min using the "dip and shoot" method.

Keywords Chemical warfare agents • Hydrolysis products • Liquid chromatography • Electrospray • Desorption electrospray • Mass spectrometry • Tandem mass spectrometry

P.A. D'Agostino (✉)
DRDC Suffield, Station Main, Medicine Hat, AB, Canada T1A 8K6
e-mail: paul.dagostino@drdc-rddc.gc.ca

J. Banoub (ed.), *Detection of Biological Agents for the Prevention of Bioterrorism*,
NATO Science for Peace and Security Series A: Chemistry and Biology,
DOI 10.1007/978-90-481-9815-3_11, © Springer Science+Business Media B.V. 2011

1   Introduction

The likelihood of battlefield use of chemical warfare agents has decreased with the ending of the Cold War and the widespread acceptance of the Chemical Weapons Convention. However there remains serious concerns world-wide that other parties may use chemical warfare agents against civilian or military targets. The Aum Shinrikyo sect used sarin, a nerve agent, during the 1995 terrorist attack on the Tokyo subway system, killing a dozen people and injuring thousands more. Public concern about the use of chemical or biological warfare agents reached a new peak following the al-Qaeda terrorist attacks of September 2001 and the subsequent delivery of anthrax letters in Washington DC. Early detection and identification of the agents used in a terrorist attack is critical and considerable effort has been expended by many nations to improve both field and laboratory based analytical methods for chemical warfare agents.

Detection and identification methods for chemical warfare agents, their degradation products and related compounds have been thoroughly reviewed with different emphases on numerous occasions.[1-10] Most methods utilize mass spectrometry for identification purposed with gas chromatography-mass spectrometry (GC-MS) being the most commonly employed technique.[9] Organic extracts of chemical warfare agents may be analysed directly by GC-MS, but the hydrolysis products of chemical warfare agents usually require derivatization prior to GC-MS analysis.[8]

The value of LC-MS as a complementary or replacement method for GC-MS, particularly for the confirmation of hydrolysis products of chemical warfare agents in aqueous extracts or samples, has been recently demonstrated.[11-23] Hydrolysis products may be analysed directly by LC-MS without the need for additional sample handling and derivatization. In addition, LC-MS may also be utilized for the determination of organophosphorus chemical warfare agents and related compounds in water, snow and aqueous extracts of soil or other samples.[19,20,23-25]

Cooks' group recently described a novel mass spectrometric method for sample ionization and analysis, and referred to it as desorption electrospray ionization (DESI).[26] During the DESI experiment charged droplets in the solvent being electrosprayed impact the surface of interest, desorbing and ionizing the analyte. Cooks recently reviewed DESI and other ambient mass spectrometry approaches,[27] including discussion on direct analysis in real time (DART),[28] a related direct analysis technique.

DESI-MS has been used for a variety of direct analyses [29] including the analysis of pharmaceutical products, [30-40] dyes on thin layer chromatography plates,[41] explosives on a variety of surfaces,[37,42-44] polymers,[45] alkaloids on plant tissue,[46] chemical warfare agents on solid phase microextraction (SPME) fibers.[25,47] DESI (and DART) allow rapid, direct sample analysis and have generated interest in the chemical defence and public security communities due to the minimal sample handling requirements and potential for rapid sample throughput.

Sarin, soman and sulfur mustard have been successfully analysed by DESI-MS from SPME fibers used to sample the headspace above chemical warfare agents and chemical warfare agents spiked onto Dacron sampling swabs and office media including office carpet, office fabrics and photocopy paper.[25,47] Sulfur mustard, a chemical warfare agent that does not produce ions during LC-ESI-MS, was also ionized during DESI-MS. It is possible that sulfur mustard and other compounds are being ionized during DESI-MS by an atmospheric chemical ionization mechanism where ionization results from gas phase proton transfer to a neutral analyte that has evaporated from the SPME surface.[42] It was also noted after removal of a highly exposed SPME fiber that the signal for an analyte remained for some time. Vaporized chemical warfare agent could be ionizing by either an ESI like mechanism where the surface is now a gas or by the atmospheric pressure chemical ionization mechanism described above. The possibility of both mechanisms occurring during DESI-MS analyses of chemical warfare agents cannot be ruled out.

Common hydrolysis products of chemical warfare agents, including methyl phosphonic acid, isopropyl methylphosphonic acid, ethyl methylphosphonic acid, pinacolyl methylphosphonic acid and thiodiglycol have also been analysed by DESI-MS.[48,49] Sufaces analysed include SPME fibers used to collect headspace samples above spiked glass surfaces and/or canola oil[49] and glass or Teflon.[48] A variant of DESI, reactive DESI using boric acid, was also employed to determine the presence of methyl phosphonic acid, isopropyl methylphosphonic acid and ethyl methylphosphonic acid at nanogram levels in urine applied to a glass surface.[48]

SPME has been used frequently for chemical sampling,[50] including direct and headspace sampling of chemical warfare agent samples and solutions.[51-59] This approach to rapid sampling and subsequent analysis has been developed for counterterrorism purposes within Canada. Direct analysis of SPME fibers by DESI-MS complements existing thermal desorption GC-MS based identification methods for chemical warfare agents and provides several advantages. The hydrolysis products of chemical warfare agents may be analysed directly without the need for derivatization procedures associated with GC-MS analyses and DESI-MS may ultimately enable higher sample throughput with less sample handling. SPME sampling and analysis can take minutes and a more rapid method of sampling for DESI-MS analysis would be desirable to increase throughput. Many of the collected samples during scenario-based counter-terrorism training exercises are liquids and a "dip and shoot" method using fused silica or stainless steel tips should reduce sampling and analysis times to less than 1 min. Unlike SPME fiber sampling, the method has application to both organic and aqueous samples and does not require fiber conditioning prior to sampling. Analysis times are much quicker as desorption is immediate, tips are disposable and useful for rapid sampling and identification of both chemical warfare agents and their hydrolysis products (without derivatization).

DESI-MS was successfully used to analyse SPME fibers used to sample the headspace above chemical warfare agents and related compounds. This technique was also evaluated for the detection and identification of common chemical warfare agents and their hydrolysis products using stainless steel and fused silica tips to rapidly sample organic and aqueous samples spiked with these compounds. This developed approach to rapid sampling and DESI-MS analysis was applied to the analysis of sulfur mustard spiked sand samples using accepted DRDC Suffield procedures for sample sterilization where chemical/biological warfare agent contamination is unknown. Thiodglycol, the hydrolysis product of sulfur mustard, was identified in the sterilized aqueous extract.

2   Experimental

2.1   Samples and Sample Handling

Standard stock solutions of chemical warfare agents and their hydrolysis products were prepared in water or dichloromethane in the 0.01–1 mg/mL range.

Spiked media samples were prepared by loading small volumes (typically 5–20 μL) of chemical warfare agent standard solutions in dichloromethane to silanized 20 mL headspace vials or onto media (e.g. Darcon swab, office carpet, sand) in the same vials. The dichloromethane solvent was allowed to evaporate and stand for 10 min. The vials were placed in a heated block at temperatures ranging from 40°C to 80°C (headspace temperature) for 10 min. Headspace sampling was typically conducted for up to 10 min using polydimethylsiloxane/divinyl benzene (PDMS/DVB) SPME fibers (65 μm film thickness, Supelco). Exposed fibers were analysed by DESI-MS and DESI-MS/MS. Blanks were handled in a similar manner.

Ottawa sand samples (3.0 g) were spiked in triplicate at the 47 μg/g level with sulfur mustard to simulate a chemical sample where the chemical and biological content is unknown. Each sample (and blank) was ultrasonically extracted with water (10 mL) for 10 min in a 20 mL scintillation vial. The lid on the scintillation vial was tightened finger tight (still allows airflow) and then autoclaved for 2 h at 121°C and 15 psi (liquid cycle) to ensure sterilization of any biological content. The sterilized aqueous extract was allowed to cool and an aliquot (1.5 mL) was removed, centrifuged for 10 min at 14,000 rpm and retained for DESI-MS and LC-ESI-MS analysis.

"Dip and shoot" sampling and analysis was performed by quickly dipping a stainless steel and/or fused silica tip (or Supelco carbowax/divinyl benzene SPME fiber for the Ottawa sand extracts) into the liquid samples. The dipped tip (sealed in a Supelco SPME manual injector) was introduced immediately through the septum port on the modified ESI interface during DESI-MS analyses.

2.2   Instrumental

Mass spectrometric data were acquired in the laboratory using a Waters (Milford, MA, USA) Q-ToF Ultima tandem mass spectrometer equipped with a Z-spray electrospray interface. The electrospray capillary was operated in the 1.5–3 kV range. The collision energy was generally maintained at 5 V for MS operation and was varied from 3 to 12 V (depending on the precursor ion selected) for MS/MS operation.

Fig. 1 DESI-MS analysis of a SPME fiber in the modified Waters Z-Spray source

RF1 was varied between 20 and 60 V during MS/MS/MS experiments to ensure product ion formation in the ESI interface. Argon was continually flowing into the collision cell at 9 psi during all analyses. Nitrogen desolvation gas was introduced into the interface (80°C) at a flow rate of 300 L/h and nitrogen cone gas was introduced at a flow rate of 50 L/h. MS data were typically acquired from 70 to 700 Da and MS/MS (product ion mass spectra) data were acquired for the protonated molecular ions of the spiked compounds (0.3–1 s). All data were acquired in the continuum mode with a resolution of 8,000 (V-mode, 50% valley definition).

During DESI-MS analyses a laboratory stand was used to hold and position the SPME manual holder so that the fiber could be introduced into the ESI plume. A plexiglass sleeve was machined for the Z-spray interface. It contained a septum port to facilitate the safe introduction of SPME fibers or tips contaminated with chemical warfare agents (Fig. 1). 50:50 acetonitrile/water (0.1% trifluoroacetic acid), were sprayed at 10 μL/min during DESI analyses.

3 Results and Discussion

DESI-MS has been applied to the analysis of numerous chemicals including common chemical warfare agents, explosives, pharmaceuticals and polymers. The technique offers some distinct advantages over traditional chromatographic-mass spectrometric methods. Analysis times are typically in the seconds to minutes time-frame as opposed to tens of minutes. The sample is easily introduced under ambient conditions into the instrument with little or no sample preparation required. Sensitivity is similar to chromatographic approaches and adduct formation may be reduced.

DRDC Suffield first reported the usefulness of DESI-MS for chemical warfare agent applications in 2005 at the ASMS[60] and has successfully used this technique for the direct analysis of SPME fibers exposed to a variety of spiked office environment and consumer product media.[25,47,49] Hydrolysis products of chemical warfare agents were also identified by DESI-MS as well as sulfur mustard and several other compounds related to sulfur mustard that do not ionize by ESI-MS.

SPME forms a cornerstone in the analytical strategy developed at DRDC Suffield for the identification of chemical warfare agents under realistic field sampling and analysis conditions. Samples (i.e., swabs, liquids, soil, materials) taken in the field would usually be contained in a septum-sealed vial, the headspace above which may be sampled by SPME using a manual holder without exposing the analyst to potentially harmful chemicals.[61] Analyses would typically be performed by fast gas chromatography-mass spectrometry (GC-MS) with DESI-MS and/or LC-ESI-MS being used to confirm identification. DESI-MS could be considered for primary analysis for large number of samples, as analysis times are typically faster than GC-MS. DESI-MS would also be preferred if identification of chemical warfare agent hydrolysis products was also required. DESI-MS does not require the derivatization step associated with GC-MS analyses, enabling identification of the actual compound with reduced sample handling and analysis time.

SPME headspace sampling can take as little as a few seconds for concentrated samples with higher volatility. For less volatile analytes, particularly at lower concentration levels, heating and sampling times of 5 or more min may be required. Direct sampling of liquid solutions with SPME fibers generally requires more time and mixing and is limited to aqueous liquids as swelling occurs following exposure to organic solutions. Many of the samples taken in the field are organic or aqueous liquids containing unknown toxic chemicals, including chemical warfare agents. Their concentration is often relatively high (mg/mL or higher), which might allow DESI-MS identification by a simple "dip and shoot" method. This would reduce sampling time, minimize the amount of analyte introduced into the MS instrument and ultimately increase throughput. Uncoated fused silica and fused silica and/or stainless steel tips associated with a SPME manual holder where the SPME coating was stripped away were investigated for sampling of organic and aqueous solutions of chemical warfare agents and their hydrolysis products at concentrations ranging from 0.01 to 1 mg/mL. The dipped tips were introduced directly into the mass spectrometer for DESI analysis.

3.1   Headspace Sampling of Munitions Grade Tabun

Terrorist use of chemical warfare agents may involve the use of crude or munitions grade chemical warfare agent that contains not only the chemical warfare agent but also related co-synthetic, degradation or other products. Identification of these additional sample components could be helpful in establishing a link between the chemical warfare agent used in the incident and a source, or provide an indication of synthetic route used to prepare the chemical warfare agent. A munitions grade sample of tabun[18] containing related organophosphorus compounds was selected to evaluate the applicability of SPME headspace sampling and DESI-MS/MS analysis for the identification purposes.

A Dacron sampling swab were placed in a headspace sampling vial and spiked with a munitions grade tabun standard at the 20 µg/g level (approximately 1–10 µg/g per sample component). The headspace above the spiked media was sampled for 10 min, with increased uptake being observed at higher temperatures. Figure 2 illustrates the DESI-MS total ion current profile (70–300 Da) and DESI-MS/MS product ion profiles collected during analysis of a Dacron sampling swab spiked with munitions grade tabun.

Tabun and a number of related organophosphorus compounds were identified by DESI-MS/MS. During DESI-MS/MS analysis, product ion data were acquired for the MH$^+$ ions of tabun and eight related organophosphorus compounds previously identified during LC-ESI-MS/MS experiments.[18] Tabun, the most abundant sample component (approximately 70% of the organic content) was easily identified on the basis of acquired DESI-MS data as well.

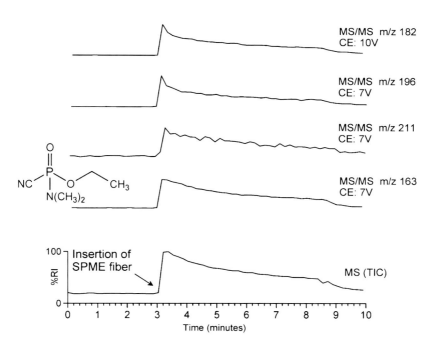

Fig. 2 DESI-MS total ion current profile (70–300 Da) and DESI-MS /MS product ion profiles for tabun (m/z 163), diisopropyl ethyl phosphate (m/z 211), ethyl isopropyl dimethylphosphoramidate (m/z 196) and diethyl dimethylphosphoramidate (m/z 182) obtained during analysis of a SPME fiber exposed to the headspace above of a Dacron swab spiked at the 20 µg/g level with munitions grade tabun (approximately 1–10 µg/g per sample component)

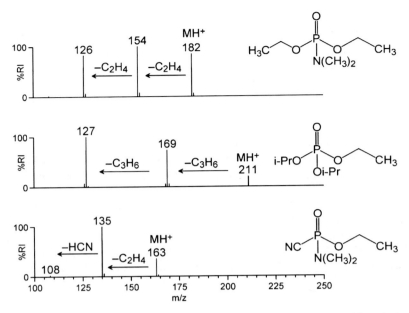

Fig. 3 Product ion mass spectra obtained for diethyl dimethylphosphoramidate (m/z 182, Collision energy: 10 V), diisopropyl ethyl phosphate (m/z 211, Collision energy: 7 V) and tabun (m/z 163, Collision energy: 7 V) during DESI-MS/MS analysis of SPME fiber exposed to the headspace above a Dacon swab spiked with munitions grade tabun

Figure 3 illustrates typical product ion mass spectra obtained for diethyl dimethylphosphoramidate, diisopropyl ethyl phosphate and tabun at collision energies that provided evidence of both the protonated molecular ion and characteristic product ions. Product ions due to the loss of the alkene associated with the alkoxy group were significant and the identity of these ions was confirmed by accurate mass measurement. Errors associated with mass measurement were typically <0.001 Da, consistent with previously acquired LC-ESI-MS/MS data.[25] Similar data were also acquired for the other office media samples spiked with munitions grade tabun, with the DESI-MS and DESI-MS/MS data being identical to that obtained during LC-ESI-MS and LC-ESI-MS/MS analyses.[47]

3.2   Headspace Sampling of Sulfur Mustard

One of the shortcomings of LC-ESI-MS for chemical warfare agent analysis has been the inability of this technique to analyse for the presence of the organosulfur chemical warfare agent, sulfur mustard, although this technique may be used for the sulfur mustard hydrolysis products.[21] Sulfur mustard (10 μg) was deposited into a 20 mL headspace sampling vial and sampled with a SPME fiber to assess the potential of DESI-MS for sulfur mustard analysis.

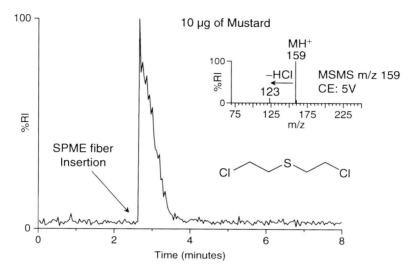

Fig. 4 DESI-MS of a SPME fiber used to sample the headspace above sulphur mustard spiked into a sealed glass vial. *Inset*: Typical mass spectrum

Figure 4 illustrates the DESI-MS/MS product ion profile acquired for product ions of m/z 159, the MH⁺ ion for sulfur mustard. The product ion mass spectrum obtained for sulfur mustard with a collision energy of 5 V contained a product ion at m/z 123 (due to loss of HCl), an ion that increased in relative intensity with increasing collision energy.

Sulfur mustard may be ionizing during DESI-MS by an atmospheric chemical ionization mechanism where ionization results from gas phase proton transfer to a neutral analyte that has evaporated from the SPME surface.[42] This implies that it would be possible to protonate analytes, including sulfur mustard, based on the ability of the analyte to accept a proton from a donor ion in the ESI plume with less proton affinity.[42] The DESI-MS/MS (and DESI-MS) data acquired for sulfur mustard appears to be consistent with this mechanism.

3.3 *"Dip and Shoot" Analyses*

"Dip and shoot" analyses involve simply dipping the tip into the liquid followed by DESI-MS analysis. DESI-MS analysis was rapid with individual analyses being completed in 1 min or less. Samples were typically screened for the presence of target compounds by DESI-MS and confirmed using this approach or by DESI-MS/MS when more specificity was required. Sarin, soman, cyclohexyl sarin, VX and tabun were all successfully analysed.

Tabun was analysed by DESI-MS/MS repeatedly over several minutes to demonstrate the ability to analyse a number of samples over a short period of time.

Figure 5 illustrates repeated analysis of a tabun sample (0.04 mg/mL in dichloromethane) using a stainless steel tip. Area measurements (arbitrary units) for repeated analysis of the tabun samples were 59 ± 20 (34%) consistent with the semi-quantitative nature of the method. In total, nine "dip and shoot" analyses were completed in less than 10 min with all the sample manipulation and analysis being done manually by the operator. The acquired DESI-MS/MS data for tabun contains a protonated molecular ion at m/z 163 and a product ion at m/z 135 (due to loss of ethylene).

Additional structural information for tabun was obtained by promoting the formation of additional characteristic ions during DESI-MS/MS/MS analyses (Fig. 6), a novel approach that makes use of higher RF1 settings.

The voltage in the RF1 region of the source was increased to promote product ion formation prior to the quadrupole. This setting was increased to 60 V from the usual setting of 20 V resulting in significant m/z 135 product ion formation (from the m/z 163 ion for tabun). The m/z 135 ion was selected with the quadrupole mass analyser and subjected to increasing collision energies (7–12 V) in the collision cell located between the quadrupole and time-of-flight mass analysers. Increasing collision energies led to increasing relative intensities of the product ions at m/z 117

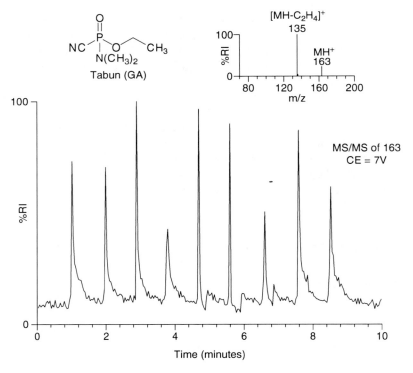

Fig. 5 Replicate DESI-MS/MS analyses of a stainless steel tip dipped into a 0.04 mg/mL tabun sample (in dichloromethane). *Inset*: Typical product ion mass spectrum

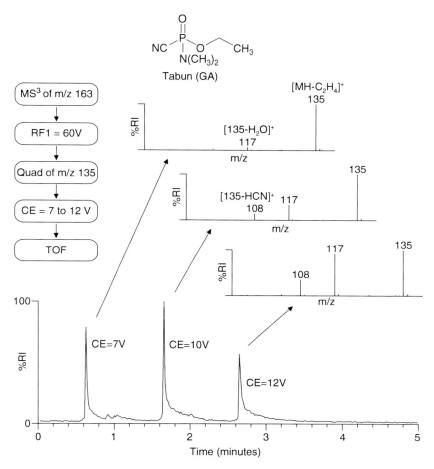

Fig. 6 DESI-MS/MS/MS analyses (vary collision energy) of a stainless steel tip dipped into a 0.2 mg/mL tabun sample (in dichloromethane)

and m/z 108, due to loss of water and HCN, respectively, from the m/z 135 ion. Acquisition of complementary MS/MS/MS data typically increases the number of ions associated with identification of a given compound. Acquisition of this additional data improves the certainty of identification and would be quite valuable for structural elucidation of unknowns.

Chemical warfare agent hydrolysis products analysed by GC-MS require an additional (undesirable) derivatization step that is not required for DESI-MS (or LC-ESI-MS) analysis. The hydrolysis products would typically be associated with aqueous samples and the "dip and shoot" method was evaluated for the hyrdolysis products of VX, sarin, soman and sulfur mustard at the 1 mg/mL level in water.

Figure 7 illustrates repeated analysis of thiodiglycol, the hydrolysis product of sulfur mustard, using a fused silica tip. Eight analyses were completed within 8

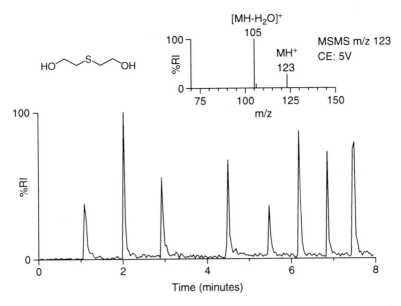

Fig. 7 Replicate DESI-MS/MS analyses of a glass tip dipped into a 1 mg/mL thiodiglycol sample (in water). *Inset*: Typical product ion mass spectrum

min. The acquired DESI-MS/MS contained the MH$^+$ ion at m/z 123 and a product ion at m/z 105 due to loss of water.

3.4 Autoclaved Soil Sample

Samples contaminated with unknown chemical and/or biological warfare agents pose a unique problem to chemical and biological specialists tasked with determining the presence of chemical or biological warfare agents. Such a sample would initially be received into biocontainment level 3 (BL-3) at DRDC Suffield where biological identification may be safely carried out. A sterility check may take up to 2 weeks and a more rapid means of analysis for chemical warfare agent content is required.

The most rapid and effective means of sterilizing a sample contaminated with biological warfare agents that allows removal of the sample from BL-3, without a sterility check, involves autoclaving the sample for 2 h. Any sample undergoing this process is necessarily exposed to water vapour at a high temperature, making the likelihood of chemical warfare agent hydrolysis high. An analytical method for chemical warfare agent identification must therefore be able to identify the principal hydrolysis products of the common chemical warfare agents. DESI-MS (and LC-ESI-MS) may be used for this purpose and have the added benefit of being able to also detect and identify intact organophosphorus chemical warfare agents and many related compounds in the aqueous sample extracts.

In the present application the actual chemical warfare agent, sulfur mustard, was spiked onto Ottawa sand at the 47 µg/g level for method evaluation. Each sample (and blank) was ultrasonically extracted with water for 10 min in a scintillation vial. The lid on the scintillation vial was tightened finger tight (still allows airflow) and then autoclaved for 2 h at 121°C and 15 psi (liquid cycle) to ensure sterilization of any biological content. The sterilized aqueous extract was allowed to cool and an aliquot (1.5 mL) was removed, centrifuged and retained for DESI-MS (and LC-ESI-MS) analysis.

A variety of Supelco SPME fibers were initially investigated for their ability to sample thiodiglycol (and any residual sulfur mustard) directly from the aqueous extract. None were effective as the thiodiglycol largely remained in the aqueous layer. However it was noted that the more polar carbowax/divinyl benzene SPME fiber did adsorb a small amount of thiodigycol from the autoclaved aqueous extract of the spiked Ottawa sand samples. This was independent of the dip time so the "dip and shoot" approach was applied to the actual SPME fiber.

Figure 8 illustrates the reconstructed-ion-current chromatograms (m/z 123) for the aqueous extract of the spiked Ottawa sand sample and blank. Thiodiglycol was not detected in the blank extract, but was readily identified in the spiked Ottawa sand sample extract (maximum concentration of thiodiglycol 14 ng/µL). The acquired

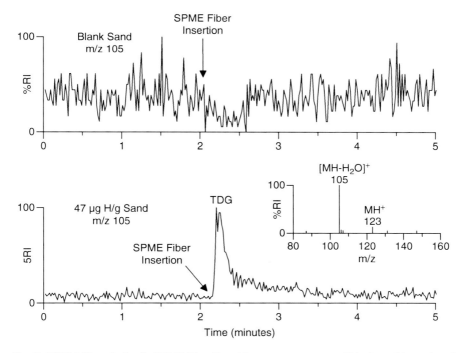

Fig. 8 DESI-MS analysis of a SPME fiber dipped into aqueous extract of blank sand (*upper*) and an aqueous extract of sand spiked at the 47 µg/g level with sulphur mustard after autoclaving (*lower* – Thiodiglycol (TDG), the hydrolysis product of sulphur mustard was detected. *Inset*: Typical mass spectrum

mass spectrum, containing the MH⁺ ion and its product due to loss of water, is inset in Fig. 8. Sulfur mustard was not detected by DESI-MS, consistent with the expectation of complete hydrolysis under the relatively harsh autoclave conditions.

4   Conclusions

DESI-MS complements existing LC-ESI-MS and GC-MS methods for chemical warfare agents. It was successfully used for the direct analysis of SPME fibers exposed to the headspace above media spiked at the μg/g level with common chemical warfare agents. Sulfur mustard, a compound that does ionize during LC-ESI-MS, was also analysed from a SPME fiber exposed to the headspace above 10 μg of sulfur mustard using the DESI-MS method.

A "dip and shoot" method was developed for the rapid sampling and DESI-MS analysis of chemical warfare agents and their hydrolysis products in liquid samples. Sampling was performed by simply dipping fused silica, stainless steel or SPME tips into the organic or aqueous samples. Replicate analyses were completed within several minutes under ambient conditions with no sample pre-treatment, resulting in a significant increase in sample throughput.

This method was applied to the determination of chemical warfare agent content in samples containing unknown chemical and/or biological warfare agents. Ottawa sand was spiked with sulfur mustard, extracted with water and autoclaved to ensure sterility. Sulfur mustard was completely hydrolysed during the extraction/autoclave step and the hydrolysis product, thiodiglycol, was identified by DESI-MS.

References

1. Witkiewicz Z, Mazurek M, Szulc J (1990) Chromatographic analysis of chemical warfare agents. J Chromatogr 503:293–357
2. Kingery AF, Allen HE (1995) The environmental fate of organophosphorus nerve agents: a review. Toxicol Environ Chem 47:155–84
3. Kientz ChE (1998) Chromatography and mass spectrometry of chemical warfare agents, toxins and related compounds: state of the art and future prospects. J Chromatogr A 814:1–23
4. Munro NB, Talmage SS, Griffin GD, Waters LC, Watson AP, King JF, Hauschild V (1999) The sources, fate, and toxicity of chemical warfare agent degradation products. Environ Health Persp 107:933–74
5. Black RM, Read RW (2000) Liquid chromatography/mass spectrometry in analysis of chemicals related to the chemicals weapons convention. In: Meyers RA (ed) Encyclopedia of analytical chemistry. Wiley, Chichester, UK, pp 1007–1025
6. Noort D, Benschop HP, Black RM (2002) Biomonitoring of exposure to chemical warfare agents: a review. Toxicol Appl Pharm 184:116–26
7. Hooijschuur EWJ, Kientz CE, Brinkman UAT (2002) Analytical separation techniques for the determination of chemical warfare agents. J Chromatogr A 982:177–200
8. Black RM, Muir B (2003) Derivatisation reactions in the chromatographic analysis of chemical warfare agents and their degradation products. J Chromatogr A 1000:253–81

9. Mesilaakso M (2005) Chemical weapons convention analysis, sample collection, preparation and analytical methods. Wiley, Chichester, UK

10. D'Agostino PA (2008) Chemical warfare agents. In: Bogusz MJ (ed) Forensic science handbook of analytical separation, vol 6. Elsevier, Amsterdam, The Netherlands, pp 839–872

11. Black RM, Read RW (1997) Application of liquid chromatography-atmospheric pressure chemical ionization mass spectrometry, and tandem mass spectrometry, to the analysis and identification of degradation products of chemical warfare agents. J Chromatogr A 759:79–92

12. D'Agostino PA, Provost LR, Hancock JR (1998) Analysis of mustard hydrolysis products by packed capillary liquid chromatography-electrospray mass spectrometry. J Chromatogr A 808:177–84

13. Black RM, Read RW (1998) Analysis of degradation products of organophosphorus chemical warfare agents and related compounds by liquid chromatography-mass spectrometry using electrospray and atmospheric pressure chemical ionization. J Chromatogr A 794:233–44

14. Read RW, Black RM (1999) Rapid screening procedures for the hydrolysis products of chemical warfare agents using positive and negative ion liquid chromatography-mass spectrometry and atmospheric pressure chemical ionization. J Chromatogr A 862:169–77

15. Mercier J-P, Morin P, Dreux M (1999) Combination of LC-MS and CE-MS analysis for the separation and the identification of phosphonic acids. Chimia 53:511–4

16. Hooijschuur EWJ, Kientz CE, Hulst AG (2000) Determination of hydrolysis products of sulfur mustard by reversed-phase microcolumn liquid chromatography coupled on-line with sulfur flame photometric detection and electrospray ionization mass spectrometry using large-volume injections and peak compression. Anal Chem 72:1199–206

17. D'Agostino PA, Hancock JR, Provost LR (2001) Determination of sarin, soman and their hydrolysis products in soil by packed capillary liquid chromatography-electrospray mass spectrometry. J Chromatogr A 912:291–9

18. D'Agostino PA, Hancock JR, Provost LR (2001) Electrospray mass spectrometry of chemical warfare agents. Adv Mass Spectrom 15:297–316

19. D'Agostino PA, Chenier CL, Hancock JR (2002) Packed capillary liquid chromatography-electrospray mass spectrometry of snow contaminated with sarin. J Chromatogr A 950:149–156

20. D'Agostino PA, Hancock JR, Chenier CL (2003) Mass spectrometric analysis of chemical warfare agents and their degradation products in soil and synthetic samples. Eur J Mass Spectrom 9:609–18

21. D'Agostino PA, Hancock JR, Chenier CL (2004) Packed capillary liquid chromatography-electrospray ionization (tandem) mass spectrometry of mustard hydrolysis products in soil. J Chromatogr A 1058:97–105

22. Liu Q, Hu XY, Xie JW (2004) Determination of nerve agent degradation products in environmental samples by liquid chromatography time-of-flight mass spectrometry with electrospray ionization. Anal Chim Acta 512:93–101

23. D'Agostino PA, Hancock JR, Provost LR (1999) Analysis of O-ethyl S-[2-(diisopropylamino) ethyl] methylphosphonothiolate (VX) and its degradation products by packed capillary liquid chromatography-electrospray mass spectrometry. J Chromatogr A 837:93–105

24. D'Agostino PA, Hancock JR, Provost LR (1999) Packed capillary liquid chromatography-electrospray mass spectrometry analysis of organophosphorus chemical warfare agents. J Chromatogr A 840:289–94

25. D'Agostino PA, Hancock JR, Chenier CL, Jackson Lepage CR (2006) Liquid chromatography electrospray tandem mass spectrometric and desorption electrospray ionization tandem mass spectrometric analysis of chemical warfare agents in office media typically collected during a forensic investigation. J Chromatogr A 1110:86–94

26. Takats Z, Wiseman JM, Gologan B, Cooks RG (2004) Mass spectrometry sampling under ambient conditions with desorption electrospray ionization. Science 306:471–473

27. Cooks RG, Ouyang Z, Takats Z, Wiseman JM (2006) Ambient mass spectrometry. Science 311:1566–1570

28. Cody RB, Laramee JA, Durst HD (2005) Versatile new ion source for the analysis of materials in open air under ambient conditions. Anal Chem 77:2297–2302
29. Takats Z, Wiseman JM, Cooks RG (2005) Ambient mass spectrometry using desorption electrospray ionization (DESI): instrumentation, mechanisms and application in forensics, chemistry and biology. J Mass Spectrom 40:1261–1275
30. Williams JP, Scrivens JH (2005) Rapid accurate mass desorption electrospray ionization tandem mass spectrometry of pharmaceutical samples. Rapid Commun Mass Spectrom 19:3643–3650
31. Chen H, Talaty NN, Takats Z, Cooks RG (2005) Desorption electrospray ionization mass spectrometry for high-throughput analysis of pharmaceutical samples in the ambient environment. Anal Chem 77:6915–6927
32. Weston DJ, Bateman R, Wilson ID, Wood TR, Creaser CS (2005) Direct analysis of pharmaceutical drug formulations using ion mobility spectrometry/quadrupole-time-of-flight mass spectrometry combined with desorption electrospray ionization. Anal Chem 77:7572–7580
33. Rodriguez-Cruz SE (2006) Rapid analysis of controlled substances using desorption electrospray ionization mass spectrometry. Rapid Commun Mass Spectrom 20:53–60
34. Leuthold LA, Mandscheff J-F, Fathi M, Giroud C, Augsburger M, Varesio E, Hopfgartner G (2006) Desorption electrospray ionization mass spectrometry: direct toxicological screening and analysis of illicit Ecstasy tablets. Rapid Commun Mass Spectrom 20:103–110
35. Kauppila TJ, Wiseman JM, Ketola RA, Kotiaho T, Cooks RG, Kostiainen R (2006) Desorption electrospray ionization mass spectrometry for the analysis of pharmaceutical s and metabolites. Rapid Commun Mass Spectrom 20:387–392
36. Williams JP, Patel VJ, Holland R, Scrivens JH (2006) The use of recently described ionization techniques for the rapid analysis of some common drugs and samples of biological origin. Rapid Commun Mass Spectrom 20:1447–1456
37. Chen H, Zheng J, Zhang X, Luo M, Wang Z, Qiao X (2007) Surface desorption atmospheric pressure chemical ionization mass spectrometry for direct ambient sample analysis without toxic chemical contamination. J Mass Spectrom 42:1045–1056
38. Venter A, Cooks RG (2007) Desorption electrospray ionization in a small pressure-tight enclosure. Anal Chem 79:6398–6403
39. Cotte-Rodriguez I, Mulligan CC, Cooks RG (2007) Non-proximate detection of small and large molecules by desorption electrospray ionization and desorption atmospheric chemical ionization mass spectrometry: Instrumentation and applications in forensics, chemistry, and biology. Anal Chem 79:7069–7077
40. Kauppila TJ, Talaty N, Kuuranne T, Kotiaho T, Kostiainen R, Cooks RG (2007) Rapid analysis of metabolites and drugs of abuse from urine samples by desorption electrospray ionization-mass spectrometry. Analyst 132:868–875
41. Van Berkel GJ, Ford MJ, Deibel MA (2005) Thin-layer chromatography and mass spectrometry coupled using desorption electrospray ionization. Anal Chem 77:1207–1215
42. Cotte-Rodriguez I, Takats Z, Talaty N, Chen H, Cooks RG (2005) Desorption electrospray ionization of explosives on surfaces: Sensitivity and selectivity enhancement by reactive desorption electrospray ionization. Anal Chem 77:6755–6764
43. Mulligan CC, Talaty N, Cooks RG (2006) Desorption electrospray ionization with a portable mass spectrometer: in situ analysis of ambient surfaces. Chem Commun 1709–1711
44. Cotte-Rodriguez I, Cooks RG (2006) Non-proximate detection of explosives and chemical warfare agent simulants by desorption electrospray ionization mass spectrometry. Chem Commun 2968–2970
45. Nefliu M, Venter A, Cooks RG (2006) Desorption electrospray ionization and electrosonic spray ionization for solid- and solution-phase analysis of industrial polymers. Chem. Commun 888–890
46. Talaty N, Takats Z, Cooks RG (2005) Rapid in situ detection of alkaloids in plant tissue under ambient conditions using desorption electrospray ionization. Analyst 130:1624–1633
47. D'Agostino PA, Chenier CL, Hancock JR, Jackson Lepage CL (2007) Desorption electrospray ionization mass spectrometric analysis of chemical warfare agents from solid-phase microextraction fibers. Rapid Commun Mass Spectrom 21:543–549

48. Song Y, Cooks RG (2007) Reactive desorption electrospray ionization for selective detection of the hydrolysis products of phosphonate esters. J Mass Spectrom 42:1086–1092
49. D'Agostino PA, Chenier CL (2007) Liquid chromatography electrospray ionization mass spectrometric (LC-ESI-MS) and desorption electrospray ionization mass spectrometric (DESI-MS) identification of chemical warfare agents in consumer products, DRDC Suffield TR 2007–074, 50 pp
50. Pawliszyn J (2002) Sampling and sample preparation for field and laboratory. Elsevier, Amsterdam, The Netherlands
51. Lakso H-A, Ng WF (1997) Determination of chemical warfare agents in natural water samples by solid-phase microextraction. Anal Chem 69:1866–72
52. Sng MT, Ng WF (1999) In-situ derivatisation of degradation products of chemical warfare agents in water by solid-phase microextraction and gas chromatographic-mass spectrometric analysis. J Chromatogr A 832:173–82
53. Schneider JF, Boparai AS, Reed LL (2001) Screening for sarin in air and water by solid-phase microextraction-gas chromatography-mass spectrometry. J Chromatogr Sci 39:420–4
54. Kimm GL, Hook GL, Smith PA (2002) Application of headspace solid-phase microextraction and gas chromatography-mass spectrometry for detection of the chemical warfare agent bis(2-chloroethyl) sulfide in soil. J Chromatogr A 971:185–91
55. Hook GL, Kimm G, Koch D, Savage PB, Ding BW, Smith PA (2003) Detection of VX contamination in soil through solid-phase microextraction sampling and gas chromatography/mass spectrometry of the VX degradation product bis(diisopropylaminoethyl)disulfide. J Chromatogr A 992:1–9
56. Hook GL, Jackson Lepage C, Miller SI, Smith PA (2004) Dynamic solid phase microextraction for sampling of airborne sarin with gas chromatography-mass spectrometry for rapid field detection and quantification. J Sep Sci 27:1017–1022
57. Zygmunt B, Zaborowska A, włatowska J, Namie nik J (2007) Solid phase microextraction combined with gas chromatography - A powerful tool for the determination of chemical warfare agents and related compounds. Curr Org Chem 11:241–253
58. Pardasani D, Kanaujia PK, Gupta AK, Tak V, Shrivastava RK, Dubey DK (2007) In situ derivatization hollow fiber mediated liquid phase microextraction of alkylphosphonic acids from water. J Chromatogr A 1141:151–157
59. Lee HSN, Sng MT, Basheer C, Lee HK (2007) Determination of degradation products of chemical warfare agents in water using hollow fibre-protected liquid-phase microextraction with in-situ derivatisation followed by gas chromatography-mass spectrometry. J Chromatogr A 1148:8–15
60. D'Agostino PA, Chenier CL, Hancock JR (2005) Development of high resolution LC-ESI-MS/MS methodology for the determination of chemical warfare agents and related compounds in an office environment. In: Proceedings of the 53rd annual conference on mass spectrometry and allied topics, June 5–9, 2005, San Antonio, TX
61. Smith PA, Jackson Lepage CR, Koch D, Haley DM, Hook GL, Betsinger G, Erickson RP, Eckenrode BA (2004) Detection of gas-phase chemical warfare agents using field-portable gas chromatography-mass spectrometry systems: instrument and sampling strategy considerations. Trends Anal Chem 23:296–306

Mass Spectrometry Applications for the Identification and Quantitation of Biomarkers Resulting from Human Exposure to Chemical Warfare Agents

J. Richard Smith and Benedict R. Capacio

Abstract In recent years, a number of analytical methods using biomedical samples such as blood and urine have been developed for the verification of exposure to chemical warfare agents. The majority of methods utilize gas or liquid chromatography in conjunction with mass spectrometry. In a small number of cases of suspected human exposure to chemical warfare agents, biomedical specimens have been made available for testing. This chapter provides an overview of biomarkers that have been verified in human biomedical samples, details of the exposure incidents, the methods utilized for analysis, and the biomarker concentration levels determined in the blood and/or urine.

Keywords Biomarkers • Biomedical samples • Chemical warfare agents • Chromatography • Mass spectrometry • Nerve agents • Sulfur mustard

1 Introduction

The Tokyo subway and Matsumoto apartment terrorist attacks in the mid-1990s by the Aum Shinrikyo cult resulted in a renewed interest in the development of methods with the capacity to verify human exposure to chemical warfare agents (CWA). The potential application of such methods is varied. Laboratory based forensic analytical methods are currently the primary use. The military, political, and legal ramifications resulting from the potential use of CWA by rogue nations or terrorist

J.R. Smith (✉)
USAMRICD, Attn: MCMR-CDT-D, 3100 Ricketts Point Road, Aberdeen Proving Ground, MD 21010-5400, USA
e-mail: John.richard.smith@us.army.mil

B.R. Capacio
US Army Medical Research Institute of Chemical Defense, 3100 Ricketts Point Road, Aberdeen Proving Ground, MD, USA

J. Banoub (ed.), *Detection of Biological Agents for the Prevention of Bioterrorism*,
NATO Science for Peace and Security Series A: Chemistry and Biology,
DOI 10.1007/978-90-481-9815-3_12, © Springer Science+Business Media B.V. 2011

organizations have generated a requirement for methods that are extremely accurate and able to measure relevant biomarkers at very low concentrations. While medical treatment used for CWA exposure is currently based on signs and symptoms of the victims, the incidents have demonstrated the critical need to identify individuals that were not exposed, the "worried well," in order to avoid unnecessary psychological stress and burden on the medical system. Pre-exposure tests, and subsequent medical surveillance, can be used to monitor potential exposure of laboratory personnel that handle CWA. Such tests may also be warranted for military and first responders who must enter or operate in a chemically contaminated environment. Low-level exposure to CWA has been implicated as a possible cause of some long-term health effects. The ability to accurately diagnose or verify exposure can be used to monitor both immediate and long-term effects of CWA exposure. Quantitative assays of CWA biomarkers have also been used to gain greater insight from animal studies on the toxicology and pre-treatment/treatment of CWA exposure.

Analysis of biomedical samples such as blood and urine for CWA biomarkers is inherently difficult due to the complex composition of the matrix and the fact that the analytes are usually present in trace quantities. Recent developments in analytical chemistry, particularly in the areas of chromatography and mass spectrometry, have aided significantly in the quest for new or improved methods. Assay techniques for CWA verification rarely target the intact agent due to the limited longevity of agents in vivo. Following exposure, many agents are rapidly converted and appear in the blood as hydrolysis products resulting from reactions with water and are excreted in the urine. Due to the rapidity of formation and subsequent urinary excretion, the use of these products as biomarkers provides a limited window of opportunity for collecting a viable sample. While the metabolites can be present in relatively high concentrations if the sample is obtained shortly after exposure, the urinary excretion of the analytes is generally rapid (hours to days). More long-lived biomarkers can be found in the blood resulting from agent interactions with macromolecules such as proteins or DNA to form adducts. The macromolecule acts as a depot for the covalently bound adduct. If the bond is stable, the residence time should be similar to the half-life of the target molecule.

The process of validating an analytical assay for human samples generally will proceed from in vitro method development to in vivo animal exposure models. The final stage is the analysis of actual patient samples. Validation of methods and biomarkers for CWA verification is limited to the analysis of biomedical specimens obtained following accidental or intentional exposure to CWA. The number of samples made available for analysis has been extremely limited. Another important aspect of method validation is the determination of background levels for the identified CWA biomarkers. Some of the biomarkers have been shown to exist in nonexposed individuals making interpretation of results in potential incidents difficult. Therefore studies to determine potential background levels and incidence of the biomarker in nonexposed human populations are essential.

This chapter will review some of most current methods used in the verification of exposure to CWAs. The review will be limited to biomarkers that have been identified in blood and urine samples obtained from human casualties of intentional or accidental exposure to sulfur mustard or nerve agents.

2 Sulfur Mustard

2.1 Background

The chemical warfare agent sulfur mustard (HD) is generally associated with World War I (WWI) when vast numbers of allied troops suffered incapacitating injuries following exposure. Since WWI, the use of sulfur mustard has been implicated in a number of conflicts, most notably during the Iran–Iraq war. In addition to the use of this chemical warfare agent in military operations or by terrorist groups, sulfur mustard poses a potential public health threat due to abandoned WWI munitions and stockpiled materials.

Sulfur mustard is classified as a vesicant chemical warfare agent and upon exposure can result in extensive damage to the skin, eyes, and lungs. It is a highly reactive, small molecular weight compound. An important reaction is the formation of a sulfonium ion produced following cyclization of an ethylene group. The sulfonium ion readily reacts with nucleophiles, such as water, or can combine with a variety of nucleophilic sites that occur in macromolecules.[1] The chemical reactions have the potential to produce a number of both free metabolites and stable adducts to macromolecules that can be exploited for analysis in urine, blood, or tissue samples.[2–4]

The laboratory analysis of biomedical samples following a suspected CWA exposure has rarely been performed, but has occurred most frequently following suspected exposure to sulfur mustard. Many of the samples made available for testing were obtained from casualties of the Iran–Iraq war in the 1980s. Some of the victims exhibiting clinical signs and symptoms consistent with sulfur mustard exposure were transported to hospitals in Europe for medical treatment. Published reports at that time on the laboratory analysis of their urine specimens utilized methods for the determination of unmetabolized sulfur mustard or for thiodiglycol (TDG), a hydrolysis product of sulfur mustard.[5–9] Fortunately, some of the urine specimens as well as some blood samples were frozen, stored as archived samples, and reanalyzed years later using more recent techniques.

Since 1995, a number of significant advances have occurred and are reflected in the published reports from that time to the present. Large strides have been made in the number of new metabolites that have been identified from sulfur mustard-exposed individuals. Developments in the field of mass spectrometry (MS) have resulted in large increases in sensitivity and selectivity. Analytical methods incorporating MS techniques have enabled the use of isotopically labelled forms of the analytes for use as internal standards during the earliest stages of sample preparation. This has resulted in greater reproducibility of the assays and made them more amenable to quantitative analysis.

In addition to biomedical samples received from the Iran–Iraq war, a small number of laboratory or field exposures to sulfur mustard have occurred where specimens were made available for verification testing. One recent case study in particular will be the primary focus of this review.

2.2 Urinary Biomarkers

There are currently five urinary metabolites that are of primary interest for verification of human exposure to sulfur mustard. Two of the metabolites, TDG and thiodiglycol sulfoxide (TDG-sulfoxide), are derived primarily from chemical hydrolysis reactions. The other three products are formed following the reaction of sulfur mustard with glutathione. Each of the five analytes has been identified in the urine of sulfur mustard-exposed individuals.

Efforts to analyze for specific biomarkers of sulfur mustard exposure in urine prior to 1995 targeted either unmetabolized sulfur mustard or TDG. Some of the urine samples obtained from casualties of the Iran–Iraq war exhibiting injuries consistent with a sulfur mustard exposure yielded positive results. Unfortunately, positive results were also found in some of the control specimens obtained from non-exposed individuals.[5–8]

More recently, a number of additional methods have been developed for the trace level analysis of TDG and/or TDG-sulfoxide in urine.[10–13] They are all gas chromatography-mass spectrometry (GC-MS) methods and require derivatization of the analyte. Some of the methods incubate the urine samples with glucuronidase with sulfatase activity to release glucuronide-bound conjugates. Some of the methods use titanium trichloride (TiCl$_3$) in hydrochloric acid to reduce TDG-sulfoxide to TDG. The strong acid also appears to hydrolyze acid-labile esters of TDG and TDG-sulfoxide. Ultimately, each of the methods will convert all target analytes into the single analyte TDG for analysis. All of the methods have similar limits of detection, approximately 0.5–1 ng/mL.

Unfortunately the TDG and TDG-sulfoxide methods all suffer from the observation that background levels have consistently been found in urine samples obtained from non-exposed individuals. In the most extensive study of background levels to date, urine samples from 105 individuals were examined for TDG and TDG-sulfoxide.[13] Quantifiable background levels were observed in 82% of the samples that were evaluated. The study indicated that the free and bound forms of the TDG-sulfoxide rather than TDG were primarily responsible for the observed background levels.

Observed background levels in control urine samples prevent the use of TDG and TDG-sulfoxide as definitive biomarkers of sulfur mustard exposure. Their analysis can still have utility on an individual level if multiple specimens over a period of time can be obtained to produce an excretion profile. A rare occurrence of this is the excretion profile generated from an individual that was the victim of an accidental laboratory exposure to sulfur mustard.[14] The erythematous and vesication areas on the individual were estimated to be less than 5% and 1% of the total body surface area, respectively. The analytical method used for the urine analysis measured both the free and the conjugated TDG.[11] A maximum TDG urinary excretion rate of 20 µg/day was observed on day 3. TDG concentrations of 10 ng/mL or greater were observed in some samples for up to a week after the exposure. A first-order elimination was calculated from days 4 through 10, and the elimination half-life of TDG was found to be approximately 1.2 days. A second example is discussed in the case study later in this chapter.

Black et al. identified a series of metabolites formed from the reaction of sulfur mustard with glutathione, a small molecular weight tripeptide that acts as a free radical scavenger.[15–17] While a large number of metabolites were identified in animal experiments, three reaction products have been verified in urine samples obtained from sulfur mustard-exposed individuals. Two of the reaction products are believed to result from the further metabolism of the sulfur mustard-glutathione conjugate by β-lyase enzyme: 1-methylsulfinyl-2-[2-(methylthio)ethylsulfonyl] ethane (MSMTESE) and 1,1′-sulfonylbis[2-(methylsulfinyl)ethane] (SBMSE). MSMTESE and SBMSE can be reduced using $TiCl_3$ and analyzed by GC/MS/MS as a single analyte: 1,1′-sulfonylbis[2-(methylthio)ethane] (SBMTE).[18,19] Black et al. reported a limit of detection of 0.1 ng/mL,[18] while Young et al. extended the lower limit of detection to 0.038 ng/mL.[19] To date, no background levels have been found in the urine of unexposed individuals including studies where urine samples from over 100 individuals were analyzed using two different methods.[13,19] Alternatively, MSMTESE and SBMSE can be analyzed individually without reducing the two analytes to a single analyte using electrospray LC/MS/MS.[20] Lower limits of detection were reported as 0.1–0.5 ng/mL for each of the analytes. The third urinary biomarker derived from the reaction of sulfur mustard with glutathione is 1,1′-sulfonylbis[2-S-(N-acetylcysteinyl)ethane]. Using solid-phase extraction (SPE) for sample cleanup and analyte concentration followed by analysis with negative ion electrospray LC/MS/MS, Read and Black were able to achieve detection limits of 0.5–1.0 ng/mL.[21]

The presence of the β-lyase metabolites has been verified in urine samples obtained from multiple incidents of sulfur mustard exposure. Several archived samples obtained from Iranian casualties were reanalyzed using the β-lyase metabolite assay. Concentrations found in the urine ranged between 0.5 and 5 ng/mL, except for one individual with a reported concentration of 220 ng/mL that died 5 days after hospital admission.[12] Two casualties of an alleged sulfur mustard attack on the Kurdish town of Halabja in 1988 also tested positive for the β-lyase metabolites.[12] Urine samples were analyzed using both GC/MS/MS[12] and LC/MS/MS.[20]

Black and Read also analyzed the urine from two individuals that were accidentally exposed to a WWI munition containing sulfur mustard using both the GC/MS/MS[15] and LC/MS/MS methods.[20] When the β-lyase metabolites were analyzed individually by LC/MS/MS, their concentrations ranged from 15 to 17 ng/mL and from 30 to 34 ng/mL for the mono-sulfoxide and bis-sulfoxide, respectively. When analyzed as the single, reduced form SBMTE using GC/MS/MS, observed concentrations were 42 and 56 ng/mL. Samples were also analyzed for the bis-(N-acetylcysteine) conjugate using LC/MS/MS.[21] While it is a major metabolite in rats exposed to sulfur mustard, it had not been previously reported to be found in urine samples from exposed individuals. The metabolite was found in urine samples from both exposed individuals, but concentrations were near the lower limit of detection of 0.5–1 ng/mL. A final incident is described in more detail in the case study below.

2.3   Case Study – Field Exposure to WWI Munition Containing Sulfur Mustard, Part 1

In 2004 an exposure incident involved two explosive ordnance technicians tasked with the destruction of a suspected WWI 75-mm munition. The munition had been discovered in a clamshell driveway and was believed to have originated from material dredged from a seafloor dumping area. Following demolition procedures, the two individuals came into contact with a brown oily liquid that was found leaking from the remnants of the munition. During the disposal operation, none of the ordnance team members complained of any eye and/or throat irritation, or breathing difficulties. The clinical sequence of events for one of the individuals has been reported, [22, 23] but will be described in brief here. Within 2 h of the munition's destruction, patient #1 noticed a tingling sensation on one arm and then showered. The next morning (approximately 14 h after the liquid contact), he had developed painful areas of the hand with noticeable reddening and the formation of small blisters. He went to a local emergency room where the blisters were observed to grow and coalesce, and he was subsequently transferred to a regional burn center. Blisters developed on the patient's arm, hand, ankle, and foot. The erythema and blistering area on the patient was estimated to be 6.5% body surface area. The patient never suffered any ocular or respiratory complications; consequently it appeared that the patient's injuries were only the result of a cutaneous liquid exposure. Patient #2 had a small, single blister and was not hospitalized. Urine samples from patient #1 were collected on days 2 through 11 and days 29, 35, and 42 after exposure and collected from patient #2 on days 2, 4, and 7 after exposure.

Hydrolysis metabolites were determined using GC/MS/MS with two different sample preparation methods. The first assay measured only free and glucuronide-bound TDG. The second assay used a reduction step to analyze free and glucuronide-bound TDG, free and glucuronide-bound TDG-sulfoxide, and acid-labile esters of TDG and TDG-sulfoxide as a single analyte.[24] The individual with the single blister (patient #2) did not have detectable levels in any of the urine samples using the assay for only TDG. Positive results were observed using the second assay, but the observed concentrations (2–4 ng/mL) fell within the range of concentrations previously observed in urine samples of unexposed individuals (approximately 20 ng/mL).[13] Using the first assay (TDG only), patient #1 had the highest observed TDG concentration of 24 ng/mL at day 2. TDG concentrations ranged between 6 and 11 ng/mL over the next 5 days and decreased to a range of 1–2 ng/mL for the following 4 days; TDG was undetected after day 11. The second assay (TDG, TDG-sulfoxide, and acid-labile esters) also had the highest observed concentration of 50 ng/mL at day 2. Only days 2, 5, and 6 produced concentrations that were greater than the highest observed background control levels, 50, 28, and 24 ng/mL, respectively. While these analytes would not be considered confirmatory due to their presence in background controls, the excretion profile provides strong evidence of a sulfur mustard exposure.

β-lyase metabolites were measured as the single analyte SBMTE using GC/MS/MS.[24] Levels for both patients decreased dramatically by day 3 after exposure. Patient #2 urine SBMTE concentrations were 2.6, 0.8, and 0.08 ng/mL for samples taken 2, 4, and 7 days after exposure, respectively. Patient #1 SBMTE concentrations decreased from 41 ng/mL at day 2 after exposure and continued to decrease to 7, 3.3, and 1.3 ng/mL over the next 3 days. For days 6–11, concentrations ranged between 0.07 and 0.02 ng/mL, and SBMTE was not detected beyond day 11. The urine from days 2 and 3 of patient #1 was also examined for the presence of the bis-(N-acetylcysteine) conjugate using LC/MS/MS. It was detected at a concentration of 3.1 ng/mL in the urine sample collected 1 day after exposure, but was not detected in the day 3 sample.

Results to date suggest that the analytes of choice for assessment of potential exposure to sulfur mustard in urine samples would be the two β-lyase metabolites. They have been verified in human exposure cases, sensitive and selective assays have been developed, and to date, no known examples of background levels have been found in the urine from unexposed individuals. The β-lyase analytes can be measured individually using LC/MS/MS or reduced to a single analyte (SBMTE) and measured using GC/MS/MS.

Assays for several other potential urinary metabolites have been developed, but the analytes have yet to be confirmed in human exposure samples. N7-(2-hydroxyethylthioethyl) guanine is a breakdown product from alkylated DNA that has been observed in animal studies. Fidder et al. developed both a GC/MS method that requires derivatization of the analyte and a LC/MS/MS method that can analyze the compound directly.[25] Other possible urinary analytes are an imidazole derivative formed from the reaction of sulfur mustard with protein histidine residues[26] and sulfur mustard adducts to metallothionien.[27]

2.4 Blood Biomarkers

Whereas urinary metabolites undergo relatively rapid elimination from the body, blood components offer biomarkers that have the potential to be used for verification of sulfur mustard exposure long after the exposure incident. Three different approaches have been utilized for blood biomarker analysis. The intact macromolecule such as protein or DNA with the sulfur mustard adduct(s) still attached can be analyzed. For proteins, an alternate approach is to enzymatically digest the proteins to produce a smaller peptide with the adduct still attached. Methods of this type have been developed for both hemoglobin and albumin. A third approach has been to cleave the sulfur mustard adduct from the macromolecule and analyze in a fashion similar to that used for free metabolites found in the urine. The latter two approaches have both been successfully utilized to verify human exposure of sulfur mustard.

Hemoglobin is an abundant and long-lived protein in human blood. Alkylation reactions between sulfur mustard and hemoglobin have been shown to occur with six histidine, three glutamic acid, and two valine amino acids of hemoglobin.[28,29] While the histidine adducts appear to be the most abundant type, their analysis using mass spectrometry techniques is problematic, and the method does not appear to be as sensitive as the method for the analysis of the N-terminal valine adducts.[30] Adducts to the N-terminal valine amino acids represent only a small fraction of the total alkylation of the macromolecule, but their location on the periphery of the molecule allows them to be selectively cleaved using a modified Edman degradation. Following isolation of the globin from the red blood cells, the globin is reacted with pentafluorophenyl isothiocyanate to form a thiohydantoin compound, which is further derivatized prior to analysis. The derivatized compound can then be analyzed using negative ion chemical ionization GC/MS.[31-33] The lower limit of detection for the assay was determined using in vitro exposures of sulfur mustard in human whole blood and was determined to be equivalent to a 100 nM exposure level.[31,33]

The sulfur mustard adduct to the N-terminal valine of hemoglobin has been verified from the blood of multiple casualties of sulfur mustard exposure.[32,34,35] Archived blood samples from two Iranian casualties believed to have been exposed to sulfur mustard 22 and 26 days prior to blood collection tested positive for the valine adduct. The same samples also tested positive for DNA sulfur mustard adducts using an ELISA method.[34] A separate set of archived blood samples from four Iranian casualties was examined for both the valine and histidine sulfur mustard adducts of hemoglobin.[32] Blood samples were collected 5 or 10 days following the suspected exposure incident. Levels of the valine and histidine adducts ranged between 0.3 and 0.8 ng/mL and 0.7 and 2.5 ng/mL, respectively. Similar values were observed following an accidental exposure of an individual with a WWI sulfur mustard munition.[32]

Following the administration of a single dose of sulfur mustard to a marmoset (4.1 mg/kg; i.v.), the valine adduct was still detected in blood taken 94 days later.[2] Intact hemoglobin with the sulfur mustard adducts attached have been examined using matrix-assisted laser desorption/ionization time-of-flight mass spectrometry, but to date the technique has only been utilized for in vitro experiments at relatively high concentrations of sulfur mustard.[27]

Human serum albumin has been found to be alkylated by sulfur mustard at the cysteine-34 position. Following isolation of the albumin from the blood, the albumin can be reacted with Pronase enzyme to digest the protein. One of the resulting peptide fragments is a tripeptide of the sequence cysteine-proline-phenylalanine, which contains the sulfur mustard alkylated cysteine-34. After SPE extraction, the tripeptide can be analyzed using LC/MS/MS.[36] The lower limit of detection for the assay is 1 nM, once again reported as an equivalent exposure level as determined using in vitro exposures of sulfur mustard in human whole blood. Recently, a modification to the isolation of the albumin from blood was reported using affinity chromatography rather than the precipitation procedure used previously.[37] The modified procedure reduced the sample preparation time significantly.

Blood samples obtained from nine Iranian casualties, collected between 8 and 9 days after a suspected exposure and previously found to be positive for the N-terminal valine sulfur mustard adduct of hemoglobin using GC/MS,[35] were analyzed for the albumin cysteine adduct using LC/MS/MS.[36] Adduct levels for both the hemoglobin valine adduct and for the albumin cysteine adduct were in very close agreement with each other, between 0.3 and 2 μM and 0.4 and 1.8 μM for the hemoglobin and albumin adducts, respectively. More recently the albumin adduct assay was used to analyze a series of blood samples collected from the accidental exposure detailed in the case study section above. The albumin adduct results along with a final blood protein assay are presented below in a continuation of the sulfur mustard exposure case study. A summary of the biomarkers that have been verified in urine and blood samples obtained from casualties of intentional or accidental exposure to sulfur mustard is provided in Fig. 1.

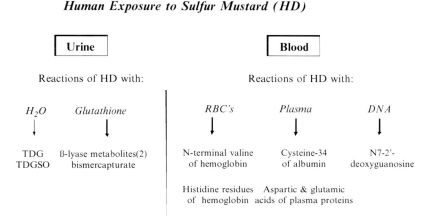

Fig. 1 Summary of biomarkers that have been verified in urine and blood samples obtained from casualties of intentional or accidental exposure to sulfur mustard

2.5   Case Study – Field Exposure to WWI Munition Containing Sulfur Mustard, Part 2

Details of an accidental exposure of two individuals to a WWI munition in 2004 were given in the urine section above. Generally, biomedical samples that are made available for verification analysis are from a single time point after the exposure. In this rare instance, the patient with the more severe injuries had blood and urine collected almost daily for a 10-day period soon after exposure and then again on days 29, 35, and 42. The blood from both individuals was analyzed using two different assays. The first assay targeted the sulfur mustard adduct to cysteine-34 of albumin as described in the section above. Based on in vitro exposures of sulfur mustard in human whole blood, concentrations of albumin adducts found in the plasma of the more severely injured patient were 350 nM on day 2 after the exposure and had decreased by 74% (90 nM) on day 42. The rate of decrease over that time period was consistent with the reported half-life of human albumin of 21 days. Albumin adduct concentrations for the patient with a single blister ranged between 16 and 18 nM over the sample collection period of 2–7 days after exposure.[38]

A second method was used for the analysis of the blood samples in which adducts to blood proteins are targeted using a more general approach. It was previously shown that sulfur mustard adducts of glutamic and aspartic acids to keratin could be cleaved using base.[39] Using a similar approach, precipitated proteins from plasma, whole blood, or red blood cells were treated with base to liberate the sulfur mustard adduct (hydroxyethylthioethyl) from the protein. Upon release, the adduct (in the form of thiodiglycol) was derivatized and analyzed using negative-ion chemical ionization GC/MS. The lower limit of detection for the assay in plasma was 25 nM as determined using in vitro exposures of sulfur mustard in human plasma.[40] This was the first reported use of this assay in the verification of a human exposure to sulfur mustard. Concentrations of the plasma protein adducts for the severely injured patient were 97 nM on day 2 and decreased by 76% (23 nM) on day 42 based on in vitro exposures of sulfur mustard in human plasma.[38] The assay could not detect plasma protein adducts in the patient with the single blister. The assay was modified slightly to lower the reported lower limit of detection of 25 nM, but the limited amount of plasma received did not permit reanalysis using the modified method. More recently, additional improvements to the assay have resulted in a lower limit of quantitation of 1.6 nM.[41]

Blood provides several options for assessing potential exposure to sulfur mustard since a variety of different metabolites have been verified in human exposure cases. Most of the assays are sensitive and selective, and the majority of the methods utilize gas or liquid chromatography combined with mass spectrometric techniques. Background levels have not been found in the blood from unexposed individuals. The time period between exposure and when the sample is collected and the severity of the injury, both of which will impact sensitivity requirements of the assay, are probably the most important factors to consider when selecting the appropriate assay.

3   Nerve Agents

3.1   *Background*

The organophosphorus (OP) nerve agents are considered to be extremely toxic compounds due to their binding and inhibition of the enzyme acetylcholinesterase (AChE). Inhibition of AChE results in excessive accumulation of acetylcholine, which in turn produces hyperstimulation of cholinergic tissues and organs. It has the potential to produce a life-threatening cholinergic crisis in humans.[42] The nerve agents most commonly referenced are tabun (GA), sarin (GB), soman (GD), cyclosarin (GF), and VX. They inhibit AChE by the formation of a covalent bond between the phosphorus atom of the nerve agent and a serine residue of the enzyme active site. That interaction results in the displacement or loss of fluorine from GB, GD, and GF. Binding of GA and VX is different in that the leaving group is cyanide or a thiol group, respectively.[43] Spontaneous reactivation of the enzyme or hydrolysis reactions with water produce corresponding alkyl methylphosphonic acids (AMPA). Alternatively, the loss of the O-alkyl group while bound to the enzyme produces a highly stable organophosphoryl-ChE bond, a process referred to as aging. Once aging has occurred the enzyme is resistant to reactivation by oximes used for post-exposure therapy or by other nucleophilic reagents.[42] The spontaneous reactivation and aging rates of the agents vary depending on the O-alkyl group. For example, VX-inhibited red blood cell (RBC) AChE reactivates at an approximate rate of 0.5–1% per hour for the first 48 h with minimal aging. On the other hand, GD-inhibited AChE does not spontaneously reactivate and has a very rapid aging rate with a half-time of approximately 2 min.[42]

Analysis of the parent nerve agents from biomedical matrices such as blood or urine is not a viable option for a diagnostic technique or retrospective detection of exposure.[2] The parent agents are relatively short-lived due to rapid hydrolysis reactions and from binding of the nerve agent to plasma/tissue proteins. The short residence time is especially profound with the G-agents relative to VX.[44,45] The initial methods developed for nerve agent verification targeted the AMPAs. Many of the more recent assays developed for exposure verification are based on the interaction of nerve agents with AChE and other cholinesterase (ChE) enzymes or with blood proteins.

3.2   *Urinary Biomarkers*

AMPAs are produced in vivo as a result of hydrolysis with water and/or split off following spontaneous regeneration of the AChE enzyme. Studies with radiolabeled parent nerve agents in animals suggest that they are rapidly metabolized/hydrolyzed in the blood and appear in the urine as their respective alkyl methylphosphonic acids.[46-48] This observation led to the development of the initial assay for AMPAs in biological samples,[49] the applicability of which was subsequently demonstrated in animals exposed to nerve agents.[50] The common products found are isopropyl methylphosphonic acid (IMPA), pinacolyl methylphosphonic acid (PMPA), cyclohexyl methylphosphonic acid (CMPA), and ethyl methylphosphonic acid (EMPA) derived from GB, GD, GF and VX, respectively. Over time, numerous variations for the analysis of AMPAs in biological fluids such as plasma and urine have been developed. These include GC separations with MS[51-53] or tandem MS (MS-MS),[51,52,54,55] and liquid chromatography (LC) with MS-MS.[56] The AMPAs require derivatization prior to GC analysis.

To date, only the presence of IMPA has been verified in urine samples following a human exposure to an OP chemical warfare nerve agent. The terrorist attacks by the Aum Shinrikyo cult in Japan utilized GB on two separate occasions. The first was in an apartment complex in Matsumoto City. Approximately 12 L of GB were released using a heater and fan resulting in 600 casualties and 7 deaths. Urine specimens were collected from one individual that was found unconscious and believed to be suffering from an inhalation and possibly a cutaneous exposure. The patient was hospitalized and found to have a depressed AChE level, but was not given oxime therapy. Nakajima et al.[57] demonstrated the presence of IMPA and the secondary hydrolysis compound methyl phosphonic acid (MPA) in the urine using GC with a flame photometric detector (FPD). Sample preparation included C18 SPE prior to derivatization. Urinary concentrations of IMPA were 0.76, 0.08 and 0.01 ug/mL for IMPA on the 1st, 3rd, and 7th day after exposure, respectively. MPA concentrations were 0.14 and 0.02 ug/mL on the 1st and 3rd day after exposure, respectively.[57]

In the second attack, GB was released into the Tokyo subway, resulting in greater than 5,000 casualties and 12 deaths. Minami et al.[58] demonstrated the presence of IMPA and MPA in the urine from two individuals collected over a 7-day period. One patient was comatose while the second patient was discharged from the hospital within 3 days. The method utilized a SPE ion-exchange cartridge prior to derivatization followed by GC-FPD analysis. Although the report on the Tokyo incident did not directly indicate urinary concentrations, calculations were made on estimates of total sarin exposure. Estimates on the two individuals involved were 0.13–0.25 mg/person in the comatose patient and 0.016–0.032 mg/person in a less severely exposed casualty.[58] These numbers are approximately tenfold less than those reported by Nakajima et al.[57] for the severely intoxicated patient. Consistent with the rapid elimination of these compounds, the maximum urine concentration was reported to be within 12 h of exposure for IMPA and between 10 and 18 h for MPA.

The AMPAs provide a convenient marker for determining exposure to nerve agents and multiple methods have been developed for their analysis. It is important though to consider the extent of exposure and time of sample collection following the suspected exposure incident. Except in cases of severe exposure, the results above indicate that the hydrolysis products would not be present for more than a few days at most following exposure.

3.3   Blood Biomarkers

AMPAs should also be present in blood samples. Noort et al., using isobutanol/toluene extraction followed by negative ion electrospray LC-MS-MS analysis, detected IMPA in the serum of 15 individuals poisoned in the Tokyo and Matsumoto incident.[56] Samples were obtained 1.5 h after the incident. IMPA serum concentrations ranged from 2 to 135 ng/mL.[56,59] For some patients a second sample was obtained at 2–2.5 h after the incident. In those samples the authors reported significantly lower IMPA concentrations consistent with the rapid elimination of these compounds.[56]

Tsuchihashi et al. demonstrated the presence of EMPA in the serum of an individual assassinated with VX in Osaka, Japan, in 1994.[51] The authors also reported the presence of diisopropyl aminoethyl methyl sulfide (DAEMS), which is believed to result from the in vivo methylation of diisopropyl aminoethanethiol (DAET) subsequent to cleavage of the P-S bond. The serum sample was subjected to both aqueous and organic extractions with EMPA in the aqueous fraction and DAEMS in the organic portion. Following the extraction, the samples were analyzed using GC-MS-MS. Reported concentrations in serum that was collected 1 h after exposure were 143 ng/mL for DAEMS and 1.25 ug/mL for EMPA.

From the standpoint of exposure verification, the relatively rapid excretion and short-lived presence of hydrolysis products imposes time restrictions for collecting a viable sample. Efforts to increase the sampling window have taken advantage of interactions of CWAs with macromolecules such as proteins. The reaction of CWAs with biomolecules provides a pool of the adducted compound that can be utilized to verify exposure. Theoretically, the longevity of the marker is consistent with the in vivo half-life of the target molecule, provided that the binding affinity is high enough that spontaneous reactivation does not occur. Binding of nerve agents to ChE or blood protein targets can be leveraged to develop assays with an increased window of opportunity for detecting exposure. Several assays have been developed based on variations of this concept.

Polhuijs et al.[60] developed an assay based on earlier observations that ChE inhibited by GB could be reactivated with fluoride ions.[61] The displacement of GB covalently bound to butyrylcholinesterase (BChE) was accomplished by incubation of inhibited plasma with fluoride to form free enzyme plus the parent agent (isopropyl methylphosphonofluoridate). Following isolation from the matrix with solid-phase extraction techniques, the agent was then analyzed using GC with MS or other appropriate detection systems. The fluoride ion regeneration procedure was used to analyze serum from exposed individuals in the Aum Shinrikyo terrorist attacks at Matsumoto and in the Tokyo subway. The same set of serum samples was previously found to be positive for IMPA in 15 of 18 samples[56] (see above). Regenerated GB was found in 12 of the serum samples. Concentrations ranged from 1.8 to 2.7 ng/mL in the Matsumoto samples and between 0.2 and 4.1 ng/mL in the Tokyo samples.[60]

The fluoride ion regeneration procedure has also been applied to VX exposure verification. For VX, the thiol group initially lost upon binding to the enzyme is replaced by fluorine, resulting in a fluorinated analog of VX (ethyl methyl-phosphonofluoridate; VX-G).[62] The assay was used to analyze biomedical samples obtained from one individual following an accidental, low-level laboratory exposure to vaporized VX. Urine specimens were collected from 1 to 6 days post-exposure and were found to be negative for EMPA using GC-MS-MS. Blood specimens were collected from 1 to 27 days post-exposure. Plasma was analyzed using GC high resolution MS, and RBCs were analyzed using positive chemical ionization GC-MS-MS. The concentration of regenerated plasma VX-G decreased from 81 pg/mL on day1 post-exposure to 7 pg/mL by day 27. The estimated half-life was 7.5 days.[63] Regenerated RBC VX-G concentrations decreased from 220 to 97 pg/mL between days 1 and 27, respectively, with an estimated half-life of 24.5 days.[64]

With regard to the ability of the fluoride ion to regenerate GD bound to BChE, it is well known that the process of aging would preclude release from the enzyme. However, studies have indicated that the fluoride ion regeneration process as applied to GD-poisoned animals has produced successful results.[65] These studies suggest that soman can be displaced from sites where aging does not play a significant role. Black et al.[66] have demonstrated that both GB and GD bind to tyrosine residues of human serum albumin. The observation that the alkyl group remained intact in particular for GD strongly argues that binding to this site does not result in aging as seen with cholinesterases.[66] Currently it is unclear as to the utility of fluoride regeneration in human exposures involving GD, but animal studies indicate that binding to other sites may offer potential for using this procedure.

An alternate approach based on OP binding to BChE has been reported by Fidder et al.[67] Using a procainamide affinity gel, BChE can be isolated from serum. Following pepsin digestion of BChE to produce nonapeptide fragments containing the serine-198 residue to which nerve agents are known to bind, the peptide fragments were analyzed using positive ion electrospray LC-MS-MS. The utility of the method was demonstrated by analyzing archived samples from two victims of the Tokyo subway terrorist attack. Concentrations of GB-inhibited BChE in the serum samples ranged between 10 and 20 pmol/mL.[67] An advantage of this technique is that aged or non-aged OP adducts should be identifiable. A limitation to the assay is that advance information on the agent identity is needed for MS analysis. For this reason, an extension of this procedure utilizing a generic approach has been developed.[68] The methodology employs the chemical modification of the phosphyl group on the serine residue to a common nonapeptide regardless of the specific agent involved. Therefore, since a common nonapeptide is the outcome, a single MS method would be employed in the analysis.[68]

Nagao et al.[69] developed a GC-MS procedure using an alkaline phosphatase digestion process to measure IMPA following its release from the GB-AChE complex. The method was used to analyze blood RBC specimens from four victims of the Tokyo subway attack. Initially they attempted to detect MPA or IMPA directly from the blood and were not successful. They were able to detect the compounds following release from erythrocyte-derived AChE, using the alkaline phosphatase digestion method. Although the authors indicated that all four RBC specimens were

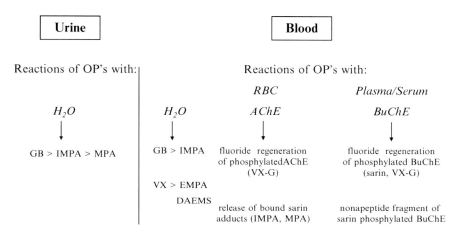

Fig. 2 Summary of biomarkers that have been verified in urine and blood samples obtained from casualties of intentional or accidental exposure to nerve agents

positive for GB adducts, they did not report MPA or IMPA concentrations. A summary of the biomarkers that have been verified in urine and blood samples obtained from casualties of intentional or accidental exposure to OP nerve agents is provided in Fig. 2.

4   Conclusions

Due to the limited number of human exposures involving sulfur mustard and nerve agents, fully ascertaining the advantages and disadvantages of the various definitive methodologies that have been developed is difficult. Several factors can influence the choice of sample and the method of analysis. One of the primary considerations is the length of time between the suspected exposure event and the time of sample collection. Human exposure results to date indicate that hydrolysis products in blood or urine undergo relatively rapid elimination. Consequently, urine specimen collection will provide a more limited opportunity to obtain a viable sample than will blood specimens. The advantage of utilizing adducts formed with macromolecules such as protein targets in the blood is a longer window of opportunity to verify exposures relative to that of hydrolysis products. Another important consideration is the route and duration of exposure. In instances of mild exposure to CWAs, the sensitivity and selectivity of the assay will become a critical factor in the verification process. Mass spectrometry and, in most instances, tandem mass spectrometry based methods are currently the methods of choice to accomplish this critical task.

Acknowledgements This work was supported by the Defense Threat Reduction Agency-Joint Science and Technology Office, Medical S&T Division. The opinions, interpretations, conclusions, and recommendations are those of the authors and are not necessarily endorsed by the U.S. Army or the Department of Defense.

References

1. Papirmeister B, Feister AJ, Robinson SI, Ford RD (1991) Medical defense against mustard gas: toxic mechanisms and pharmacological implications. CRC Press, Boca Raton, FL, pp 91–122
2. Noort D, Benschop HP, Black RM (2002) Biomonitoring of exposure to chemical warfare agents: a review. Tox Appl Pharmacol 184:116–126
3. Black RM, Noort D (2005) In: Mesilaakso M (ed) Chemical weapons convention chemicals analysis: sample collection, preparation, and analytical methods. Wiley, Chichester, West Sussex, England, pp 403–431
4. Noort D, Black RM (2005) In: Mesilaakso M (ed) Chemical weapons convention chemicals analysis: sample collection, preparation, and analytical methods. Wiley, Chichester, West Sussex, England, pp 433–451
5. Vycudilik W (1985) Detection of mustard gas bis(2-chloroethyl)-sulfide in urine. Forensic Sci Int 28:131–136
6. Vycudilik W (1987) Detection of bis(2-chloroethyl)-sulfide (yperite) in urine by high resolution gas chromatography-mass spectrometry. Forensic Sci Int 35:67–71
7. Wils ERJ, Hulst AG, de Jong AL, Verweij A, Boter HL (1985) Analysis of thiodiglycol in urine of victims of an alleged attack with mustard gas. J Anal Toxicol 9:254–257
8. Wils ERJ, Hulst AG, van Laar J (1988) Analysis of thiodiglycol in urine of victims of an alleged attack with mustard gas, part II. J Anal Toxicol 12:15–19
9. Drasch G, Kretschmer E, Kauert G, von Meyer L (1987) Concentrations of mustard gas [bis(2-chloroethyl)sulfide] in the tissues of a victim of vesicant exposure. J Forensic Sci 32:1788–1793
10. Black RM, Read RW (1988) Detection of trace levels of thiodiglycol in blood, plasma and urine using gas chromatography-electron-capture negative-ion chemical ionisation mass spectrometry. J Chromatogr 449:261–270
11. Jakubowski EM, Woodard CL, Mershon MM, Dolzine TW (1990) Quantification of thiodiglycol in urine by electron ionization gas chromatography-mass spectrometry. J Chromatogr 528:184–190
12. Black RM, Read RW (1995) Improved methodology for the detection and quantitation of urinary metabolites of sulphur mustard using gas chromatography-tandem mass spectrometry. J Chromatogr B 665:97–105
13. Boyer AE, Ash D, Barr DB et al (2004) Quantitation of the sulfur mustard metabolites 1, 1'-sulfonylbis[2-(methylthio)ethane] and thiodiglycol in urine using isotope-dilution gas chromatography-tandem mass spectrometry. J Anal Toxicol 28:327–332
14. Jakubowski EM, Sidell FR, Evans RA et al (2000) Quantification of thiodiglycol in human urine after an accidental sulfur mustard exposure. Toxicol Method 10:143–150
15. Black RM, Read RW (1995) Biological fate of sulphur mustard, 1, 1'-thiobis(2-chloroethane): identification of β-lyase metabolites and hydrolysis products in human urine. Xenobiotica 25:167–173
16. Black RM, Brewster K, Clarke RJ, Hambrook JL, Harrison JM, Howells DJ (1992) Biological fate of sulphur mustard, 1, 1'-thiobis(2-chloroethane): isolation and identification of urinary metabolites following intraperitoneal adminstration to rat. Xenobiotica 22:405–418
17. Black RM, Hambrook JL, Howells DJ, Read RW (1992) Biological fate of sulfur mustard, 1,1'-thiobis(2-chloroethane). Urinary excretion profiles of hydrolysis products and β-lyase metabolites of sulfur mustard after cutaneous application in rats. J Anal Toxicol 16:79–84

18. Black RM, Clarke RJ, Read RW (1991) Analysis of 1, 1'-sulphonylbis[2-(methylsulphinyl) ethane] and 1-methylsulphinyl-2-[2-(methylthio)ethylsulphonyl]ethane, metabolites of sulphur mustard, in urine using gas chromatography-mass spectrometry. J Chromatogr 558: 405–414

19. Young CL, Ash D, Driskell WJ et al (2004) A rapid, sensitive method for the quantitation of specific metabolites of sulfur mustard in human urine using isotope-dilution gas chromatography-tandem mass spectrometry. J Anal Toxicol 28:339–345

20. Read RW, Black RM (2004) Analysis of β-lyase metabolites of sulfur mustard in urine by electrospray liquid chromatography-tandem mass spectrometry. J Anal Toxicol 28:346–351

21. Read RW, Black RM (2004) Analysis of the sulfur mustard metabolite 1, 1'-sulfonylbis[2-S-(N-acetylcysteinyl)ethane] in urine by negative ion electrospray liquid chromatography-tandem mass spectrometry. J Anal Toxicol 28:352–356

22. Carroll LS (2005) Images in clinical toxicology: sulfur mustard cutaneous exposure. Clin Toxicol 1, 55

23. Newmark J, Langer JM, Capacio BR, Barr JR, McIntosh RG (2007) Liquid sulfur mustard exposure. Mil Med 2:196–198

24. Barr JR, Pierce CL, Smith JR et al (2008) Analysis of urinary metabolites of sulfur mustard in two individuals after accidental exposure. J Anal Toxicol 32:10–16

25. Fidder A, Noort D, de Jong LPA, Benschop HP, Hulst AG (1996) N7-(2-hydroxyethylthioethyl)-guanine: a novel urinary metabolite following exposure to sulphur mustard. Arch Toxicol 70:854–855

26. Sandelowsky I, Simon GA, Bel P, Barak R, Vincze A (1992) N^1-(2-hydroxyethylthioethyl)-4-methylimidazole (4-met-1-imid-thiodiglycol) in plasma and urine: a novel metabolite following dermal exposure to sulphur mustard. Arch Toxicol 66:296–297

27. Price EO, Smith JR, Clark CR, Schlager JJ, Shih ML (2000) MALDI-TOF/MS as a diagnostic tool for the confirmation of chemical agent exposure. J Appl Toxicol 20:S193–S197

28. Noort D, Verheij ER, Hulst AG, de Jong LPA, Benschop HP (1996) Characterization of sulfur mustard induced structural modifications in human hemoglobin by liquid chromatography-tandem mass spectrometry. Chem Res Toxicol 9:781–787

29. Black RM, Harrison JM, Read RW (1997) Biological fate of sulphur mustard: *In vitro* alkylation of human haemoglobin by sulphur mustard. Xenobiotica 27:11–32

30. Noort D, Hulst AG, Trap HC, de Jong LPA, Benschop HP (1997) Synthesis and mass spectrometric identification of the major amino acid adducts formed between sulphur mustard and haemoglobin in human blood. Arch Toxicol 71:171–178

31. Fidder A, Noort D, de Jong AL, Trap HC, de Jong LPA, Benschop HP (1996) Monitoring of *in vitro* and *in vivo* exposure to sulfur mustard by GC/MS determination of the N-terminal valine adduct in hemoglobin after a modified Edman degradation. Chem Res Toxicol 9:788–792

32. Black RM, Clarke RJ, Harrison JM, Read RW (1997) Biological fate of sulphur mustard: identification of valine and histidine adducts in haemoglobin from casualties of sulphur mustard poisoning. Xenobiotica 27:499–512

33. Noort D, Fidder A, Benschop HP, de Jong LPA, Smith JR (2004) Procedure for monitoring exposure to sulfur mustard based on modified Edman degradation of globin. J Anal Toxicol 28:311–315

34. Benschop HP, van der Schans GV, Noort D, Fidder A, Mars-Groenendijk RH, de Jong LPA (1997) Verification of exposure to sulfur mustard in two casualties of the Iran-Iraq conflict. J Anal Toxicol 21:249–251

35. Benschop HP, Noort D, van der Schans GV, de Jong LPA (2000) Diagnosis and dosimetry of exposure to sulfur mustard: development of standard operating procedures; further exploratory research on protein adducts. Final report for contract DAMD17-97-2-7002, ADA381035

36. Noort D, Hulst AG, de Jong LPA, Benschop HP (1999) Alkylation of human serum albumin by sulfur mustard *in vitro* and *in vivo*: mass spectrometric analysis of a cysteine adduct as a sensitive biomarker of exposure. Chem Res Toxicol 12:715–721

37. Noort D, Fidder A, Hulst AG, Woolfitt AR, Ash D, Barr JR (2004) Retrospective detection of exposure to sulfur mustard: improvements on an assay for liquid chromatography-tandem mass spectrometry analysis of albumin/sulfur mustard adducts. J Anal Toxicol 28:333–338
38. Smith JR, Capacio BR, Korte WD, Woolfitt AR, Barr JR (2008) Analysis for plasma protein biomarkers following an accidental human exposure to sulfur mustard. J Anal Toxicol 32:17–24
39. Noort D, Fidder A, Hulst AG, de Jong LPA, Benschop HP (2000) Diagnosis and dosimetry of exposure to sulfur mustard: development of a standard operating procedure for mass spectrometric analysis of haemoglobin adducts: Exploratory research on albumin and keratin adducts. J Appl Toxicol 20:S187–S192
40. Capacio BR, Smith JR, DeLion MT et al (2004) Monitoring sulfur mustard exposure by gas chromatography-mass spectrometry analysis of thiodiglycol cleaved from blood proteins. J Anal Toxicol 28:306–310
41. Lawrence RJ, Smith JR, Boyd BL, Capacio BR (2008) Improvements in the methodology of monitoring sulfur mustard exposure by gas chromatography-mass spectrometry analysis of cleaved and derivatized blood protein adducts. J Anal Toxicol 32:31–36
42. Sidell FR (1997) In: Sidell FR, Takafuji ET, Franz DR (eds) Medical aspects of chemical and biological warfare. Office of The Surgeon General, Department of the Army, Washington, DC, pp 129–179
43. Degenhardt CEAM, Pleijsier K, van der Schans MJ et al (2004) Improvements of the fluoride reactivation method for the verification of nerve agent exposure. J Anal Toxicol 28:364–371
44. Benschop HP, De Jong LPA (2001) In: Somani SM, Romano JR (eds) Chemical warfare agents: toxicity at low levels. CRC Press, Boca Raton, FL, pp 25–81
45. van der Schans MJ, Lander BJ, van der Wiel H, Langenberg JP, Benschop HP (2003) Toxicokinetics of the nerve agent (±)VX in anesthetized and atropinized hairless guinea pigs and marmosets after intravenous and percutaneous administration. Tox Appl Pharmacol 191:48–62
46. Harris LW, Braswell LM, Fleisher JP, Cliff WJ (1964) Metabolites of pinacolyl methylphosphonofluoridate (soman) after enzymatic hydrolysis in vitro. Biochem Pharmacol 13:1129–1136
47. Reynolds ML, Little PJ, Thomas BF, Bagley RB, Martin BR (1985) Relationship between the disposition of [³H]soman and its pharmacological effects in mice. Toxicol Appl Pharmacol 80:409–420
48. Little PJ, Reynolds ML, Bowman ER, Martin BR (1986) Tissue disposition of [³H]sarin and its metabolites in mice. Toxicol Appl Pharmacol 83:412–419
49. Shih ML, Smith JR, McMonagle JD, Dolzine TW, Gresham VC (1991) Detection of metabolites of toxic alkylmethylphosphonates in biological samples. Biological Mass Spectrom 20:717–723
50. Shih ML, McMonagle JD, Dolzine TW, Gresham VC (1994) Metabolite pharmacokinetics of soman, sarin, and GF in rats and biological monitoring of exposure to toxic organophosphorus agents. J Appl Toxicol 14:195–199
51. Tsuchihashi H, Katagi M, Nishikawa M, Tatsuno M (1998) Identification of metabolites of nerve agent VX in serum collected from a victim. J Anal Toxicol 22:383–388
52. Fredriksson S-Å, Hammarström L-G, Henriksson L, Lakso H-Å (1995) Trace determination of alkyl methylphosphonic acids in environmental and biological samples using gas chromatography/negative-ion chemical ionization mass spectrometry and tandem mass spectrometry. J Mass Spectrom 30:1133–1143
53. Miki A, Katagi M, Tsuchihashi H, Yamashita M (1999) Determination of alkylmethylphosphonic acids, the main metabolites of organophosphorus nerve agents, in biofluids by gas chromatography-mass spectrometry and liquid-liquid-solid-phase-transfer-catalyzed pentafluorobenzylation. J Anal Toxicol 23:86–93
54. Driskell WJ, Shih ML, Needham LL, Barr DB (2002) Quantitation of organophosphorus nerve agent metabolites in human urine using isotope dilution gas chromatography-tandem mass spectrometry. J Anal Toxicol 26:6–10

55. Barr JR, Driskell WJ, Aston LS, Martinez RA (2004) Quantitation of metabolites of the nerve agents sarin, soman, cyclosarin, VX and Russian VX in human urine using isotope-dilution gas chromatography-tandem mass spectrometry. J Anal Toxicol 28:371–378

56. Noort D, Hulst A, Platenburg DHJM, Polhuijs M, Benschop HP (1998) Quantitative analysis of O-isopropyl methylphosphonic acid in serum samples of Japanese citizens allegedly exposed to sarin: estimation of internal dosage. Arch Toxicol 72:671–675

57. Nakajima T, Sasaki K, Ozawa H, Sekijima Y, Mortia H, Fukushima Y, Yanagisawa N (1998) Urinary metabolites of sarin in a patient of the Matsumoto incident. Arch Toxicol 72:601–603

58. Minami M, Hui D-M, Katsumata M, Inagaki H, Boulet CA (1997) Method for the analysis of methylphosphonic acid metabolites of sarin and its ethanol-substituted analogue in urine as applied to the victims of the Tokyo sarin disaster. J Chromatogr B 695:237–244

59. Polhuijs M, Langenberg JP, Noort D, Hulst AG, Benschop HP (1999) In: Sohns T, Voicu VA (eds) NBC risks: current capabilities and future perspectives for protection. Kluwer, Dordrecht, The Netherlands, pp 513–521

60. Polhuijs M, Langenberg JP, Benschop HP (1997) New method for retrospective detection of exposure to organophosphorus anticholinesterases: Application to alleged sarin victims of Japanese terrorists. Toxicol Appl Pharmacol 146:156–161

61. Albanus L, Heilbeonn E, Sundwall A (1965) Antidote effect of sodium fluoride in organophosphorus anticholinesterase poisoning. Biochem Pharmacol 14:1375–1381

62. Jakubowski EM, Heykamp LS, Durst HD, Thompson SA (2001) Preliminary studies in the formation of ethyl methylphosphonofluoridate from rat and human serum exposed to VX and treated with fluoride ion. Anal Lett 34:727–737

63. Solano MI, Thomas JD, Taylor JT et al (2008) Quantification of nerve agent VX-butyrylcholinesterase adduct biomarker from an accidental exposure. J Anal Toxicol 32:73–77

64. McGuire JM, Taylor JT, Byers CE, Jakubowski EM, Thomson SA (2008) Determination of VX-G analogue in red blood cells via gas chromatography-tandem mass spectrometry following an accidental exposure to VX. J Anal Toxicol 32:73–77

65. Adams TK, Capacio BR, Smith JR, Whalley CE, Korte WD (2004) The application of the fluoride reactivation process to the detection of sarin and soman nerve agent exposures in biological samples. Drug Chem Toxicol 27:77–91

66. Black RM, Harrison JM, Read RW (1999) The interaction of sarin and soman with plasma proteins: the identification of a novel phosphonylation site. Arch Toxicol 73:123–126

67. Fidder A, Hulst D, Noort D et al (2002) Retrospective detection of exposure to organophosphorus anti-cholinesterases: mass spectrometric analysis of phosphylated human butyrylcholinesterase. Chem Res Toxicol 15:582–590

68. Noort D, Fidder A, van der Schans MJ, Hulst AG (2006) Verification of exposure to organophosphates: generic mass spectrometric method for detection of human butyrylcholinesterase adducts. Anal Chem 78:6640–6644

69. Nagao M, Takatori T, Matsuda Y, Nakajima M, Iwase H, Iwadate K (1997) Definitive evidence for the acute sarin poisoning diagnosis in the Tokyo subway. Toxicol Appl Pharmacol 144:198–203

Applications of Mass Spectrometry in Investigations of Alleged Use of Chemical Warfare Agents

Robert W. Read

Abstract Chemical warfare agents were used extensively throughout the twentieth century. Many such uses are well documented; however some allegations of use of chemical warfare agents were not easily confirmed. During the early 1980s interest developed into investigation of alleged use by analytical techniques, particularly mass spectrometry. Since that time, many combined chromatographic – mass spectrometric methods have been developed, both for application to the analysis of environmental and biomedical samples and for investigation of physiological interactions of chemical warfare agents. Examples are given of some of the investigations in which the author has been involved, including those into Yellow Rain and uses of chemical warfare agents in Iraq and Iran. These examples illustrate the use of combined chromatographic-mass spectrometric methods and emphasise the importance of controls in analytical investigations.

Keywords Chemical warfare agents • Chromatography • Mass spectrometry • Yellow Rain • Protein adducts • Environmental samples

1 Introduction

Before World War 1 efforts were made to outlaw the use of chemicals in warfare. The Hague Declaration on the Use of Projectiles the Object of Which is the Diffusion of Asphyxiating or Deleterious Gases[1] was signed on July 29, 1899 and entered into force on Sept 4, 1900. The Hague Convention on Laws and Customs of War on Land[1] signed on October 18, 1907, forbade employment of poison or poisoned weapons, and entered into force on Jan 26, 1910. Despite these efforts the use of chemical warfare during World War 1 was widespread.

R.W. Read (✉)
Defence Science and Technology Laboratory, Porton Down, Salisbury, Wiltshire, SP4 0JQ, UK
e-mail: RWREAD@mail.dstl.gov.uk

J. Banoub (ed.), *Detection of Biological Agents for the Prevention of Bioterrorism*,
NATO Science for Peace and Security Series A: Chemistry and Biology,
DOI 10.1007/978-90-481-9815-3_13, © British Crown 2011

The first chemicals used were irritants. Ethyl bromoacetate was used by France in 1914 and Germany attempted to use xylyl bromide in January 1915 at Bolimow in Poland. The latter was unsuccessful due to the low temperature and consequent lack of volatility of the chemical. Various industrial chemicals were later employed as weapons. Germany used chlorine for the first time during the second Battle of Ypres in April 1915. Britain also used chlorine at Loos in September 1915. Phosgene was widely deployed and used by both sides. The first use of a chemical specifically designed for warfare was that of the vesicant, sulphur mustard, by Germany at Ypres in 1917.

Following World War 1, further efforts were made to ban the use of chemical weapons. The Geneva Protocol to the 1907 Hague Convention, the Protocol for the Prohibition of the Use in War of Asphyxiating Gas, and of Bacteriological Methods of Warfare,[1] was signed on June 17, 1925 and entered into force on Feb 8, 1928. However, uses of chemical weapons continued.

The Rif rebellion of 1921–1927 in Morocco had already seen use of sulphur mustard, phosgene and the irritant, chloropicrin, by Spain. Sulphur mustard was also used by the Soviet Union in Xinjiang, China, against Japanese forces in 1934 and 1936–1937, and by Italy in Abyssinia (Ethiopia) in 1935–1940. Sulphur mustard and the arsenical vesicant lewisite were used in the second Sino-Japanese War of 1937–1945.

There were no deliberate uses of chemical weapons during World War 2 despite the possession of significant stockpiles and great concern that they might be used against both military forces and civilian populations. Plans were made in Britain for use of sulphur mustard to contaminate beaches and other terrain in the event of an invasion. However there were some mistaken uses of sulphur mustard by Germany against Polish and Russian forces, and by Poland in an isolated incident against German forces. The only significant event involving chemical weapons was at Bari harbour, Italy, in 1943. The US Liberty ship John Harvey, which was carrying a cargo of munitions containing sulphur mustard, was destroyed in an air raid, resulting in spillage of liquid sulphur mustard into the sea. This incident was covered up for several years and the number of casualties, both military and civilian, is disputed.

Post World War 2, uses of chemical weapons continued. Mustard and phosgene were used by Egypt in North Yemen during the civil war of 1963–1967. There were no confirmed uses of chemical weapons during the Vietnam War of 1963–1975 but the irritant o-chlorobenzalmalononitrile (CS) was extensively used for clearing Viet Cong forces from tunnels.

In 1975–1983 allegations were made of use of a chemical warfare agent in South East Asia. This so-called Yellow Rain became the subject of the first major scientific investigation of an alleged use and is described in more detail below. Allegations were also made of the use of the same material in Afghanistan following the Soviet invasion of 1979.

There was widespread use of sulphur mustard during the Iraq-Iran war of 1983–1988, confirmed by chemical analysis of samples collected from the scenes of alleged uses (see below). The use of nerve agents was strongly suspected, but no analytical confirmation was made. Uses of sulphur mustard and the nerve agent sarin by Iraq against its Kurdish population, during the Anfal campaign of 1986 –1989, were confirmed by chemical analysis (see below).

Nerve agents were used by the Aum Shinrikyo cult in Japan in the 1990s. Sarin was disseminated in Matsumoto City, June 27–28, 1994, and in the Tokyo subway, March 20, 1995. VX was used in an assassination in 1994. These terrorist uses were confirmed by chemical analyses, mostly carried out in Japan and The Netherlands.

2   Yellow Rain

In the late 1970s and early 1980s allegations were made of use of chemical weapons by Vietnamese and Pathet Lao forces against the Hmong tribes in northern Laos. These allegations concerned attacks from the air with an agent in the form of a mist of yellow droplets – Yellow Rain. Following these attacks victims reported symptoms including burning skin, vomiting, eye pain, headache, dizziness, chest pain, coughing fits, breathing difficulties, diarrhea, blistering of skin, subcutaneous bleeding, bleeding from nose and gums, blindness and seizures. Domestic animals and birds were also reported to have been killed. The agent was also said to be toxic to plants with holes appearing in leaves and crops dying within 2 weeks. Similar attacks were also reported in Cambodia between 1978 and 1981, and in Afghanistan following the Soviet invasion in 1979.

The symptoms were assessed as being similar to those resulting from exposure to trichothecene mycotoxins. These mycotoxins are produced by fusarium fungi which occur naturally in grain stored in cold, damp conditions. Their toxicity is due to inhibition of protein synthesis causing damage to bone marrow, skin and the gastro-intestinal tract. When contaminated grain is consumed, a chronic condition known as alimentary toxic aleukia (ATA) results. Trichothecene mycotoxins were not known to be naturally occurring in South East Asia.

Samples of vegetation and yellow material collected from the scenes of alleged attacks were analysed in US university laboratories and the presence of trichothecene mycotoxins reported. However an investigation of the yellow material by electron microscopy at the then Chemical Defence Establishment (CDE), Porton Down, UK, revealed that it consisted of partially digested pollen grains. This finding supported a theory that the material was bee faeces and the reported mist of yellow droplets was produced by bees defaecating as a swarm. Sensitive gas chromatography - selected ion monitoring negative ion chemical ionization mass spectrometry methods were developed at CDE in collaboration with the Admiralty Research Establishment.[2,3] Trichothecenes are detected by these methods as their pentafluoropropionyl or heptafluorobutyryl ether derivatives. High sensitivity was necessary owing to the very small size (a few milligrams) of many samples. A large number of samples, both environmental and biomedical, were analysed with strict attention to controls. Positive and negative (glassware blank) controls were analysed alongside each sample or set of samples. No trichothecenes were detected in any sample associated with alleged uses of Yellow Rain. Participation in two round-robin exercises resulted in correct identification of trichothecenes spiked into blood and urine at ng/ml levels with no false positives.[2,3]

In 1987 samples of foodstuffs collected in Thailand were obtained. These samples were analysed at CDE and two samples of sorghum, known to be old and mouldy, found to be contaminated with trichothecenes – nivalenol, scirpenetriol and 15-monoacetoxyscirpenol – at low ng/g levels. These analyses confirmed that trichothecenes do occur naturally in South East Asia.[4] The major lesson learned from these investigations was the absolute necessity of robust controls in analytical protocols.

An informative account of the investigations into Yellow Rain is given by Tucker.[5]

3   OPCW Analytical Requirements

The Convention on the Prohibition of the Development, Production, Stockpiling and Use of Chemical Weapons and on their Destruction (Chemical Weapons Convention, CWC)[1,6] was signed on Jan 13, 1993 and entered into force on April 29, 1997. The implementing body of the convention is the Organisation for the Prevention of Chemical Weapons (OPCW) based in The Hague, The Netherlands.

The Verification Annex to the CWC allows for the collection and analysis of samples during inspection activities and investigations of alleged use. Any such samples would be analysed by at least two of a worldwide group of Designated Laboratories according to stringent criteria laid down by the Technical Secretariat of the OPCW. Laboratories are designated on the basis of their performance in proficiency tests held each April and October. Analytical criteria require identification of chemicals present as defined by the Schedules of Chemicals to the CWC, by at least two separate techniques, one of which must be spectrometric. Spectrometric identification may be based on interpretation of spectra, comparison with library spectra, or comparison with authentic standards. Positive and negative control samples prepared by the OPCW are provided alongside actual samples. All chemicals present in positive control samples must be correctly identified and no false positives reported. All samples and controls are provided to the Designated Laboratories with no information to identify them as such. An analytical report must be produced according to strict criteria within 15 calendar days of receipt of samples.

Would these requirements have proved successful in the investigations into Yellow Rain? Firstly the requirements only apply to amounts of chemicals that can be detected using scanning techniques, e.g. full scan mass spectrometry, infra-red and nuclear magnetic resonance spectroscopy. In proficiency test samples the concentration of spiking chemicals is usually in the 1–10 µg/g (or µg/ml) range. Many samples of Yellow Rain were of such a small size that only selective, trace analysis methods capable of detecting nanogram amounts, or less, were applicable. Secondly, trichothecenes are not included in the schedules of chemicals and so would not normally be included in analytical methods used by Designated Laboratories. Finally, the 15 day timescale required is unrealistic; the Yellow Rain investigations took several years in total. It should be noted however, that the emphasis on control samples in the OPCW criteria is at least partly due to the experience gained during these investigations.

4 Blood Protein Adducts of Sulphur Mustard

When sulphur mustard was used on a large scale during the Iran-Iraq war of 1983 −1988, attention was focused on the development of methods for verification of exposure in casualties. Mass spectrometric methods, predominantly gas chromatography-mass spectrometry, were available for analysis of environmental samples. Sensitive methods for analysis of biomedical samples, blood and urine, were not available. In order to address this problem, work was carried out to elucidate the metabolism of sulphur mustard and to identify adducts with blood proteins. Animal studies using rats were required for identification of urinary metabolites.[7]

Blood protein adducts were determined from in-vitro studies[9] and later confirmed in samples from human casualties.[11] Blood was incubated with a mixture of $^{13}C_4/^{32}S$ and $^{12}C_4/^{35}S$ mustard. The ^{35}S radio-label allowed use of liquid chromatography (LC) with radioactivity detection for semi-preparative fractionation; the $^{13}C_4$ mass-label allowed the detection of $^{12}C_4$ (M) and $^{13}C_4$ (M+4) doublets in mass spectra of semi-preparative scale radio-LC fractions. Haemoglobin, or globin, was isolated from incubates using a standard method and digested with the enzyme Pronase E.

Digests were fractionated by radio-LC and the resulting fractions analysed by gas chromatography-mass spectrometry (GC-MS), as t-butyldimethylsilyl derivatives, and by liquid chromatography-mass spectrometry (LC-MS). Amino acids alkylated on their side chains by sulphur mustard were detected by the presence of doublet ions due to a mass labelled 2-[(hydroxyethyl)thio]ethyl, $HOCH_2CH_2SCH_2CH_2$ (HETE), moiety. Amino acid residues were easily identified by GC-MS by assuming derivatisation of the amino, carboxylic acid and HETE moieties and simple subtraction of the relevant masses from the mass of the molecular ion.

The identity of alkylated amino acids was confirmed by LC-MS (Fig. 1) and LC-tandem MS (LC-MS/MS) of radio-LC fractions. The use of atmospheric pressure chemical ionisation (APCI) was found superior to electrospray ionisation (ESI) for this purpose, with the Finnigan TSQ700 triple quadrupole instrument available at that time. Detection by LC-MS was based on the presence of mass-labelled quasi-molecular ion (MH+) doublets (Fig. 2). Precursor ion scans of m/z 105 were also found useful for screening fractions for the presence of molecules containing a HETE moiety (Fig. 3). The amino acid residue was then identified using product ion scans of the quasi-molecular ion(s). Alkylated amino acids

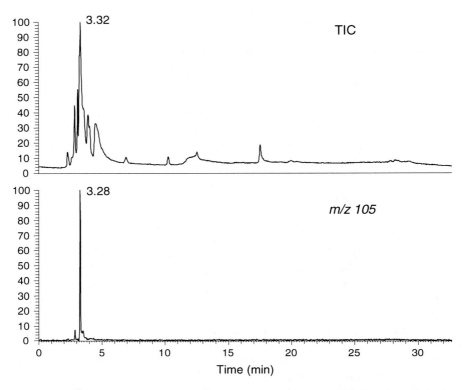

Fig. 1 LC-MS APCI total ion and extracted ion (*m/z* 105) chromatograms from a radio-LC fraction from a Pronase E digest of human haemoglobin isolated from blood incubated with sulphur mustard

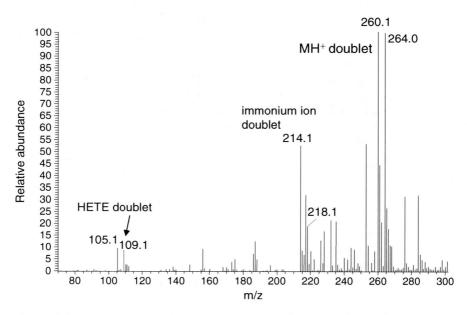

Fig. 2 APCI spectrum of the peak at retention time of approx. 3.3 min in Fig. 1, identified as HETE-histidine

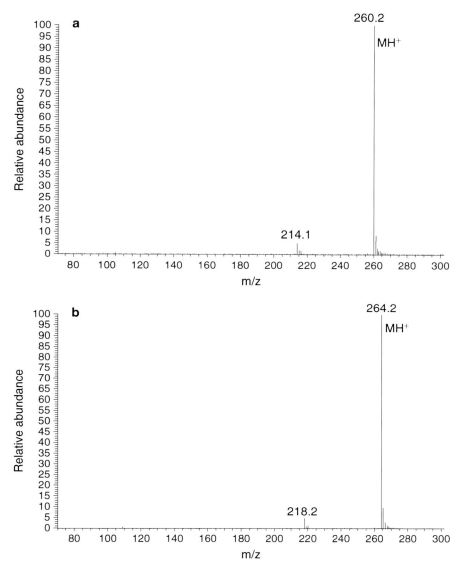

Fig. 3 APCI precursor ion spectra of *m/z* 105 (**a**) and *m/z* 109 (**b**) for the adduct identified as HETE-histidine

detected by these methods were histidine (His), cysteine (Cys), aspartic acid (Asp), glutamic acid (Glu), lysine (Lys) and tryptophan (Trp). Although Pronase E should digest proteins to individual amino acids, the dipeptides valyl leucine (Val-Leu), valyl histidine (Val-His) and histidyl leucine (His-Leu) were also found.

In order to determine the positions of the alkylation sites in haemoglobin, further digestions of isolated haemoglobin were performed using the enzyme trypsin.

The protein of normal human haemoglobin consists of two identical α-globin chains and two identical β-globin chains, the amino acid sequences of which are known. Masses of fragment peptides produced by digestion with trypsin may be determined knowing that the enzyme cleaves the protein at the C-terminal side of lysine and arginine. Masses of alkylated tryptic peptides may be determined by addition of 104 Da (HETE = 105 Da) to the mass of the native peptide. Alkylated peptides were thus detected by searching electrospray LC-MS chromatograms, from analysis of radio-LC fractions of trypsin digests, for the presence of doublets at masses corresponding to the alkylated peptides. Singly and doubly charged quasi-molecular ion doublets were detected in most cases (Fig. 4).

Alkylation sites were identified using product ion scans of doubly or singly charged quasi-molecular ions and comparison of sequence ions with those predicted using Finnigan Bioworks™ software. This software allowed the addition of a mass modification of 104 Da (or 109 Da for $^{13}C_4$ label) to amino acids in a peptide sequence and prediction of the resulting ESI-MS/MS product (sequence) ion series. Most spectra obtained showed good coverage of y (C-terminal fragment) and b (N-terminal fragment) series sequence ions derived from both doublet ions used as precursors, and allowed confirmation of alkylation sites (Fig. 5).

Sites identified were α-chain Val 1, His 20, His 45 and His 50, and β-chain Val 1, Glu 22, Glu 26, Glu 43, His 97 and His 146. Interestingly no cysteine, aspartic

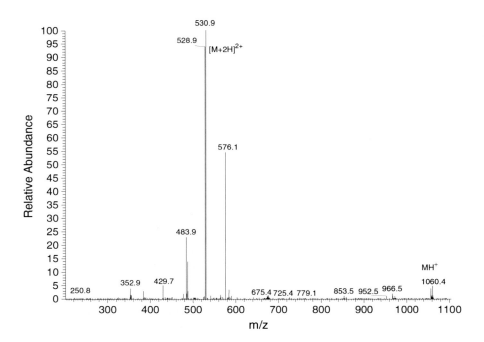

Fig. 4 LC-MS ESI spectrum of the HETE-alkylated globin β-chain T1 tryptic peptide

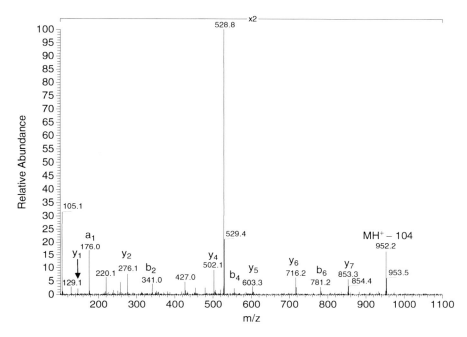

Fig. 5 LC-MS/MS ESI product ion spectrum of the m/z 528.8 doubly charged quasi-molecular ion from HETE- alkylated globin β-chain T1 tryptic peptide, (HETE)Val-His-Leu-Thr-Pro-Glu-Glu-Lys

acid, lysine or tryptophan alkylation was detected, despite these alkylated amino acids being detected in Pronase E digests. The adduct with cysteine may originate from glutathione present as an impurity in isolated haemoglobin. The complete results obtained from Dutch[8] and UK[9] work on sulphur mustard adducts of haemoglobin are shown in Fig. 6.

5 Haemoglobin Adducts in Human Casualties

Following identification of alkylation sites, attention turned to development of sensitive methods for verification of exposure. Initially work focused on N-terminal valine adducts. A method for analyzing N-alkylated valine residues in haemoglobin for monitoring exposure to environmental carcinogens had been developed by Törnquist et al.[10] This method selectively cleaves the N-terminal valine residue using a modified Edman reagent, pentafluorophenyl isothiocyanate (PFPITC), to produce a pentafluorophenyl thiohydantoin (PFPTH). The method was adapted for trace analysis of valine alkylated by sulphur mustard with an additional derivatisation of the HETE hydroxyl group with heptafluorobutyric anhydride. The resulting derivative was analysed by GC-MS with selected ion monitoring at a resolution of α-chain

T1**Val**-Leu-Ser-Pro-Ala-Asp-Lys-T2Thr-Asn-Val-Lys-T3Ala-Ala-Trp-Gly-Lys-T4**Val-Gly-
Ala-His-Ala-Gly-Glu-Tyr-Gly-Ala-Glu-Ala-Leu-Glu-Arg**-T5Met-Phe-Leu-Ser-Phe-Pro-
Thr-Thr-Lys-T6**Thr-Tyr-Phe-Pro-His-Phe-Asp-Leu-Ser-His-Gly-Ser-Ala-Gln-Val-Lys**-
T7Gly-His-Gly-Lys-T8Lys-T9**Val-Ala-Asp-Ala-Leu-Thr-Asn-Ala-Val-Ala-His-Val-Asp-
Asp-Met-Pro-Asn-Ala-Leu-Ser-Ala-Leu-Ser-Asp-Leu-His-Ala-His-Lys**-T10Leu-Arg-
T11Val-Asp-Pro-Val-Asn-Phe-Lys-T12Leu-Leu-Ser-His-Cys-Leu-Leu-Val-Thr-Leu-Ala-Ala-
His-Leu-Pro-Ala-Glu-Phe-Thr-Pro-Ala-Val-His-Ala-Ser-Leu-Asp-Lys-T13 Phe-Leu-Ala-Ser-
Val-Ser-Thr-Val-Leu-Thr-Ser-Lys-T14Tyr-Arg

β-chain

T1**Val-His-Leu-Thr-Pro-Glu-Glu-Lys**-T2Ser-Ala-Val-Thr-Ala-Leu-Trp-Gly-Lys-T3**Val-Asn-
Val-Asp-Glu-Val-Gly-Gly-Glu-Ala-Leu-Gly-Arg**-T4Leu-Leu-Val-Val-Tyr-Pro-Trp-Thr-
Gln-Arg-T5**Phe-Phe-Glu-Ser-Phe-Gly-Asp-Leu-Ser-Thr-Pro-Asp-Ala-Val-Met-Gly-Asn-
Pro-Lys**-T6Val-Lys-T7Ala-His-Gly-Lys-T8Lys-T9**Val-Leu-Gly-Ala-Phe-Ser-Asp-Gly-Leu-
Ala-His-Leu-Asp-Asn-Leu-Lys**-T10**Gly-Thr-Phe-Ala-Thr-Leu-Ser-Glu-Leu-His-Cys-
Asp-Lys**-T11**Leu-His-Val-Asp-Pro-Glu-Asn-Phe-Arg**-T12Leu-Leu-Gly-Asn-Val-Leu-Val-
Cys-Val-Leu-Ala-His-His-Phe-Gly-Lys-T13Glu-Phe-Thr-Pro-Pro-Val-Gln-Ala-Ala-Tyr-Gln-
Lys-T14Val-Val-Ala-Gly-Val-Ala-Asn-Ala-Leu-Ala-His-Lys-T15**Tyr-His**

Fig. 6 Identified alkylated tryptic fragments (in bold) of α- and β- globin chains of haemoglobin. Alkylation sites are underlined. Superscripts show N-termini of tryptic fragments. Alkylated α-chain T9, β-chain T10 and T10-S-S-T12 fragments were also identified but alkylation sites not established

approximately 7,500 on a double focusing magnetic sector instrument. The method was used to detect this adduct in samples from five human casualties, four Iranian and one accidental exposure.[11]

A second method was developed for the detection of alkylated histidine. This residue was chosen as a target analyte as five histidines in haemoglobin had been found to be alkylated by sulphur mustard. The method involved rather tedious sample preparation. Globin was isolated from red blood cells (or whole blood in the case of samples received frozen) by precipitation. A portion of the isolated globin was digested in 6M hydrochloric acid under vacuum at 110 °C overnight. The digest was dried in a centrifugal evaporator and the residue subjected to solid phase extraction on a propyl sulphonic acid (PRS) cation exchange cartridge. The eluate was concentrated to dryness and the residue derivatised with fluorenylmethyl chloroformate (FMOC) in pH 8 borate/acetone. The resulting FMOC derivative was analysed by electrospray LC-MS/MS with multiple reaction monitoring (mrm). In spite of the complex sample preparation, surprisingly good linear calibrations were obtained from incubates of blood with mustard. The method was sensitive enough to detect adduct in samples from the above human casualties.[11]

6   Blood Protein Adducts of Nerve Agents

The physiological target of organpophosphorus nerve agents is the active site serine residue in the serine hydrolase enzyme acetylcholinesterase. The corresponding residue in butyrylcholinesterase is also a target but this enzyme has no known physiological function. LC-MS/MS detection of an active site nonapeptide following pepsin digestion of isolated butyrylcholinesterase is the basis of the method, developed by Fidder et al., for confirmation of nerve agent exposure.[12] This method is not unequivocal for agents, for example soman, that age by loss of the O-alkyl moiety, or V-agents (which give the same adducts as the corresponding G-agents). They may also not be applicable following oxime therapy, although this requires further investigation. An alternative adduct with tyrosine in albumin has recently been identified using radio- and mass-labelled agents and similar methods to those described for sulphur mustard. In this case, deuterated (D_3) and ^{14}C labels on the P-methyl group of sarin and soman were used. Phosphylated tyrosine was detected in protease digests of human plasma incubated with sarin and soman[13] (Figs. 7 and 8).

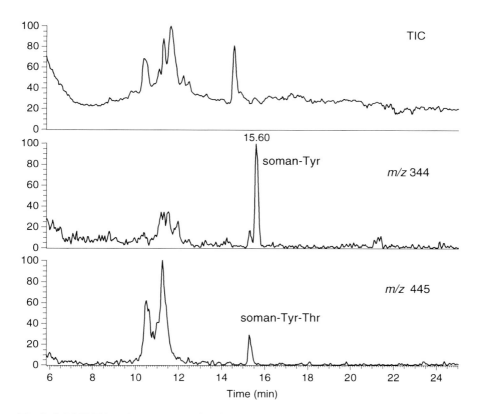

Fig. 7   LC-MS ESI total and extracted ion chromatograms from a radio-LC fraction of a Pronase E digest from a human plasma incubate with soman, showing detection of adduct with tyrosine, MH$^+$ = m/z 344. The peak in the m/z 445 extracted ion chromatogram was tentatively identified as the phosphonylated dipeptide Tyr-Thr

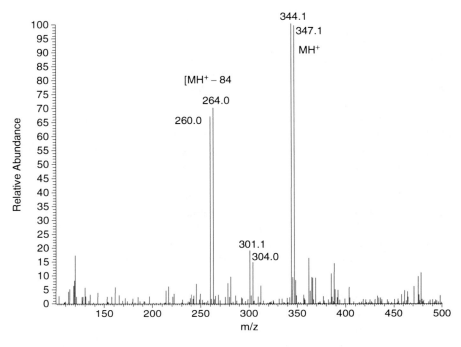

Fig. 8 ESI mass spectrum of the soman adduct with tyrosine from data in Fig. 7

Analysis of tryptic digests of incubates showed the presence of a phosphylated tripeptide, Tyr-Thr-Lys, consistent with phosphylation of tyrosine-411 in albumin. Analysis of samples from guinea pigs exposed percutaneously or subcutaneously to sarin, soman, cyclosarin and tabun showed the presence of adducts with all four agents.[14] There was no indication of ageing with soman and the adduct with tabun retains complete structural integrity. The phosphylation is probably a chemical reaction and applicable only to moderate exposures. Adducts have been detected ex vivo in marmoset plasma up to 23/24 days after agent doses of $2 \times LD_{50}$ in the presence of oxime therapy[15] (Fig. 9). No butyrylcholinesterase adducts were detected in the same animals, except for soman and tabun in animals that died within a few hours of exposure. These adducts were aged (soman) or de-amidated during sample preparation (tabun). Albumin tyrosine adducts should be applicable to verification of exposure to the G-series nerve agents in the presence of therapies containing oximes.

7 Environmental Samples from Iran and Iraq

There were many uses of chemical warfare agents during the Iraq-Iran war of 1983–1988, several of which were confirmed by analysis of environmental samples. One example concerned the analysis at the then Chemical and Biological

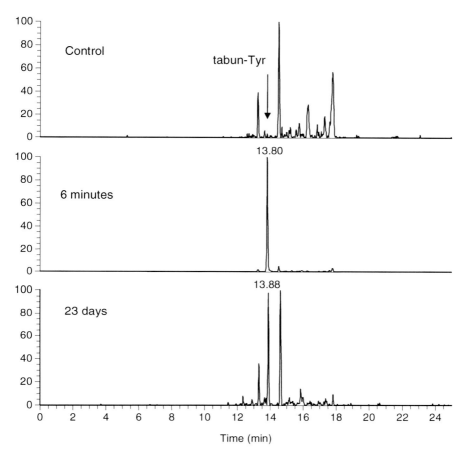

Fig. 9 LC-MS/MS multiple reaction monitoring chromatograms for the transition m/z 317 $(MH^+) \rightarrow m/z$ 198, from a plasma sample taken from a control (therapy only) animal, and samples taken post mortem 6 min and 23 days after a $2 \times LD_{50}$ intra-muscular dose of tabun together with oxime therapy

Defence Establishment (CBDE), Porton Down, of some contaminated equipment from a film crew who had recorded the decanting of the contents of a chemical filled artillery shell in Iran in 1987. Some of the members of this crew later developed blisters symptomatic of sulphur mustard exposure. Analysis of samples taken from a video camera, headphones and a plastic handled corkscrew (essential equipment for a film crew) confirmed the presence of sulphur mustard.

The following year, some samples were collected by a journalist investigating alleged use of chemical warfare agents in a Kurdish area of northern Iraq. The samples from a munition casing, soil and what appeared to be sheep's wool, were initially sent to a commercial laboratory in the UK which identified 1,4-dithiane, 1,4-thioxane and divinyl sulphide, decomposition products of sulphur mustard. Further analysis at

CBDE confirmed the presence of those decomposition products. Application of a variety of analytical techniques identified a total of twenty five compounds in four samples. The compounds identified included intact sulphur mustard and the explosives tetryl and TNT, all easily confirmed using full scan GC-MS methods.[16]

7.1   Environmental Samples from Birjinni, N Iraq

In 1992 CBDE was asked to analyse samples from the Kurdish village of Birjinni in Iraq.[17] The samples were collected in June 1992 by a team of forensic scientists from the scene of an alleged chemical attack on August 25, 1988. The samples consisted of soil, insect pupae and clothing from exhumed remains of two victims, and soil and metal fragments from four of a total of twelve bomb craters. It was initially considered unlikely that any useful data would be obtained from samples collected almost 4 years after the alleged attack. The analytical protocol followed emphasized the use of stringent controls to eliminate any chance of contamination and ensure that any traces of agent detected were real. Analyses of the samples from exhumed remains using sensitive selected ion monitoring GC-MS methods were negative for the presence of chemical warfare agents. However analyses of samples of soil and metal fragments produced some unexpected results. Extracts of samples from two of the four craters were found to contain sulphur mustard and its decomposition products, thiodiglycol, 1,4-dithiane and 1,4-thioxane, together with the explosive tetryl (Figs. 10 and 11). Concentrations of these compounds present in extracts were sufficient to allow the acquisition of full scan electron ionization (EI) mass spectra. Sulphur mustard was detected in three soil samples at estimated concentrations of 600 ng/g to 10 µg/g.

Even more surprisingly, extracts of samples from the other two craters were found to contain methylphosphonic acid (MPA) and i-propyl methylphosphonic acid (iPMPA), the main hydrolysis products of sarin, detected as their t-butyldimethylsilyl (tBDMS) derivatives (Fig. 12). Estimated concentrations of MPA present ranged from 60 ng/g to 60 µg/g, and iPMPA from 6 ng/g to 200 ng/g.

Analysis of extracts from a green painted metal fragment associated with one of these soil samples detected a trace amount of intact sarin. Because of this unexpected finding, a series of confirmatory analyses were carried out. Extracts were analysed using two different phase GC columns and three ionization modes, EI (Fig. 13) and chemical ionisation (CI) with both methane and ammonia as reagent gas (Fig. 14), with both selected ion and MS/MS (multiple reaction monitoring). Each analysis gave a positive result for sarin, and all associated controls showed that no contamination had occurred in the laboratory. The paint on the metal fragment was analysed by infra-red spectroscopy and shown to be of a military alkyd type.

Later analysis using LC-MS also confirmed the presence of the agent hydrolysis products, thiodiglycol, MPA and iPMPA. Ion chromatography showed the presence of greatly elevated levels of fluoride ion in samples associated with sarin. Probably uniquely in this investigation, samples from the craters containing mustard residues could be used as controls for those containing sarin residues, and vice

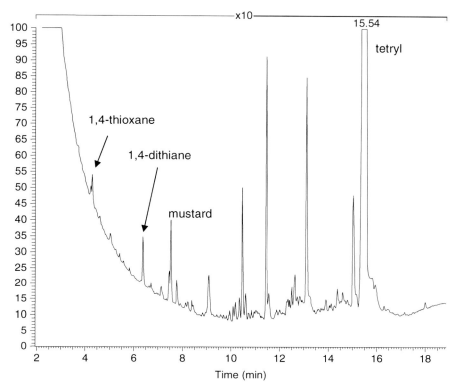

Fig. 10 GC-MS EI chromatogram from a dichloromethane extract of a soil sample from a munition crater, showing detection of sulphur mustard, breakdown products and tetryl

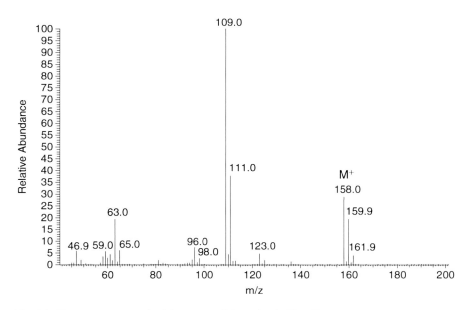

Fig. 11 EI mass spectrum of sulphur mustard from data in Fig. 10

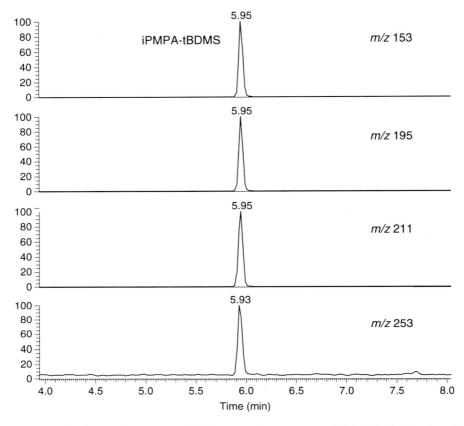

Fig. 12 GC-MS EI selected ion monitoring mass chromatograms of the tBDMS derivative of i-propyl methylphosphonic acid (iPMPA) in an aqueous extract of a soil sample from a munition crater

versa. A complete audit trail was also maintained from sample collection to analysis. The analytical results provided the first confirmation of use of a nerve agent by Iraq against its Kurdish population, and it remains the only such confirmation. The results of this investigation were published in early 1994, shortly before the terrorist use of sarin in the Tokyo subway.

8 Conclusion

Over the last 25 years there have been several analytical investigations using mass spectrometric methods into alleged military and terrorist uses of chemical weapons. The scientific community involved in this type of analysis has generally taken heed of the lessons learned from the Yellow Rain investigations. The use of positive and negative controls is taken seriously, particularly by those working in OPCW

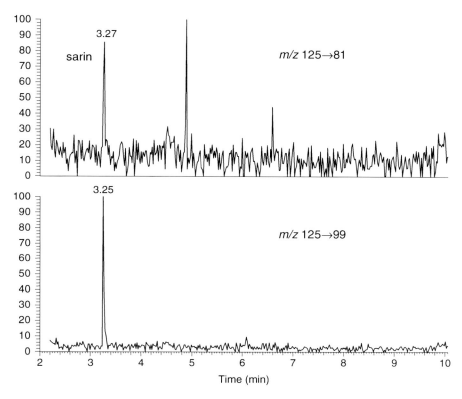

Fig. 13 GC-MS/MS EI chromatograms from a dichloromethane extract of a painted metal fragment showing the detection of sarin. Multiple reaction monitoring using a BPX5 non-polar GC column

designated laboratories. The impact of those lessons in the biological warfare agent area has yet to be seen. With the exception of several examples of methods for the forensic identification by mass spectrometry of proteinaceous toxins, particularly ricin and botulinum, little if any emphasis has been put on the importance of controls. The Biological and Toxin Weapons Convention (Convention on the Prohibition of the Development, Production and Stockpiling of Bacteriological (Biological) and Toxin Weapons and on Their Destruction),[1] which entered into force in 1975, contains no requirement for verification, and is currently unlikely ever to do so. Work on identification of biological agents by mass spectrometry, and other techniques, appears to be focussed on homeland security requirements rather than on any future applications to treaty verification. Work on toxins is also focussed on homeland security and food contamination, with the exception of ricin which, together with the marine algal toxin saxitoxin, is included in Schedule 1 of the Chemical Weapons Convention.

Everyone carrying out investigations into alleged uses of chemical, or biological, agents should be aware that their results need to withstand national and international scrutiny to forensic standards. The consequences of false positive results, either legal or military, would likely be severe.

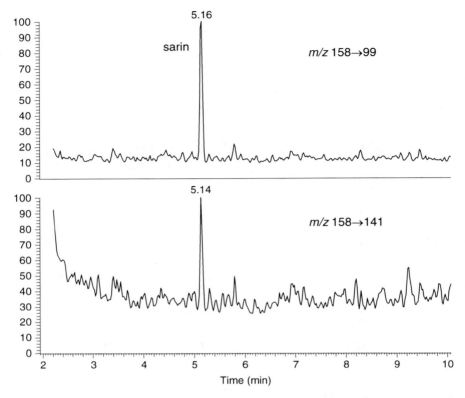

Fig. 14 GC-MS/MS ammonia CI chromatograms from a dichloromethane extract of a painted metal fragment showing the detection of sarin. Multiple reaction monitoring using a DBWax polar GC column

References

1. http://avalon.law.yale.edu/subject_menus/lawwar.asp
2. Begley P, Foulger BE, Jeffery PD, Black RM, Read RW (1986) Detection of trace levels of trichothecenes in human blood using capillary gas chromatography – electroncapture negative ion chemical ionisation mass spectrometry. J Chromatogr 367:87–101
3. Black RM, Clarke RJ, Read RW (1986) Detection of trace levels of trichothecene mycotoxins in human urine by gas chromatography-mass spectrometry. J Chromatogr 367:103–115
4. Black RM, Clarke RJ, Read RW (1987) Detection of trace levels of trichothecene mycotoxins in environmental residues and foodstuffs using gas chromatography with mass spectrometric or electron-capture detection. J Chromatogr 388:365–378
5. Tucker JB (2001) The 'Yellow Rain' controversy: lessons for arms control compliance. The Nonproliferat Rev 8(1):25–42
6. http://www.opcw.org/chemical-weapons-convention/
7. Black RM, Brewster K, Clarke RJ, Hambrook JL, Harrison JM, Howells DJ (1992) Biological fate of sulphur mustard, 1, 1'-thiobis(2-chloroethane): isolation and identification of urinary metabolites following intraperitoneal administration to rat. Xenobiotica 22:405–418

8. Noort D, Verheij ER, Hulst AG, de Jong LPA, Benschop HP (1996) Characterization of sulphur mustard induced structural modification in human hemoglobin by liquid chromatography – tandem mass spectrometry. Chem Res Toxicol 9:781–787

9. Black RM, Harrison JM, Read RW (1997) Biological fate of sulphur mustard: in vitro alkylation of human haemoglobin by sulphur mustard. Xenobiotica 27:11–32

10. Törnquist M, Mowrer J, Jensen S, Ehrenberg L (1986) Monitoring of environmental cancer initiators through hemoglobin adducts by a modified Edman degradation method. Anal Biochem 154:255–266

11. Black RM, Clarke RJ, Harrison JM, Read RW (1997) Biological fate of sulphur mustard: identification of valine and histidine adducts in haemoglobin from casualties of sulphur mustard poisoning. Xenobiotica 27:499–512

12. Fidder A, Hulst AG, Noort D, de Ruiter R, van der Schans MJ, Benschop HP, Langenberg JP (1992) Retrospective detection of exposure to organophosphorus anti-cholinesterases: mass spectrometric analysis of phosphylated human butyrylcholinesterase. Chem Res Toxicol 15:582–590

13. Black RM, Harrison JM, Read RW (1999) The interaction of sarin and soman with plasma proteins: the identification of a novel phosphonylation site. Arch Toxicol 73:123–126

14. Williams NH, Harrison JM, Read RW, Black RM (2007) Phosphylated tyrosine in albumin as a biomarker of exposure to organophosphorus nerve agents. Arch Toxicol 81:627–639

15. Read RW, Riches JR, Stevens JA, Stubbs SJ, Black RM Biomarkers of organophosphorus nerve agent exposure: comparison of phosphylated butyrylcholinesterase and phosphylated albumin after oxime therapy. Arch Toxicol (in press). doi: 10.1007/s00204-009-0473-4

16. Black RM, Clarke RJ, Cooper DB, Read RW, Utley D (1993) Application of headspace analysis, solvent extraction, thermal desorption and gas chromatography-mass spectrometry to the analysis of chemical warfare samples containing sulphur mustard and related compounds. J Chromatogr 637:71–80

17. Black RM, Clarke RJ, Read RW, Reid MTJ (1994) Application of gas chromatography-mass spectrometry and gas chromatography-tandem mass spectrometry to the analysis of chemical warfare samples, found to contain residues of the nerve agent sarin, sulphur mustard and their degradation products. J Chromatogr A 662:301–321

Mass Spectrometry for the Analysis of Low Explosives

Bruce McCord, Megan Bottegal, and John Mathis

Abstract It is extremely important to be able to characterize a wide variety of organic and inorganic materials which may be of potential use in improvised explosives manufacture. As a result our research group has become increasingly interested in the development of mass spectrometric methods for the analysis of various classes of low explosives. We present in this document, the development of ion chromatographic/mass spectrometric methods for the analysis of inorganic explosives and HPLC-ESI-MS for the analysis of smokeless powders.

1   Introduction

The recent increase in terrorist activities has renewed interest in the development of methods for the analysis and characterization of explosives. Explosives can be used as a weapon of terror and/or as an agent for the dispersal of biological or chemical agents. Thus it is important for anyone involved in the analysis of terrorist incidents to understand the methodology used in explosive analysis. Generally speaking, analysis of explosives can occur pre-blast or post blast. In pre-blast analysis, the goal is to characterize the materials used in a bomb or to detect traces of explosives on the hands or clothing of a bombmaker. In post blast analysis the goal is to quickly determine the identity of the explosive based on trace levels of residues remaining on the ground or on shrapnel from the bomb. In both situations a chromatographic separation followed by mass spectral analysis can provide the specific information on the type of explosive used and enable the forensic investigator to more quickly identify potential perpetrators, forestalling explosive attacks or preventing future explosive incidents.

B. McCord (✉), M. Bottegal, and J. Mathis
Department of Chemistry, Florida International University, University Park, Miami, FL, 33199
e-mail: mccordb@fiu.edu

J. Banoub (ed.), *Detection of Biological Agents for the Prevention of Bioterrorism*,
NATO Science for Peace and Security Series A: Chemistry and Biology,
DOI 10.1007/978-90-481-9815-3_14, © Springer Science+Business Media B.V. 2011

Explosives can be defined as substances, alone or in combination, which are capable of undergoing rapid exothermic chemical reaction or decomposition without the participation of atmospheric oxygen. There are two major types of explosives, high explosives and low explosives. The difference between them is in their respective reaction rates. Low explosives react slower than the speed of sound. They are said to deflagrate, meaning they burn rapidly. Low explosives are commonly used as propellants in which the release of large amounts of gas during deflagration can be used to propel rockets or fire projectiles. To do damage, low explosives must be held in some sort of container, and it is the disruption of the container due to the buildup of these gasses that results in the explosion. Chemically, low explosives can be further divided into two classes, inorganic pyrotechnics which consist of mixtures of oxidizing agents and carbon or metal based fuels, and smokeless powders, which are based on nitrocellulose with a variety of additives to control burn rates.

High explosives react faster than the speed of sound and do not need to be contained to produce damage. Instead, the rapid detonation of a high explosive produces a shock wave that propages outward from the seat of the blast and results in a violent pressure pulse that creates damage. High explosives can be classified as primary or secondary explosives depending on their sensitivity to friction and shock. Common high explosives include nitrate esters such as PETN and nitroglycerine, nitramines such as RDX, and aromatic nitrates like TNT and DNT. A third class of high explosives are so insensitive to shock that they are more commonly referred to as blasting agents. Ammonium nitrate is an example of this type of material. Due to its safety, it is by far the most commonly used explosive in the mining and construction industry with over three million metric tons used in the USA alone.[1]

As a result of the increasing use of blasting agents in mines and the increasing control over access to military and other high explosives, it has become correspondingly difficult for terrorist elements to obtain high explosives. As a result, such groups are drawn towards the use of improvised low explosives such as commercial and improvised pyrotechnic mixtures and smokeless powders. Recipes for manufacturing devices from such materials are easily found on the internet. It is interesting that these web sites often lack specific safety or chemical instructions, (caveat emptor) resulting in a sort of chemical Darwinism that can be a nightmare for a forensic investigator interested in making a clandestine explosives laboratory safe. In any case, because such recipes exist, the investigator and chemical analyst must keep up to date on current improved explosive methods in order to be able to recognize the method and means of preparation. An additional source of energetic materials, for persons interested in improvised explosive manufacture are commercial propellants such as black powder, black powder substitutes and smokeless powders. These products are commonly used by hunters, sportsmen, and those individuals interested in ancient weapons and theatrical reenactment of warfare.

Thus it becomes important to be able to characterize a wide variety of organic and inorganic materials which may be of potential use in improvised explosives manufacture. In addition, with the expansion in the capabilities of the forensic analysis there is an increasing expectation that highly specific methods such as mass spectrometry should be used to determine absolute identities of substances. In the USA, the technical working group on the analysis and detection of explosives (TWGFEX) has developed a series of guidelines for explosive analysts, emphasizing this point.[2]

As a result our research group has become increasingly interested in the development of mass spectrometric methods for the analysis of various classes of low explosives. As mentioned above, there are two basic chemical preparations for which an active explosives laboratory interested in low explosives must be prepared to analyze, inorganic based pyrotechnics and organic smokeless powders. Thus for this paper we have focused on the development of ion chromatographic/mass spectrometric methods for the analysis of inorganic explosives and HPLC/MS for the analysis of smokeless powders.

2   IC/MS Techniques for Inorganic Explosives

Inorganic explosives include black powder, black powder substitutes, flash powders and various improvised mixtures. These materials generally consist of an inorganic oxidizer such as potassium nitrate, potassium chlorate, or potassium perchlorate and a fuel such as a carbon source (charcoal, ascorbic acid, sodium benzoate etc.), sulfur or powdered metal. Black powder, consisting of sulfur, potassium nitrate and charcoal is by far the oldest explosive, invented in China some hundreds of years ago.[3] Flash powders contain an oxidizing agent (nitrate, chlorate or perchlorate) combined with finely powdered metals as fuels. These mixtures are commonly used in pyrotechnic displays producing bright lights and sound. Both black powder and flash powder are somewhat sensitive to ignition by friction or electrical discharge and as a result, sporting goods manufacturers have developed alternative compositions for use in antique weapons such as muskets which are less dangerous to ship and/or less fouling. These materials are known as black powder substitutes and may substitute a different oxidizer or fuel when compared to the standard black powder composition. Common black powder substitutes include Pyrodex which contains potassium perchlorate, dicyandiamide and sodium benzoate in addition to potassium nitrate, sulfur and charcoal and Jim Shokey's Gold powder which contains potassium perchlorate, potassium nitrate and ascorbic acid.[4] Improvised mixtures may substitute any number of different oxidizers and fuels ranging from powdered sugar to zinc to walnut hulls. Upon deflagration these ingredients react to produce an even wider range of inorganic ions with up to 60% of the weight of these powders remaining as inorganic salts.[5] For example black powder produces nitrate, nitrite, sulfate, sulfide, carbonate, thiocyanate and cyanate upon burning. Pyrodex will produce these ions as well as chloride, benzoate and chlorate. Proper analysis requires the detection of a wide range of inorganic ions and our laboratory and others have developed a wide a variety of ion chromatographic and electrophoretic methods for their analysis.[5–8]

Table 1 The range of inorganic ions found in these types of devices.

Explosive	Uninitiated	Observation	Post-blast/burn
Black powder	KNO_3, C, S	Sulfur smell, white smoke, visible grey residue	SO_4^{2-}, NO_2^-, CO_3^{2-}, SCN^-, OCN^-, S^{2-}, K^+, $S_2O_3^{2-}$
Pyrodex	KNO_3, C, S, DCDA, SB, $KCIO_4$	Less sulfur smell and residue than black powder	SO_4^{2-}, CO_3^{2-}, Cl^-, benzoate, DCDA, OCN^-, NO_2^-, K^+
Triple seven	KNO_3, C, DCDA, SB, $KCIO_4$, 3-NBA	No sulfur smell	CO_3^{2-}, Cl^-, K^+, NO_2^-, benzoate, DCDA, OCN^-
Black canyon/ clean shot pioneer/ golden powder	KNO_3, C, S, ascorbic acid	Sulfur smell, white smoke, visible grey residue	SO_4^{2-}, NO_2^-, CO_3^{2-}, S,K, SCN^-, OCN^-, $S_2O_3^{2-}$
Flash powder	Fuel [Al, Mg, S] and oxidizer	Silvery residue	CIO_4^-, CIO_3^-, metals, oxides

Table 1 illustrates the range of inorganic ions found in these types of devices. It should also be noted that these procedures are also applicable in the analysis of fertilizer based high explosives, urea nitrate and ammonium nitrate, where ammonium and nitrate ions may be detected.[9]

In the past the identification of these ions in explosive residue were performed by running aqueous extracts using two orthogonal chromatographic procedures such as ion chromatography and capillary electrophoresis.[6] Recently we have been working on the development of ion chromatography/mass spectrometric methods for their analysis. Previous work had demonstrated that IC/MS methods could be used for the determination of explosives residue.[10] Unfortunately these methods utilized a harsh mobile phase that destroyed ascorbic acid prior to detection. As a result an improved procedure was developed consisting of a 10 mM ammonium bicarbonate (pH10) gradient with acetonitrile at a flow rate of 0.5 mL/min.[4] Exact mass capability and MS^2 detection was provided using a Waters Premier Quadrupole time of flight system operated at a source temperature of 120 °C and a desolvation temperature of 350 °C. Figure 1 shows the result of a bulk analysis of Jim Shockey's Gold powder demonstrating the detection of nitrate, chlorate, perchlorate and ascorbic acid. In this project the exact mass capability of the system proved useful in determining the compositon of the large nitrate cluster at 209.9405 amu which was determined to contain the ions [NO_3^- NO_3^- NO_2^- H^+ K^+]. This system was also able to determine the structure of the ascorbate fragments at m/z 115.0026, and 87.0083 which were determined to be [$C_4H_3O_4$]$^-$ and [$C_3H_3O_3$]$^-$ based on exact mass measurements and chemical analysis. The procedure was tested on a variety of inorganic explosives and showed good specificity for the determination of precursor and reactant ions. It should prove to be an excellent confirmation procedure for the presence of inorganic ions following a prescreen by ion chromatography.

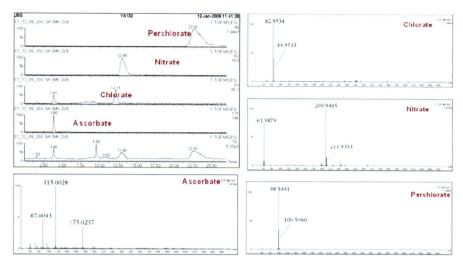

Fig. 1 The analysis of a black powder substitute (Jim Shockey's Gold) by electrospray IC/MS. The procedure involves a gradient with ammonium bicarbonate and acetonitrile using a Waters IC-Pak Anion HR (4.6×250 mm) coupled to a Waters Premier QTOF system. The figure shows an extracted ion chromatogram followed by individual high resolution mass spectra of the various components present in the powder[4]

3 Analysis of Smokeless Powders by HPLC/MS

A second class of low explosives, utilized in the preparation of improvised explosive devices, are the smokeless powders. Smokeless powders consist of nitrocellulose combined with various additives and stabilizers used to improve shelf life and control burn rates. The powders are used as propellants for pistols, rifles and shotguns and are classified as single or double base depending on whether nitroglycerine has been added to the base nitrocellulose. The name smokeless refers to the relative lack of residue produced following the explosion when compared to the large amount of residue produced by black powder. Smokeless powder can be manufactured as rods, disks, flakes or balls with an approximate particle size of 0.2–1.0 mm in diameter. Both morphology and chemical composition are important in its characterization. In chemical analysis it is the additive package that is key to identifying the powder.[11] These additives include energetic materials such as nitroglycerine and DNT, stabilizers such as ethyl centralite and diphenyl amine, and plasticizers such as dibutyl phthalate. Table 2 lists the various components present in smokeless powders and their application. The stabilizers are of particular interest as they are designed to minimize decomposition produced by the release of nitric acid. This decomposition results in the gradual formation of nitroso and nitrodiphenyl amines which can be used to help individualize the powder to a particular lot number or age.[12]

Table 2 Common components in smokeless powders and their application[11]

Nitrocelluose, nitroglycerine, dinitrotoluene	Energetic Material
Diphenylamine, nitrodiphenyl amine, methyl centralite, ethyl centralite	Stabilizer
Dibutyl phthalate, diethyl phthalate, dioctyl phthalate	Plasticizers and burn rate deterrents
Dyes, carbon black and inorganic salts	For identification, reduction of static sensitivity and flash reduction, respectively

Following the explosion of a device in which smokeless powder has been used, it is common to find partially burned and unburned particles expelled from the device in the surroundings. In addition shrapnel formed in the blast may be coated with residue permitting recovery and detection of the additives and stabilizers present in the powder. Because the particles contain polymeric nitrocellulose which can contaminate HPLC and GC systems, it is common to extract away the additives in the powder by soaking the particles overnight in methylene chloride.[13] Nitrocellulose is insoluble in this solvent and the supernatant can then be analyzed directly by GC/MS or evaporated to dryness and reconstituted in an HPLC compatible solvent such as methanol or acetonitrile.[14]

GC/MS methods for smokeless powder must be performed with specialized systems to minimize decomposition of the nitroglycerine and nitrosodiphenyl amines present in the sample. It is also important to provide sufficient resolution to detect various geometrical isomers of nitrodiphenyl amines and nitrotoluenes. Typically lower injection temperatures are used with larger bore columns at high flow rates to minimize decomposition in the column and injector. Sometimes a post column splitter is used to minimize the gas input to the mass spectrometer. Analysis is generally performed using electron impact in the positive ion mode although higher sensitivity for NG and other explosive compounds can be obtained using negative ion chemical ionization.[16,17] LC/MS methods have also been developed for the analysis of smokeless powders.[15] It is common for these procedures to be used in the positive ion mode to detect the additives, however detection of energetic materials such as nitroglycerine and dinitrotoluene is best performed in the negative ion mode.[16] A potential advantage of HPLC/MS procedures is their potential application in quantitative analysis for determination of lot to lot variation.[14] Figure 2 demonstrates the analysis of a standard containing a mixture of various additives present in smokeless powders.[14,15] The sample analysis was performed using gradient elution on a Restek pinnacle C8 column with 1 mM ammonium acetate and methanol. Electrospray MS analysis was performed using a Bruker Esquire ion trap system in the positive ion mode.[15]

Detection of nitroglycerine in these samples requires the use of negative ion electrospray MS analysis in the presence of an ammonium salt to enhance adduct ion formation. In a series of experiments with various high explosives in the presence of different adduct ions, we have demonstrated that optimal detection of nitroglycerine can be obtained using an isocratic eluent consisting of 50% methanol/50% aqueous mixture of ammonium salts.[16] The importance of proper selection of the adduct ion is shown in Fig. 3. This figure illustrates the effect of mixing a variety of ammonium

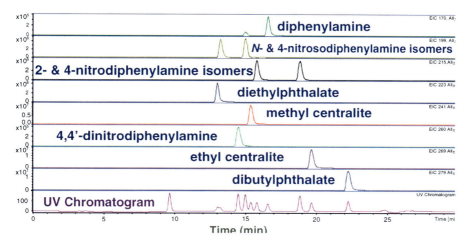

Fig. 2 The analysis of a standard containing various components present in smokeless powders analyzed by electrospray mass spectrometery using a Bruker Esquire ion trap system. The analysis was performed using methanol/water gradient with a 2.1 mm id Restek pinnacle C8 column[15]

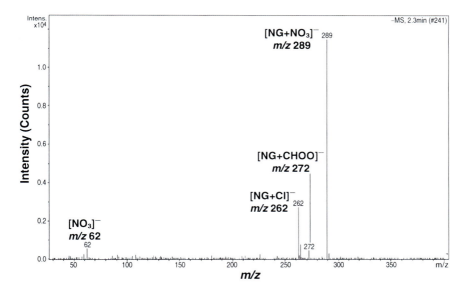

Fig. 3 Negative ion mass spectrum of NG collected using a Bruker Esquire ion trap system, which illustrates detection of NG-chloride, m/z 262, NG-formate, m/z 272, and NG-nitrate, m/z 289 adducts. The analysis was performed using a Hewlett-Packard C18 column (2.1×100 mm) in the isocratic mode using 50% methanol/50% aqueous solution containing 0.05 mM ammonium nitrate, 0.1 mM ammonium chloride, 0.1 mM ammonium formate, and 0.1 mM ammonium acetate[16]

salts into the buffer when analyzing nitroglycerine. As seen in the figure, the base peak is the nitrate adduct. Other products formed include formate and chloride ion adducts of NG.[16]

4   Conclusions

This report describes the application of various mass spectrometric procedures for the detection of organic and inorganic low explosives. For inorganic explosives, ion chromatography/mass spectrometry is used to detect the presence of oxidizers and their decomposition products in aqueous extracts. Proper selection of eluents is necessary to avoid decomposition of ascorbic acid which is present in certain commercial formulations. For the analysis of smokeless powders electrospray mass spectrometery offers certain improvements over GC/MS in terms of its capability to detect thermally labile components such as nitrosodiphenyl amine and nitroglycerine. Both positive and negative ion MS procedures must be used to detect all components present in the powder. Overall these procedures increase specificity and confidence in the analysis and detection of explosive residue.

Acknowledgements   The author would like to thank the following individuals without whom this paper would not be possible. Their scientific contributions and friendship are gratefully acknowledged here and in the papers I have referenced. These individuals include Janet Doyle, Kelly Mount, Mark Miller, Lisa Lang, Chad Wissinger, Olivier Collin, Maximilien Blas, and Kristy Lahoda. Financial and material support is gratefully acknowledged from the FBI, BATF, TSWG, DHS, and the National Institute of Justice, grant #2009-DN-BX-K251. Points of view in the document are those of the authors and do not necessarily represent the official view of the US Department of Justice.

References

1. http://www.reportlinker.com/p046420/explosives.html. Accessed 4/24/09
2. http://www.ncfs.ucf.edu/twgfex/docs/Guide%20for%20identification%20of%20intact%20 explosives.pdf. Accessd 4/24/10
3. Davis, T (June 1972) The chemistry of powder and explosives, Angriff Press, Las Vegas, ISBN-10: 0913022004
4. Bottegal M, Lang GL, Miller M, McCord B (2010) Analysis of ascorbic acid-based black powder substitutes by high performance liquid chromatography-electrospray ionization-quadrupole time-of-flight mass spectrometry (HPLC-ESI-QToFMS), Rapid communications in mass spectrometry 24(9):1377–1386
5. McCord BR, Bender E (1998) Chromatography of explosives. In: Beveridge AD (ed) Forensic investigations of explosives. Taylor & Francis, London, pp 231–265
6. McCord BR, Hargadon K, Hall K, Burmeister S (1994) Forensic analysis of explosives using ion chromatographic methods. Anal Chim Acta 288:43–56
7. Doyle JM, Miller ML, McCord BR, McCollam DA, Mushrush GW (2000) A multicomponent mobile phase for ion chromatography applied to the separation of anions from the residue of low explosives. Anal Chem 72(10):2303–2307
8. Doyle JM, McCord BR (1998) Novel electrolyte for the analysis of cations in low explosive residue by capillary electrophoresis. J Chromatogr B 714:105–111
9. http://www.ncfs.ucf.edu/twgfex/doc   s/Guidelines%20for%20Forensic%20Ident%20 of%20 PB%20Explosive%20Residues.pdf, accessed 4/24/10
10. Lang GL, Boyle K (2009) The analysis of black powder substitutes containing ascorbic acid by ion chromatography/mass spectrometry. J Forensic Sci 54(6):1315–1322

11. Heramb R, McCord B (2002) Smokeless powders and their analysis, a brief review. Forensic Sci Commun 4(2):1–7
12. Stine GY (1991) An investigation into propellant stability. Anal Chem 63(8):475A–478A
13. Martz RM, Lasswell LD (1983) Identification of smokeless powders and their residues by capillary column gas chromatography/mass spectrometry. In: Proceedings of the International Symposium on the Analysis and Detection of Explosives. U.S. Government Printing Office, Washington, DC, pp 245–254
14. Wissinger CE, McCord BR (2002) A reversed phase HPLC procedure for smokeless powder comparison. J Forensic Sci 47(1):168–174
15. Mathis J, McCord B (2005) Mobile phase influence on electrospray ionization for the analysis of smokeless powders by gradient reversed phase liquid chromatorgraphy-EIMS. Forensic Sci Int 154:159–166
16. Mathis J, McCord B (2005) The analysis of high explosives by liquid chromatography electrospray ionization mass spectrometry: multiplexed detection of negative ion adducts. Rapid Commun Mass Spectrom 19(2):99–104
17. Miller ML (1994) GC/MS analysis of unburned/burned smokeless powders and pipe bomb residues. In: Presented 46th Annual Meeting AAFS San Antonio, TX February 14–19, 1994 Abstract # B83

Identification of the Bacterial Cellular Lipid Fraction by Using Fast GC × GC-MS and Innovative MS Libraries

Luigi Mondello, Peter Quinto Tranchida, Giorgia Purcaro, Chiara Fanali, Paola Dugo, Erminia La Camera, and Carlo Bisignano

Abstract The bacteria fatty acid (FA) profile has been extensively studied for taxonomic classification purposes, since bacteria, in general, contain particular and rare fatty acids, compared to animal and plant tissues. In the last few years, the concern about pathogenic microorganisms used as bioterrorist agents has increased; therefore, rapid methods for the characterization of bacteria are necessary. In the present research, a half-an-hour procedure, to analyze bacteria, was developed: a 2-min one-step sample preparation step, was followed by a relatively fast comprehensive 2D GC-MS separation (25 min). Furthermore, dedicated mass spectrometry libraries were constructed for bacteria and FA identification. Finally, data-processing was carried out with the support of novel comprehensive 2D GC software.

Keywords Bacterial fatty acid methyl esters • Comprehensive two-dimensional gas chromatography • Split-flow • Mass spectrometry • Linear retention indices

L. Mondello (✉), P.Q. Tranchida, G. Purcaro, and P. Dugo
Dipartimento Farmaco-chimico, Facoltà di Farmacia,
Università di Messina, viale Annunziata, 98168 Messina, Italy
e-mail: lmondello@unime.it; ptranchida@pharma.unime.it; giopurcaro@libero.it; pdugo@pharma.unime.it

C. Fanali
Istituto di Biochimica e Biochimica Clinica, Università Cattolica del Sacro Cuore, Rome, Italy
e-mail: c.fanali@unicampus.it

L. Mondello and C. Fanali
Campus-Biomedico, Via Alvaro del Portillo, 21, 00128 Rome, Italy
e-mail: lmondello@unime.it; c.fanali@unicampus.it

E. La Camera and C. Bisignano
Dipartimento Farmaco-biologico, Università Campus Bio-medico, viale Annunziata, 98168 Messina, Italy
e-mail: elacamera@unime.it; bisignano@pharma.unime.it

J. Banoub (ed.), *Detection of Biological Agents for the Prevention of Bioterrorism*, 231
NATO Science for Peace and Security Series A: Chemistry and Biology,
DOI 10.1007/978-90-481-9815-3_15, © Springer Science+Business Media B.V. 2011

1   Introduction

1.1   Principles of Comprehensive Two-Dimensional Gas Chromatography

The introduction of comprehensive two-dimensional gas chromatography (GC×GC) dates back to 1991, and can be certainly considered as one of the most important inventions in the GC field.[1] The separation power gain, compared to conventional GC, is probable greater than that between the packed and open-tubular capillary column.

A typical GC×GC system consists of a primary and secondary column, linked in series, and with independent selectivities. A transfer device, defined as modulator, is the most important part of any GC × GC system, and is (normally) located at the head of the second dimension. The function of the modulator is to cut, re-concentrate and launch chromatography bands from a primary conventional column (e.g., 30 m×0.25 mm ID×0.25 μm d_f), onto a short micro-bore column (e.g., 1 m×0.10 mm ID×0.10 μm d_f). The modulation process is carried out continuously throughout the GC×GC experiment, has a duration of typically 4–8 s, which corresponds to the time-window of each analysis in the second dimension.

In the field of GC×GC, it is very common to use a primary apolar column, which achieves separation on a boiling-point basis. Isovolatile constituents are then subjected to a fast 2D analysis, generally on a polar capillary.[2] Such an internal diameter combination enables slow and fast peak production in the first and second dimension, respectively. In an ideal GC × GC analysis, the total peak capacity is equivalent to the product of the peak capacities relative to each dimension. Although such a value exceeds the "real" peak capacity, comprehensive 2D GC is certainly the most powerful tool today-available for the analysis of complex volatile mixtures.[2]

Although several modulation systems have been developed and are currently employed, the principles of the process have remained essentially unaltered. An example of how (twin-stage) thermal modulation was initially achieved is shown in Fig. 1. In step A, a narrow band, in this case containing two co-eluting compounds, is formed at the head of the modulator, which is maintained at a sufficiently low temperature to generate a primary re-concentration effect. In step B, a heating pulse (ΔI) is directed to the first segment of the modulator, causing band re-mobilization. In step C, the released band hits a second cold spot, and is again re-concentrated; at the same time, volatiles begin to undergo compression at the modulator head, which has rapidly cooled down. In step D, a further heating pulse (ΔII) is directed to the second segment of the modulator, and the narrow band is launched onto the second dimension. In step E, the two analytes are subjected to a fast GC separation and reach the detector at different times. During the second-dimension analysis, the subsequent fraction is subjected to modulation.[2]

Ideally, each first-dimension peak must be subjected to minimum three modulations, in order to maintain the resolution achieved on the primary column. An important requisite is that all compounds reach the detector within the modulation

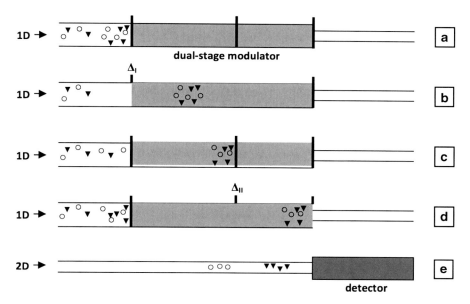

Fig. 1 Schematic of a dual-stage modulation process

time-window; a phenomenon defined "wrap-around" occurs if second-dimension retention times exceed the modulation period.

Thermal modulation has a beneficial effect on sensitivity: band re-concentration produces a signal-to-noise increase in the 10–50 factor range. Furthermore, thermal modulation generates very narrow and rapid peaks; hence fast detectors (a sampling frequency of at least 50 Hz is necessary) are mandatory for correct peak re-construction.[2]

If a 3,600 s GC×GC analysis is considered, with a 4-s modulation period, then nine hundred 4-s 2D chromatograms, stacked side-by-side, will form a "raw" GC × GC chromatogram. Dedicated software is mandatory to visualize the raw monodimensional data in a bidimensional format (contour plot): the single rapid GC chromatograms, positioned at 90° with respect to an x-axis, are characterized by first-dimension t_R values that are expressed in minutes. The compounds separated in the second dimension, aligned along a y-axis, are characterized by an oval shape and with t_R values that are expressed in seconds. The colour and dimension of each blob are directly related to detector response. Dedicated software is also required for GC × GC quantitation: the peak areas relative to the same compound in each fast 2D chromatogram are summed.

The advantages of comprehensive two-dimensional gas chromatography, over conventional GC, are

– Selectivity (two separation dimensions, related to volatility and polarity)
– Sensitivity (band compression)
– Separation power

- Structure (formation of group-type patterns on the 2D plane)
- Speed (comparable to ultra-fast GC experiments, if the number of peaks resolved per unit of time is considered)

If a third MS dimension is added to a GC×GC instrument (GC × GC-MS), then the most powerful tool today-available for volatile analysis is generated.[2]

1.2 Fatty Acid Characterization by Using GC×GC

Comprehensive 2D GC has been applied successfully in many research areas. Historically, the most common GC×GC application has been on petrochemical samples, one of the most complex sample-types known to analytical chemists. GC × GC has also been widely employed for the unravelling of complex food matrices,[3] in many instances for the analysis of fatty acid methyl esters (FAMEs). In particular, enhanced sensitivity and formation of structured chromatograms are two GC × GC features which are of great help in the elucidation of FAME profiles. Fish and vegetable oils,[4,5] milk fat,[6] plasma,[7] and micro-algae and aquatic meiofauna species,[8] have been characterized by using GC×GC.

Recently, David and co-workers exploited GC×GC for the study of bacteria FAs.[9] Precious information was attained through highly-structured chromatograms, and the data derived was compared with that attained by using the standardized Sherlock MIDI system (Sasser M., MIDI, Technical Note 101, 1990, see www.midi-inc.com). The latter process is based on the careful control of growth conditions (24 h), before harvesting the bacterial cells, and processing them through saponification, methylation and extraction (the procedure takes about 1 h). Bacterial fatty acid methyl esters (BAMEs) were identified on the 2D space plane by using two parameters, namely ECL (equivalent chain length) in the first dimension and RPV (relative polarity value) in the second dimension.

In general, bacteria are characterized by different FA profiles, compared to animal and plant tissues. Appreciable amounts of C_{14} to C_{18} straight-chain saturated and monounsaturated FAs are present; however, the latter differ from the common plant ones: for example, the main C_{18} monoenoic FA is not oleic acid, but *cis*-vaccenic acid ($C_{18:1\omega7}$). Furthermore, bacteria are characterized by odd-chain, branched-chain (mainly iso- and anteiso-), cyclopropane and hydroxy fatty acids. These different constituents can be easily identified in the GC×GC contour-plot, if a group-type separation is obtained.

Since 1963, when Abel and co-workers first-described the GC analysis of BAMEs, fatty acid profiles have been widely-used for the taxonomic classification of microorganisms.[10-12] It is important to note that the FA composition, both in qualitative and quantitative terms, is markedly affected by the nature of the medium, by the growth conditions, as well as by the age of the culture when harvested. Therefore, the knowledge and the standardization of these conditions are essential for valid and repeatable studies.

The reliable analysis of BAMEs is a useful tool for bacteria identification. Furthermore, following a series of international terrorist attacks, concern about the use of pathogenic microorganisms, as biological agents, has increased exponentially. A biological event could spread out in time and space, before the realization that a bioterrorism attack has occurred. Therefore, very rapid and reliable methods for bacteria characterization are required.

The aim of the present research is the development of a rapid approach to identify bacteria, through specific BAME profiles. A single-step, rapid methylation procedure (2–3 min), using a trimethylsulfonium hydroxide (TMSH) methanol solution, was employed,[13] followed by a relatively rapid GC×GC–MS analysis.

The employment of a split-flow, twin-oven GC×GC-MS system enabled the use of a micro-bore column set [apolar (0.1 mm ID)×polar (0.05 mm ID)], under close-to-optimum gas linear velocity conditions. Furthermore, dedicated GC×GC-MS libraries were developed to obtain unambiguous bacteria and BAME identification. Data-processing was carried out with the support of recently-developed comprehensive chromatography software.

2   Material and Methods

2.1   Samples and Sample Preparation

Five different strains of bacteria were studied in this work, namely *P. fluorescens*, *B. subtilis*, *S. aureus*, *E. coli*, and *P. aeruginosa*. The bacteria were grown on R2A Agar at 28°C for 24 hours. Bacteria were then processed according to Müller and co-workers.[13] Briefly, a few colonies of bacteria were harvested and suspended in 10 μL of distilled water, then 30 μL of methanolic trimethylsulfonium hydroxide (TMSH) solution (0.25 M) were added. The reaction mixture was dried under nitrogen stream and then re-dissolved in 200 μL of *tert*-butyl-methyl ether (MTBE)/methanol (MeOH) mixture (10/1 *v/v*) and directly injected in the GC×GC-MS system.

All the reagents employed were purchased from Sigma Aldrich (Milano, Italy).

A BAMEs mixture was kindly provided by Supelco (Bellafonte, PA, USA) and was used to optimize the GC×GC-MS method. The fatty acid methyl esters used to construct the GC × GC BAME library were from Supelco and Larodan (Malmö, Sweden).

2.2   Instrumentation

The GC×GC-MS applications were carried out on a Shimadzu GC × GC-MS system, consisting of two independent GC2010 gas chromatographs, connected

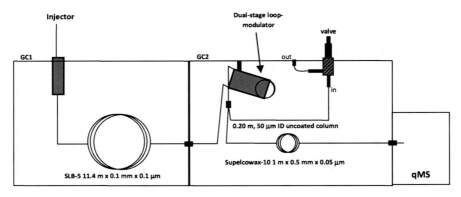

Fig. 2 Scheme of the split-flow, twin-oven GC×GC-MS instrument

through a heated (280°C) transfer line, and a QP2010 Plus quadrupole mass spectrometer (Shimadzu, Kyoto, Japan) linked to the primary GC through a cable extension (due to the presence of the second oven).

The system was equipped with an AOC-20i auto-injector, and a split-splitless injector (250°C). The primary column (situated in GC1), an SLB-5ms (silphenylene polymer) 11.4 m×0.1 mm ID×0.1 μm d_f, was connected to a custom-made Supelcowax-10 1 m×0.05 mm ID×0.05 μm d_f [poly(ethyleneglycol)] capillary and to a 0.20 m×0.05 mm ID uncoated capillary (both columns, provided by Supelco, were located in GC2), by using a fixed outlet capillary column splitter (SGE, Ringwood, Victoria, Australia). The uncoated capillary was connected to a manually-controlled valve, namely an OSS-2 outlet splitter system (SGE).

A scheme of the split-flow, twin-oven GC×GC-MS instrument is reported in Fig. 2.

The final part of the primary column was used to create a double loop for the cryogenic dual-stage, loop-type modulator (under license from Zoex Corporation, Houston, TX, USA). The modulation time applied was 4 s and the duration of the hot pulse (325°C) was 375 ms.

Optimized conventional GC×GC-MS: GC1 temperature program: 115°C–280°C at 5°C/min. GC2 temperature program: 190–280°C at 5°C/min. Initial He pressure (constant linear velocity): 680.2 kPa. Injection volume: 0.5 μL; split ratio: 5:1.

The MS transfer line and ion source were maintained at 250°C and 200°C, respectively. The scan range was of 40–360 m/z, at a scan speed of 10,000 amu/s, and a sampling frequency of 25 spectra/s. MS ionization mode: electron ionization (70 eV).

Data were collected by the GCMS solution software; bidimensional visualization was carried out by using the ChromSquare v.1.0® software (Chromaleont, Messina, Italy). The MS libraries used for spectral matching were laboratory-constructed.

3 Results and Discussion

3.1 GC × GC-MS Method Optimization

Comprehensive 2D GC method optimization can be a painstaking issue and is, probably, one of the main reasons behind the still rather limited employment of the technique (another is probably a natural, though often exaggerated affection towards well-established methods). The scenario is much more complex, if compared with conventional GC, because the two dimensions are intimately related. Apart from modulation parameters (period and temperature), the main operational conditions that must be considered are the stationary phase chemistries, capillary column dimensions, gas flow, the temperature program(s), outlet pressure conditions, and the detector settings.

In the present research, a relatively-fast GC × GC-MS method was optimized for the analysis of BAMEs. A 0.1 mm ID apolar capillary was used in the first dimension, while a 0.05 mm ID polar column was employed as second dimension. If selectivity is considered, the present stationary phase combination has proved to be nicely suited to the analysis of FAMEs. In fact, some highly-structure 2D chromatograms have been reported in the literature.[4,5,7]

A discussion on GC × GC gas flows and on the split-flow configuration herein used (2.2), is necessary for reasons of clarity. In recent years, it has been shown that the potential of GC × GC is only partially expressed when using the 0.25 mm ID + 0.10 mm ID column combination, because gas velocities are generally near-to-ideal in the first dimension, while secondary ones are far from optimum.[13] The main consequence of non-ideal GC × GC chromatography conditions is that a substantial amount of chromatogram retention-time space is not occupied.

The GC × GC flow drawback, due to the fact that a single inlet pressure is applied to a dual-column set, with differing internal diameters, has been recently circumvented: a novel method, defined as "split-flow" GC × GC, enabled the generation of greatly improved H_2 velocities.[14] A first-dimension apolar 30 m × 0.25 mm ID capillary was linked to a 1 m × 0.10 mm ID polar one, and to a 30 cm × 0.10 mm ID uncoated capillary, by using a Y press fit. The uncoated segment was connected to a manually-operated split valve (located on top of the GC oven), while the polar column was passed through a cryogenic modulator, and then connected to a flame ionization detector (FID). Gas flows in both dimensions were improved simply by manually regulating the split valve; the advantages of the innovative approach were shown using a single-oven GC × GC system, under split-flow and non split-flow conditions.

In 2009, a split flow, twin-oven, GC × GC-FID system was used for the analysis of coffee volatiles.[15] The columns used were: a 30 m × 0.25 mm ID polar first dimension linked, through a Y-union, to a 1 m × 0.05 mm ID apolar column, and to a 0.20 m × 0.05 mm ID uncoated capillary segment. The main objective of the experiment was to demonstrate that the split-flow approach enabled the optimum exploitation of a "high resolution" 0.05 mm ID second dimension, in a H_2-based application.

In a further split flow research, the dual-oven GC×GC-FID system was exploited (again, H$_2$ was used as mobile phase) for the use of a 1 m×0.10 mm ID and 1 m × 0.05 mm ID secondary polar column (the primary apolar column was the same in both cases), under ideal operational conditions.[16] A direct comparison was made between the results attained, highlighting the considerable advantages of using a 0.05 mm ID capillary.

The most recent split flow GC×GC research was focused on the optimized use of a 1 m×0.05 mm ID secondary GC×GC column, employing He as carrier gas, and a rapid-scanning quadrupole mass spectrometer as detector.[17] The column configuration consisted of: a 30 m (0.25 mm ID) apolar primary capillary connected to a 1 m×0.05 mm ID polar analytical column, and to a 0.20 m × 0.05 mm ID uncoated capillary. A series of experiments were carried out on a commercial perfume sample, under split-flow and non split-flow GC×GC-MS conditions. Data-processing was carried out by using recently-developed comprehensive chromatography software.

In the present split-flow, twin-oven GC×GC-MS experiment, a 11.4 m×0.1 mm ID apolar capillary was connected to a 1 m × 0.05 mm ID polar column and to a 20 cm segment of 0.05 mm ID uncoated capillary. Various split-flow experiments were carried out by manually-regulating the split valve at different stages. The best operational conditions were attained with the split-valve completely opened: linear velocities of *circa* 45 and 70 cm/s were calculated in the first and second dimension, respectively. It must be emphasized that these calculations, due to non-ideal gas behaviour, must be considered only as approximations. The relatively high (but ideal) first-dimension He velocity was combined with a relatively accelerated temperature program (115–280°C at 5°C/min), to produce a rather fast GC×GC separation (the last compounds of interest eluted within 25 min). However, the operational conditions applied generated rather narrow primary column chromatography bands (6–9 s) and, hence, a short modulation period (4 s) was necessary to achieve a near-to-sufficient number of cuts per peak.

Initially, a split-flow GC×GC-MS experiment was carried out on a standard solution, containing 26 methyl esters and a C$_7$–C$_{30}$ alkane series (the contemporary presence of the hydrocarbons will be later explained), using the same temperature program in both ovens (data not shown). Although the GC × GC separation performance was good, wrap-around occurred, being due to the low primary column elution temperatures. Consequently, a series of positive temperature offsets were tested in GC2, with the best result attained through the application of +75°C. A bidimensional chromatogram expansion, relative to the optimized split-flow GC×GC-MS application, is shown in Fig. 3.

As can be observed, the contour plot is characterized by an ordered structure: saturated BAMEs (C$_{11:0}$–C$_{20:0}$) are located in the non polar zone of the chromatogram, and are aligned along a distinct band; methyl esters with one bond (C$_{16:1}$–C$_{19:1}$) are also aligned horizontally and are slightly more retained than the saturated counterparts. The only methyl ester with two double bonds (C$_{18:2}$) is, as expected, slightly more retained than the monosaturates. The hydroxy BAMES, characterized by intense H-bond interactions with the poly(ethyleneglycol) phase, are situated in the

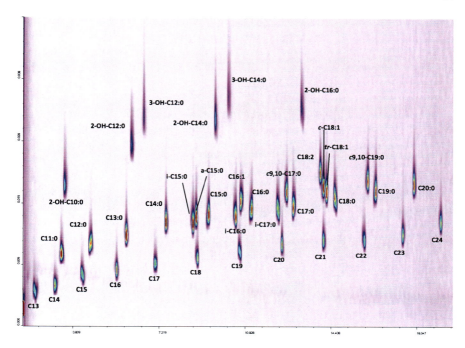

Fig. 3 GC×GC-MS chromatogram of a standard mixture of BAMEs and C_7–C_{30} alkanes. Abbreviations: i=iso; a=anteiso; tr=trans; c9, 10-C17:0=methyl cis-9,10-methylenehexadecanoate

most polar zone of the chromatogram, while the alkanes are characterized by the lowest second-dimension elution times.

The rapid-scanning quadrupole mass spectrometer was operated using a 40–360 m/z mass range, a 10,000 amu/s scan speed and a 25 Hz sampling frequency. Such MS parameters enabled the attainment of average 6–8 spectra per peak, enabling reliable peak identification and near-to-satisfactory peak reconstruction.

3.2   Construction of the GC×GC-MS Libraries

Once optimized the split-flow GC×GC-MS method, the same was used to construct two dedicated MS libraries. The first, named GC × GC BAME library, contained pure MS spectra relative to fatty acids (methyl esters) commonly found in bacterial lipid extracts, and linear retention index (LRI) values. Initially, mixtures of standard fatty acid methyl esters were subjected to GC × GC-MS analysis and the spectra attained were included in the library. It must be emphasized that the use of GC × GC enabled the isolation of the chemical noise, through the re-concentration effect of cryogenic modulation. Hence, the MS spectra were characterized by a high degree of purity. After, the LRI values of the methyl esters were automatically calculated by using the comprehensive chromatography and GC-MS softwares, on the basis of the retention times of a C_7–C_{30} alkane series, co-analyzed with the standard compounds. At present, the GC × GC BAME library contains MS spectra and LRI values relative to *circa* 80 BAMEs.

The GC-MS software used (2.2), enables the application of a twin-filter: (1) spectral similarity, and (2) linear retention index range. The primary filter eliminates GC×GC BAME library matches with a spectral similarity (expressed in%) lower than a minimum value set by the analyst; the other filter deletes GC × GC BAME library spectra, characterized by an LRI value outside a pre-defined range, and with respect to the LRI value of the unknown compound.

The other MS library, named as bacteria library, is in a very early stage of development. The library was constructed by subjecting real bacterial fatty acid samples to the fast GC×GC-MS method, by deriving a single averaged spectrum, comprising all the methyl esters in the retention time range, and by finally subtracting the compressed chemical the noise at three points across the chromatogram. The bacterial spectrum attained can be considered as a sample fingerprint, and is altogether comparable (in terms of concept) to direct electron ionization-MS analysis. The averaged spectrum was derived from the untransformed GC × GC-MS chromatogram, using the instrumental GC-MS software. At present, the library contains only five bacterial spectra (see Section 2.1), because it is currently in the testing stage. Consequently, the results attained using the GC × GC bacterial library are to be considered as preliminary. As an example, the GC × GC-MS derived averaged spectrum of *P. fluorescens* is shown in Fig. 4.

3.3   GC × GC-MS Analysis of Bacteria Samples

Fatty acid methyl esters, relative to five bacteria samples, were prepared by using a very rapid sample preparation method (2.1). At the same time, a C_7–C_{30} alkane series was subjected to GC×GC-MS analysis and, before the end of application, the bacteria samples were ready. The real-world BAME samples were then subjected to GC × GC-MS analysis. A bidimensional GC × GC-MS chromatogram expansion, relative to a bacterial sample, is shown in Fig. 5. As can be observed, the complexity of the chromatogram justifies the use of a bidimensional methodology, but excludes the contemporary analysis of the C_7–C_{30} alkane series. It must be noted that there are slight secondary-column retention time differences, in comparison to the standard BAMEs chromatogram illustrated in Fig. 3. It was concluded that such slight elution time deviations were due to matrix effects. The averaged spectrum of the untransformed GC × GC-MS chromatogram was subjected to library matching, and the bacterium was identified as *E. coli* with a 98% similarity (Fig. 6). The considerable difference, between the *E. coli* spectrum and the other bacteria library spectra, can be deduced observing the second best match provided by the GC-MS software (66% similarity).

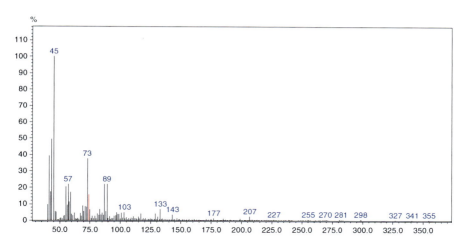

Fig. 4 Averaged MS spectrum of *P. fluorescens*

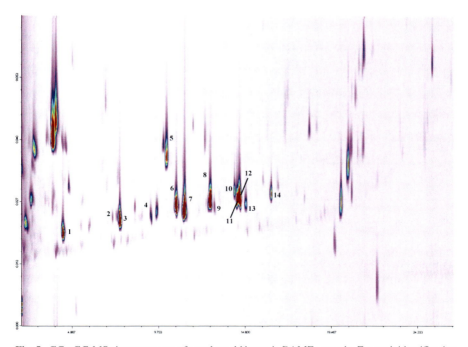

Fig. 5 GC×GC-MS chromatogram of a real-world bacteria BAMEs sample. For peak identification refer to Table 1

Once the bacterium was correctly identified, then, the individual FAs can be identified. Peak assignment was automatically achieved by using a recently-developed comprehensive chromatography (ChromSquare) program, in combination

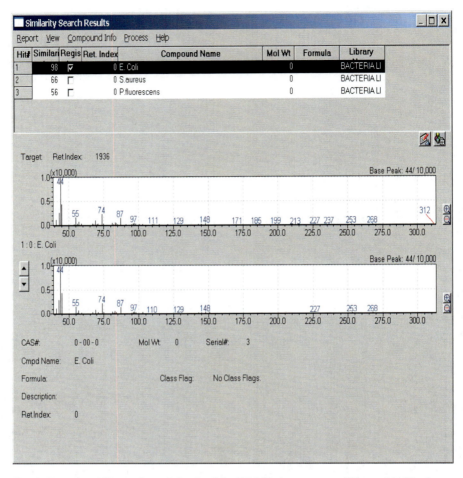

Fig. 6 Bacterium MS search result for the GC×GC-MS chromatogram illustrated in Fig. 5

with the GC-MS software, and the GC×GC BAME library. As aforementioned, the GC-MS software enabled the application of a dual-filter, allowing the elimination of all matches with an: (1) MS similarity below a pre-defined limit (90% in this case); (2) LRI value outside a pre-defined range, and with respect to the unknown compound (± 20 LRI units in this case). The 14 methyl esters identified in the *E. coli* sample are reported in Table 1, along with retention times, LRI and MS similarity values. Some characteristic BAMEs were identified, such as hydroxyl (3-OH $C_{14:0}$) and cyclopropanic (*cis*-9,10-$C_{19:0}$) FA. As can be observed, there was rather good agreement between experimental and library LRI values, with a maximum difference of 17 LRI units for $C_{17:1\omega7}$. Considering the complexity of the chromatogram illustrated in Fig. 5, the number of identified methyl esters is certainly low and is dependent on the early-development stage of the BAME library

Table 1 BAMEs identified in the GC×GC-MS chromatogram illustrated in

Peak/Compound	1D t_R (min)	2D t_R (s)	LRI Exp.	LRI Lib.	% Simil.
(1) $C_{12:0}$	4.353	1.172	1526	1522	96
(2) $C_{14:1\omega5}$	7.157	1.414	1706	1710	94
(3) $C_{14:0}$	7.556	1.374	1729	1726	96
(4) $C_{15:0}$	9.356	1.374	1826	1830	95
(5) 3-OH $C_{14:0}$	10.236	2.182	1880	1870	96
(6) $C_{16:1\omega7}$	10.759	1.535	1910	1903	95
(7) $C_{16:0}$	11.225	1.495	1936	1925	95
(8) $C_{17:1\omega7}$	12.694	1.616	2018	2001	92
(9) $C_{17:0}$	12.892	1.535	2030	2020	92
(10) $C_{18:2\omega6}$	14.029	1.737	2095	2091	97
(11) $C_{18:1\omega9}$	14.16	1.616	2103	2097	95
(12) $C_{18:1\omega7}$	14.294	1.657	2110	2104	93
(13) $C_{18:0}$	14.693	1.576	2133	2125	97
(14) cis-9,10-$C_{19:0}$	16.095	1.697	2215	2208	95

and on the fact that many constituents were not BAMEs. For example, several spectra were characterized by a typical phthalate profile, and were not positively identified because these compounds are not present in the BAME library.[17]

4 Conclusions

Although the results attained in the present research are satisfactory, they must be considered as preliminary. The objective of this first phase of research was to determine whether an idea, namely the development of a fast GC×GC-MS method supported by two novel MS libraries, was suitable or not for the rapid characterization of bacteria. Although the outcome of the experiments reported is certainly a positive one, a series of analytical aspects must be studied in much more depth. A limited number of fatty acids were identified in all five bacterium samples, a factor related to the low number of spectra present in the MS library. Furthermore, standardized bacteria growth and collection conditions must be developed, because reproducible GC × GC-MS profiles must be attained. As a consequence, the second phase of this research will be devoted to the expansion of the MS libraries, and, more importantly, to method validation.

References

1. Liu Z, Phillips JB (1991) Comprehensive two-dimensional gas chromatography using an on-column thermal modulator interface. J Chromatogr Sci 29:227–231
2. Mondello L, Tranchida PQ, Dugo P, Dugo G (2008) Comprehensive two-dimensional gas chromatography-mass spectrometry: a review. Mass Spectrom Rev 27:101–124

3. Tranchida PQ, Dugo P, Dugo G, Mondello L (2004) Comprehensive two-dimensional chromatography in food analysis. J Chromatogr A 1054:3–16
4. de Geus H-J, Aidos I, de Boer J, Luten JB, Brinkman UATh (2001) Characterization of fatty acids in biological oil samples using comprehensive multidimensional gas chromatography. J Chromatogr A 910:95–103
5. Mondello L, Casilli A, Tranchida PQ, Dugo P, Dugo G (2003) Detailed analysis and group-type separation of natural fats and oils using comprehensive two-dimensional gas chromatography. J Chromatogr A 1019:187–196
6. Hyötyläinen T, Kallio M, Lehtonen M, Lintonen S, Peräjoki P, Jussila M, Riekkola M-L (2004) Comprehensive two-dimensional gas chromatography in the analysis of dietary fatty acids. J Sep Sci 27:459–467
7. Tranchida PQ, Costa R, Donato P, Sciarrone D, Ragonese C, Dugo P, Dugo G, Mondello L (2008) Acquisition of deeper knowledge on the human plasma fatty acid profile exploiting comprehensive two-dimensional gas chromatography. J Sep Sci 31:3347–3351
8. Akoto L, Stellaard F, Irth H, Vreuls RJJ, Pel R (2008) Improved fatty acid detection in microalgae and aquatic meiofauna species using a direct thermal desorption interface combined with comprehensive gas chromatography–time-of-flight mass spectrometry. J Chromatogr A 1186: 254–261
9. David F, Tienpont B, Sandra P (2008) Chemotaxonomy of bacteria by comprehensive GC and GC-MS in electron impact and chemical ionisation mode. J Sep Sci 31:3395–3403
10. Abel K, De Schmertzing H, Peterson JI (1963) Classification of microorganisms by analysis of chemical composition. I. feasibility of utilizing gas chromatography. J Bacteriol 85:1039–1044
11. Miller LT (1982) Single derivatization method for routine analysis of bacterial whole-cell fatty acid methyl esters, including hydroxy acids. J Clin Microbiol 16: 584–586
12. Buyer JS (2002) Identification of bacteria from single colonies by fatty acid analysis. J Microbiol Methods 48:259–265
13. Müller K-D, Husmann H, Nalik HP, Schomburg G (1990) Trans-esterification of fatty acids from microorganisms and human blood serum by trimethylsulfonium hydroxide (TMSH) for GC analysis. Chromatographia 30:245–248
14. Tranchida PQ, Casilli A, Dugo P, Dugo G, Mondello L (2007) generation of improved gas linear velocities in a comprehensive two-dimensional gas chromatography system. Anal Chem 79:2266–2275
15. Tranchida PQ, Purcaro G, Conte LS, Dugo P, Dugo G, Mondello L (2009) Enhanced resolution comprehensive two-dimensional gas chromatography applied to the analysis of roasted coffee volatiles. J Chromatogr A 1216:7301–7306
16. Tranchida PQ, Purcaro G, Conte LS, Dugo P, Dugo G, Mondello L (2009) Optimized use of a 50 μm ID secondary column in a comprehensive two-dimensional gas chromatography system. Anal Chem 81:8529–8537
17. Tranchida PQ, Purcaro G, Fanali C, Dugo P, Dugo G, Mondello L (2010) Optimized use of a 50 μm ID secondary column in comprehensive two-dimensional gas chromatography-Mass spectrometry. J Chromatogr A 1217:4160–4166

Endotoxins in Environmental and Clinical Samples Assessed by GC-Tandem MS

Bogumila Szponar

Abstract Bacteria appeared on the Earth millions years before us and human evolution was triggered by the constant presence of pathogenic and symbiotic microorganisms in our surroundings. Interplay occurred between higher organism and microbial consortia residing in the host organs and on the epithelial surfaces; another natural space of bacteria–human interaction is the indoor environment where we spend the majority of our lifetime. Indoor microbial exposure affects our well-being and can result in respiratory symptoms, such as allergies and asthma, since both dead and live microorganisms and their cell constituents, including lipopolysaccharides (LPS, endotoxins), interact with our immune system. Thus, there is a demand for robust tools for qualitative and quantitative determination of the microbial communities that we are exposed to.

This work described the reproducible approach of the Gram-negative bacteria and endotoxins assessment by their specific chemical markers, 3-hydroxy fatty acids. Gas chromatography-tandem mass spectrometry proved to be an excellent means for specific and selective detection of bacteria/endotoxin markers in the complex matrices like indoor bioaerosol and clinical samples: blood, saliva or feces. Using this method, epidemiological studies were conducted in the field of indoor air quality research, as well as in clinical investigations when bacterial consortia were involved: in Crohn's disease, periodontitis, and newborn gut microbial colonisation in association with allergy development.

B. Szponar (✉)
Department of Laboratory Medicine, Division of Medical Microbiology,
Lund University, Lund, Sweden
and
Ludwik Hirszfeld Institute of Immunology and Experimental Therapy,
Polish Academy of Sciences, Wroclaw, Poland
e-mail: bogusia02@yahoo.com; szponar@iitd.pan.wroc.pl

J. Banoub (ed.), *Detection of Biological Agents for the Prevention of Bioterrorism*, 245
NATO Science for Peace and Security Series A: Chemistry and Biology,
DOI 10.1007/978-90-481-9815-3_16, © Springer Science+Business Media B.V. 2011

1 Introduction

Lipopolysaccharide (LPS, endotoxin) is a bacterial macromolecule representing potential that had influenced vast number of biological processes and bacteria–host interactions in mammals. Gram-negative bacteria have existed on the Earth for millions of years and due to co-habitation during vertebrate evolution, it is assumed that several biomechanisms developed to conquer and defend in both parties: microbial and vertebrate. One more way of such co-existence is symbiosis, or mutualism that occurs in microbiota of the gastrointestinal tract in higher vertebrates.[1]

Lipopolysaccharides are exposed on Gram-negative bacteria surface – three to four millions of LPS molecules are present on each cell, which is ca. 3% of total dry bacterial biomass, and under certain circumstances they can also be released to the surrounding environment. LPS possesses toxic potential and is responsible for many pathogenic properties of bacteria, from stimulating the human immune system to inflammatory reaction and antibody production. Innate immunity mechanisms recognize endotoxins by one of the receptors specialized in binding macromolecules representing PAMP – pathogen-associated molecular pattern called Toll-like receptors (TLR). The receptor complex MD-2-TLR4 is designated for LPS recognition, binding, and initialization of signal transduction leading to inflammation that eliminates bacterial pathogen.

2 3-Hydroxy Fatty Acids, Chemical Marker of Gram-Negative Bacteria Lipopolysaccharides

3-Hydroxy fatty acids are compounds of specificity limited exclusively to lipopolysaccharide conservative component – lipid A,[29] and therefore ideal to search and quantify by gas chromatography coupled with mass spectrometry – GC-MS.[20,35] Indeed, environmental samples, e.g. house dust, air filtrates, or tobacco smoke successfully pass the simple chemical procedures prior to reproducible instrumental analysis by GC-MS. Today the method is established and applied in case studies as well as in multi-parameter cohort investigations aiming LPS quantitative assessment in different matrices.[13,14,19,41,44]

3 Microorganisms in Indoor Environment

The indoor environment is one of the natural sites of interaction between bacteria and humans. Houses and other buildings of public use are often affected by dampness, and even the outdoor working environment can have high exposure rates to microorganisms (farming, timber industry etc.). Several authors described how the microbial exposure affects the health causing, for example: upper airway inflammation; chronic headaches; allergies; asthma. Indoor microbial exposure studies focus on live and dead microorganisms, on the presence of bacterial and fungal biomass in settled dust and building materials, and on bioaerosols. These materials may also contain cell wall fragments, macromolecules known for their biological activity (like bacterial peptidoglycan and lipopolysaccharides), and also toxic microbial products, such as fungal mycotoxins, may be are present.

An assessment and measurement of the microbial exposure is considered to be a difficult issue. Culture methods detect only a small part of all microorganisms in indoor environments since more than 95% are dead; DNA-based methods are sensitive, and can detect also dead microorganisms. When aiming a quantitation of the microbial compounds that actually affect our health, it is crucial to choose a reliable approach. In complex matrices such as dust, building materials, air filtrates, or biological fluids and tissues, chemical microbial markers, unique for a group of bacteria or fungi, may be identified by analytical methods with the use of gas chromatography-mass spectrometry. For endotoxin studies this method provides a quantitative alternative for Limulus assay (LAL), which sometimes caused problems due to unsatisfactory interlaboratory reproducibility and selectivity.

3.1   Indoor Air Microbial Load Assessment by GC-Tandem MS

Dust suspended in the air is a bioaerosol that sediments on surfaces and building materials, thereafter persisting within the breathing zone. Sampling of such matter is performed by air filtration of a given air volume or by collecting a settled dust deposed on surfaces within a space of the breathing zone.

An integrated methodology for characterizing the microbial communities of indoor environments has been developed for determination of specific microbial markers in inhalable dust particles and building materials by using mass-spectrometry-based methods.[17,33] This approach represents a robust base in research aiming to relate microbial exposure indoors to human well-being and health. The sample is subjected to hydrolysis to release the unique, marker compounds of microbes: 3-hydroxy fatty acids, constituents of lipidic part of endotoxin, muramic acid, part of peptidoglycan present in cell wall of all bacteria, and ergosterol from the fungal cell wall. Free compounds are subjected to further derivatisation prior to obtaining trimethylsilyl (TMS) methyl esters and quantitative analysis on GC-MS/MS. We used an ion trap instrument and triple quadruple instrument, in electron impact (EI) ionization mode. The target ion for isolation was m/z 175, characteristic for the TMS derivative of 3-OH FAs, which was subsequently fragmented and the analysis of its product ion at m/z 131 was performed.

Based on this approach, several works dealt with diversity of microbial load in indoor environments, and microbial markers potential to characterize damp and non-damp buildings,[26,28] bed, shelf, and basement in the same house,[32] and schools in different geographical regions.[41] In the latter, dust was collected in 30 class-rooms in ten schools in Shanghai; among the participating 1,414 pupils of mean age 13 years, a positive correlation was found between dust levels of ergosterol (a fungal biomass marker) and the prevalence of respiratory infections ($p < 0.01$). Negative correlation between 3-hydroxy fatty acids, endotoxin markers, and symptoms including respiratory infections ($p < 0.05$), current asthma ($p < 0.05$), and daytime breathlessness ($p < 0.001$) indicated a possible protective function of bacterial LPS against respiratory diseases; the epidemiological studies are continued.

4   Lipopolysaccharide Sensing by Host as a Determinant of Disease

The sensory mechanism in vertebrates that is designed for lipopolysaccharide rec-ognition is a member of a Toll-like receptors family. Namely it consists of MD-2, an LPS binding molecule, and TLR4, a signal transduction portion. Many of patho-genic and commensal bacteria sensed by MD-2-TLR4 complex exist on the mucosal surfaces of the upper respiratory and gastrointestinal tracts. During infec-tion they are usually etiological agents of local inflammation but rarely invade the bloodstream. Contrary, Gram-negative bacteria, responsible for systemic infections in humans, produce LPS molecules which are poorly sensed by MD-2-TLR4, and therefore they escape recognition by the innate defense system.[8] Nevertheless, an innate immunity is successful when infection is stopped on the epithelium level, without systemic spread-out.

Interesting, TLR4 is not expressed on lumenal surfaces of gastrointestinal tract; it is not very numerous in normal gastrointestinal epithelium either. The most prevalent localization of the microbial pattern-recognition receptor are the defense cells: macrophages, dendritic cells, neutrophils, which are in movement to guard the respiratory and gastrointestinal submucosa, aiming to sense not all the Gram-negative bacteria present on the epithelium surface, but only those which managed to transit the epithelium and reach the submucosal space. This way represents the most straight-forward threat for systemic bacteriemia and sepsis.[25]

5   Endotoxin in Clinical Analyses

Endotoxic shock results from invasion of the pathogenic microflora or its excretory products and stimulates rapid development and diverse clinical manifestations demanding urgent help in hospital intensive care units. The attempts to identify endotoxins in biological fluids, for the early recognition of sepsis have been ongoing for a long time and produced two main approaches. One is the biological test, originally based on horseshoe crab (*Limulus polyphemus*) lymph that coagulated when exposed to lipopolysaccharide, which nowadays is commercially offered as biotechnological variants – the Limulus assay (LAL). Another way applied microbial chemical markers determined by analytic methods in a patient's sample by use a gas chromatography-mass spectrometry. Two markers are currently applied: keto-octulosonic acid (Kdo), a compound present in the inner core of LPS,[30] and 3-hydroxy fatty acids, from lipid A.[31,37]

Direct determination of LPS in human blood in patients with sepsis is, however, complicated and has many obstacles; as expressed by "the dangers of contamination, lack of precision and accuracy, and both false positive and false negative results". Endotoxaemia is present in the blood of about 30% of patients with bacteraemia, but endotoxaemia does not predict either Gram-negative bacteraemia or Gram-negative infection, nor does it predict survival from sepsis. There is some correlation with severity of sepsis, but the level of precision is poor. In particular, the positive predictive value of the test is insufficiently high to be of clinical use. It may be that the LAL assay, or one of the newer developments, may be more useful in excluding Gram-negative infection, but that remains to be shown".[3,5]

Nevertheless, several other pathological conditions exist whose course is much less dramatic than sepsis or severe systemic diseases, where Gram-negative bacteria and/or endotoxins have a significant role to play. Determination of these bacterial products in blood and physiological specimens of the patient may be used as prognostic factor of the disease or as a tool for monitoring the progress of therapy.

Endotoxaemia developing during cardiosurgery as elevated endotoxin concentrations in patient's serum may prevail over 24 h after an operation. A major reason is thought to be the increased gut permeability resulting in endotoxin and bacterial leakage.[18] In these studies we measured endotoxin markers in samples obtained during and after cardiovascular procedures, and compared them with clinical observations and routine laboratory test results (blood morphology, urine, bilirubin, kidney parameters, clotting parameters, gasometry).

Changes in the level of 3-OH FAs were measured in the serum collected from a total of 16 patients in the course of cardiosurgery. The results of these patients showed five patients (group I) had increased serum 3-OH FA, and 11 patients (group II) did not show any change in 3-OH FA 24 h after operation. All patients in group I revealed leukocytosis, and post-operative anemia (Fig. 1).

It was concluded that cardiosurgery may strongly promote gut endotoxin translocation to the blood in some patients, and prolonged leukocytosis, deep anemia, and increased liver dysfunction markers may indicate the need for observation of possible endotoxaemia development. Therefore it has been recommended that the endotoxin levels and/or endotoxaemia markers are monitored in cardiosurgery patients.

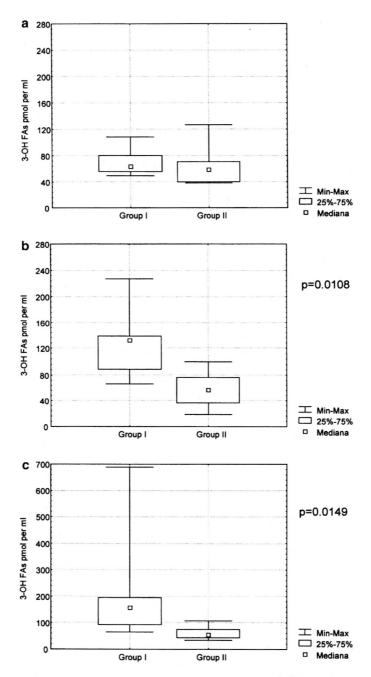

Fig. 1 Changes in the level of 3-hydroxy fatty acids (average of four acids), markers of endotoxin, in serum collected from patients in the course of cardiosurgery: (**a**) at 24 h before operation, (**b**) during the operation, (**c**) at 24 h after operation (Reprinted from Krasnik et al.,[18] with permission)

6   Endotoxins in Gastrointestinal Tract

Gram-negative commensals normally reside in the epithelium of the upper respiratory and gastrointestinal tracts and in pathological conditions they can move through disrupted tissue to the submucosal space and multiply there, causing inflammation. However, there is rather weak evidence that such bacteraemia *per se* can lead to systemic organ dysfunction. Blood-borne commensal Gram-negative bacteria may be a marker of tissue infection, but the real threat for sepsis depends more on the location of the tissue damage than on the specific microbe cultured from the blood.[4] As it is discussed recently,[25] diversification of the structure of lipid A, the most conservative part of lipopolysaccharide, is a crucial factor for the host recognition of LPS and further pathological consequences of Gram-negative bacterial diseases. An innate immune system, represented by the MD-2-TLR4 complex, best detects lipid A attributed with six saturated fatty acid chains of 12 or 14 carbon atom length (so-called "hexaacyl LPS"). This is the most popular arrangement for bacteria that naturally populate soil and water as well as anaerobic microbiome of the gastrointestinal tract. Also, it is present in aerobic and facultative anaerobic Gram-negatives having their habitat in the mucosal surfaces of the respiratory tract and gut, regardless whether commensals (i.e. *Klebsiella pneumonia*, *Enterobacter cloacae*) or pathogen (like *Salmonella* sp. and *Shigella* spp.). Other types of lipid A architecture ("non-hexaacyl LPS") are produced by most bacterial systemic pathogens that normally inhabit soil, water, insects, and vertebrates, and they attack the host's epithelial barriers through other than mucosal routes: inhalation, ingestion, cuts or bites. Examples of these are: *Yersinia pestis* (tetraacyl LPS in mammalian host); *Bordetella pertussis* (pentaacyl LPS); or *Brucella abortus* (heptaacyl LPS). The absence of the inflammation-inducing hexaacylated LPS could permit the disease development.[24] In this perspective studying lipopolysaccharide lipidic moiety composition by spectrometric approach in biological specimens generates an emerging field of research.

6.1   *Newborn Flora*

Conventional cultivation techniques for studying babies gut microbiota were extensively applied for decades. The initial bacterial colonization creates reducing conditions favorable to the development of anaerobes, mainly represented by *Bifidobacterium* spp. With time, the bacterial consortium is diversified and represented by e.g. *Bifidobacterium*, *Bacteroides*, *Streptococcus* and *Clostridium*, and is rather unstable getting gradually a more stable final pattern. Normal colonization by the human intestinal commensal microbes stimulates a range of important functions, i.e. postnatal intestinal maturation, maintenance of the mucosal barrier and nutrient absorption.[22] Proper knowledge of the types of microorganisms as well as the events that influence the timing of colonization, may provide opportunities to modulate the microbiota when modulation is necessary to enhance these functions.

Cellular fatty acids in stool samples were used to assess initial gut colonization and identify the main source of the intestinal microflora.[38] Gas chromatography-mass spectrometry employed for the analysis of volatile derivatives of fatty acids is a rapid and reliable tool. This method of bacterial identification has advantages over conventional culture, as it covers both cultivable and uncultivable microorganisms. Several clinical conditions may be diagnosed and monitored with the help of a fatty acid profile.

Development of necrotizing enterocolitis (NEC) is a condition that occurs to preterm infants and early bacterial colonization most likely has a role in the multi-factorial pathogenesis of this severe disease. Therefore it is important to characterize intestinal microbiota. Björkström et al.[2] found that NEC case in their studies had an early colonization of lactic acid bacteria, besides, prior to onset of the disease a high count of non-*E. coli* Gram-negative species was found: high count of *Klebsiella* spp., *Pseudomonas* spp., *Proteus* spp.; it was assumed that these species may be a risk factor for NEC. Early colonization by the beneficial lactic-acid bacteria alone did not prevent necrotizing enterocolitis, despite the promising results from some clinical trials with probiotics for its prevention.[2]

In addition, to complete the spectrum of bacteria involved in different pathological conditions, anaerobic Gram-negative *Bacteroides* may be included to the fatty acids profile of intestinal microbiota. Vael and co-workers in a cohort study related gut flora to a clinically relevant factor for developing asthma,[39] have found that the early gut colonization with *Bacteroides fragilis* was associated with a positive Asthma Predictive Index (API) at age of 3 years. Authors suggest that the modulation of composition of the gut flora could contribute to easing of the symptoms of asthma and the desirable modification was identified as limiting of the *Bacteroides fragilis* flora.

Studies of the gut flora of wheezing infants at different ages are needed for the better understanding of the intestinal flora abnormalities significance in association with asthma. An analytical tool relaying not only on the cultivation of stool samples, but on analytically determinate microbial chemical markers provide an opportunity to monitor status of big group of individuals and get results that may would be associated to clinically important records.

6.2 3-Hydroxy Fatty Acids and Chronic Inflammatory Bowel Diseases: Markers in Feces

Anaerobic microbiota is predominant in the distal ileum and colon and represents a complex variety of bacteria that closely interact with the host's epithelial cells and mucosal immune system. Crohn's disease and ulcerative colitis are thought to be a result of continuous microbial stimulation of pathogenic immune responses as a consequence of host genetic defects in mucosal barrier function or immunoregulation.[7] In the pathogenesis of Crohn's disease failure of the innate immune system could be involved via reduced or impaired defense against Gram-negative bacteria, but also genetic polymorphism in the LPS receptor TLR4 was associated with impaired LPS signaling and increased susceptibility to Gram-negative organisms in Crohn's disease and ulcerative colitis.[23]

Altered microbial composition and function in inflammatory bowel diseases results in increased immune stimulation, epithelial dysfunction, or increased mucosal permeability. Traditional pathogens probably are not responsible for these disorders, but increased virulence of commensally bacterial species, among these Gram-negative bacteria promote their adverse properties, like mucosal attachment, invasion and intracellular persistence, altogether stimulating pathogenic immune responses. Also, host genetic polymorphisms interact with functional microbiota modifications that consequently stimulate aggressive immune responses and lead to chronic tissue injury.

Conventional culture methods and methods based on ribosomal RNA and DNA were applied on studies of the microbiology of intestine; because more than 50% of intestine microbes are dead, the first method has obvious limitations. So far most of the research is focused on the detection of microbes using 16S rRNA-based approaches; currently the main problem is actually an inability to convert the mass of metagenomics data into biological meaningful information, i.e. by using systemic approaches. Nevertheless, those results have demonstrated that in feces of healthy adult humans: (i) majority (54–75%)[45] of the sequences are derived from Gram-positive bacteria; (ii) fecal samples do not necessarily represent the bacterial community in the other parts of the gastrointestinal tract, and (iii) mucosa-associated bacteria seemed to be uniformly distributed along the complete colon,[21] (iv). part of colonic microbes are not in direct contact with the mucosa and not a significant difference was found between colonic biopsies and feces.[40]

Bearing in mind the above, it is accepted that alternative approaches to characterize intestinal microflora would complete the picture.

Gram-negative bacteria are a minority in the overall microbial load of the intestine, but reports about a significant increase in *Enterobacteriaceae* in the feces of patients with both active and inactive colonic Crohn's disease[34] stimulated a study of a non-invasive diagnostic approach, applying 3-hydroxy fatty acids analysis by GC-MS/MS. Each individual harbors a unique microbiota pattern; therefore we aimed to characterize the Gram-negative bacteria load of each patient (hospitalized patients with Crohn's disease, patients with ulcerative colitis and healthy controls) and investigate possible correlations between Gram-negative bacteria population in feces and clinical status.

We studied the diversity in 3-OH FA composition of feces between the three groups of individuals that indicated differences in Gram-negatives community[6]. Nine saturated straight-chain 3-OH FAs of 10–18 carbon chain lengths and six *iso*- and *anteiso*-branched chain 3-OH FAs of 15–18 carbon chain lengths were detected. Three of the four most abundant 3-OH FAs, i.e. 3-OH nC16:0, 3-OH nC17:0 and 3-OH isoC17:0, are constituents of the lipid A of *Bacteroides* spp., whereas 3-OH n-C14:0, is found in *Enterobacteriaceae*; 3-OH C14:0 was higher in Crohn's disease (CD) than in ulcerative colitis (UC) and controls.

In the case of one patient, samples were collected in periods of active colonic CD and in remission. In feces of this patient three 3-OH FAs (3-OH nC13:0, 3-OH *iso*-C18:0 and 3-OH nC18:0) were found only during remission; the relative amount of 3-OH nC14:0 in feces was higher during active CD than in remission, whereas the relative amounts of the other major 3-OH FAs, i.e. 3-OH nC16:0 and 3-OH *iso*-C17:0, were lower during active CD than during remission.[6]

7 Monitoring the Oral Cavity Microbiota

The oral cavity is a microbial ecosystem populated by hundreds of species, settled in different habitats with ability to translocate to the other niches. After eruption of teeth, the dental plaque biofilm appears on non-shedding surfaces and since then bacterial species responsible for caries and periodontitis are ready to initiate the disease. Knowledge on the pattern of early oral colonization patterns would provide an explanation of biofilm development and help in monitoring progress of prevention and/or treatment of the illness.[27] Materials considered to be representative for the oral cavity microbiota are saliva samples, subgingival plaque sample, tongue sample, soft tissue sample and total supragingival plaque sample, of which early colonizers like streptococci dominate in soft tissue samples and saliva. The supra- and subgingival tissues biofilm is complex, e.g. *Actinomyces* spp. are very numerous. Between initial and late colonizers of the hard tissues emerges *Fusobacterium nucleatum*, acting as a bridge, especially important because of its ability to adhere to mammalian tissues and to coaggregate other microorganisms. Plaque biofilm is formed by *Bacteroides* spp. and related bacteria, among others *Porphyromonas gingivalis*, *Tannerella forsythia*, and *Treponema denticola*, all active in a process of peridontitis.[43]

Different markers of periodontal disease in saliva have been proposed, originating from the host (locally produced proteins, steroid hormones, cytokines, immunoglobulins) and from bacteria (cultivation of intact microorganisms, detection of bacterial products). Saliva is regarded as microbial reservoir and can serve as a carrier for bacterial transmission, it reflects presence of certain bacterial species on the tongue, within dental plaque and periodontal pockets. Noteworthy, bacterial biomarkers transit from gingival crevicular fluid to saliva and therefore saliva is considered as a relevant medium for monitoring periodontal status.

7.1 3-Hydroxy Fatty Acids as Periodontitis Markers Assessed in Saliva

Periodontitis is an inflammatory condition that leads to destruction of the periodontum, resorption of the alveolar bone, and, frequently, to tooth loss; one of the most prevalent diseases in humans and affects mainly individuals above 35 years of age.

The pathogenesis is complex and largely unknown although microorganisms are thought to play an important role, and Gram-negative bacteria appear to be essential in the process. Chronic periodontitis is associated mainly with strictly anaerobic species such as *Porphyromonas gingivalis*, *Tanerella forsythia* (formerly *Bacteroides forsythensis*) and *Treponema denticola*. Together with Gram-positive facultative anaerobe *Actinobacillus actinomycetemcomitans*, these microbes are strongly associated with aggressive periodontitis. They are recognized as putative periodontopathogens belonging to the hundreds of species that compose the microbial population of the mouth.

Diagnosis of periodontitis usually relies on clinical data, but microbiological studies are useful in establishing the etiology of the disease and in controlling of the treatment of the patients. The endotoxins of the anaerobic Gram-negative bacteria are associated with periodontitis, and therefore straight and branched-chain 3-hydroxy fatty acids were used as specific bacterial components and periodontitis markers upon analysis of saliva by a GC-MSMS chemical–analytical method.[9]

Among two groups of individuals: with periodontitis and healthy individuals, in the 3-OH FA patterns the major compounds were 3-OH-C14:0 and 3-OH-*i*C17:0, and the latter dominated in the periodontitis cases (Fig. 2).

The periodontitis cases showed many similarities with those of the control cases, i.e. 3-OH C12:0 and 3-OH C14:0 levels, likely due to *Neisseria* and *Haemophilus* spp. known to be predominant in healthy individuals saliva; 3-OH C12:0, 3-OH C13:0 and 3-OH C14:0 were present in *Campylobacter*, *Haemophilus*, *Fusobacterium*, *Neisseria* and *Veillonella*, all abundant in the oral cavity; 3-OH C16:0 is found in bacteria belonging to *Bacteroides-Porphyromonas-Prevotella* and dominates over 3-OH *i*C17:0 in *Tannerella forsythia*.[42]

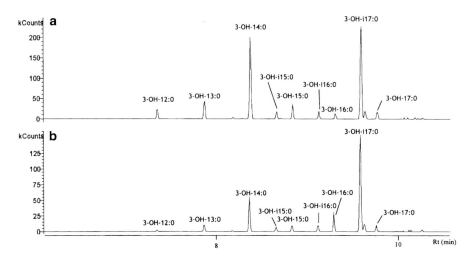

Fig. 2 3-Hydroxy fatty acid profiles of saliva of (**a**) a healthy individual, and (**b**) a periodontitis patient. Fatty acids are identified by chain length and possible branching (Reprinted from Ferrando et al.,[9] with permission from Elsevier)

The levels of 3-OH *i*C17:0 separated periodontitis cases from controls: strictly anaerobic species of *Prevotella* and *Porphyromonas*, and *Tannerella forsythensis* are implicated in chronic periodontitis; *Capnocytophaga* spp. where 3-OH-Ci17:0 is the predominant 3-hydroxy fatty acid, also plays a role in the pathogenesis of the disease. Discriminate analysis correctly classified 95.6% of the samples when taking into account 3-OH C12:0, 3-OH C14:0, 3-OH *i*C17:0 and 3-OH-C17:0. The diagnostic capacity of the described method was the following: sensitivity, 0.92; specificity, 1.00; positive predictive value, 1.00; negative predictive value, 0.90. Clearly, this analytical approach may be useful tool included in a diagnostic laboratory's battery of tools in further studies of periodontitis.

8 Diastereoisomers of 3-Hydroxy Fatty Acids in Biological Samples

The derivatization and other manipulations of biological materials are usually required for the applied analytical approach, these manipulations induce the release of lipidic and protein compounds which thus increase the chemical complexity of the sample. To overcome these problem, some modifications were found: (i) releasing the markers as monomers by methanolysis and producing the methyl esters of the 3-hydroxy fatty acids; (ii) saponification, extraction, and hydrolysis; (iii) derivatization to volatile tri-methysilyl esters, suitable for gas chromatography; (iv) derivatization to chiral forms of 3-hydroxy fatty acids, which permits to unequivocally, distinguish compounds of bacterial origin and the endogenous derivatives obtained from the acyl-coenzyme A participating in this β-oxidation.

The latter was developed and extensively used in the diagnosis of some inborn metabolic diseases,[15,16] as free 3-hydroxy fatty acids are involved as intermediates during mitochondrial fatty acid oxidation pathways. These compounds (e.g. as *S*-3-OH palmitoylCoA) when present in blood, serum or tissue samples turn into derivatives that are identical with bacterial 3-hydroxy palmitic acid. The difference between bacterial and mitochondrial 3-hydroxy fatty acids lies in the type of optical isomerisation: bacterial are of *R*-form, whereas mitochondrial 3-hydroxy fatty acids-coenzyme A complexes are of *S*-forms.

Our previous studies showed that rats which have been injected with *E. coli* lipopolysaccharide revealed high levels of endotoxin markers in the studied tissues, but in fact we were not able to determine the actual migration of these markers.[36] Therefore, to solve the question of the origin of the detected 3-hydroxy fatty acids in the body, lipopolysaccharides of unique chemical composition that were isolated from the bacteria *Pectinatus cerevisiiphilus* needed to be used.[11] Lipid A of *P. cerevisiiphilus* lipopolysaccharide is characterised by a high content of amide linked *R*-3-hydroxy tridecanoic acid and ester-linked undecanoic acid with minor amounts of ester linked tridecanoic acid and *R*-3-hydroxy undecanoic acid.[12]

It is very important to distinguish 3-hydroxy fatty acids of bacterial from non-bacterial origin. Therefore, we prepared chiral derivatives that were assessed in GC-MS/MS, EI in selected ion monitoring mode (SIM). As an authentic standard we used lipopolysaccharide of this unique chemical composition isolated from *Pectinatus cerevisiiphilus*, considering characteristic ions m/z 120 and m/z 347 (3-OH C13). To exemplify the bacterial origin of 3-hydroxy fatty acids in biological specimens of human origin, we took saliva samples and newborn feces samples, from healthy individuals, and prepared in parallel with chemical standards of 3-OH FAs and LPS of *P. cerevisiiphilus*.

The phenylethylamide derivatives of 3-hydroxy fatty acid are used in order to determine of the optical configuration by GC-MS. This method allowed the recognition of compounds of bacterial origin, having the R-configuration exclusively. This method was developed in our laboratory for measuring environmental samples and allowed us to unequivocally determine the bacterial origin of 3-OH FAs in organic dust.[10] Samples which revealed the highest levels of 3-OH FAs in the previous experiment (trimethylsilyl derivatives) were selected for determination of absolute configuration.

8.1   *Pectinatus cerevisiiphilus Lipopolysaccharide*

The major component of lipid A in *P. cerevisiiphilus* is an unbranched R-3-hydroxy tridecanoic acid. Amide derivatives of R- and S-diastereoisomers of 3-OH FAs are shown on Fig. 3.

The chromatogram shown on Fig 3a represents *R*-3-OH 13:0 from *P. cerevisiiphilus*; following figures are external standards: phenylethylamide derivatives of 3-OH 10–18, recorded in selected ion monitoring (SIM) mode according to m/z 120, the common ion for these derivatives (Fig 3b), and according to m/z 347 which is a molecular ion (M+) of 3-OH 13:0.

Spectrum of the phenylethylamide derivative of 3-OH 13:0 contains ions typical for 3-hydroxy acids (m/z 105 and m/z 120), and a molecular ion (M+) m/z 347 (Fig. 4). All standard acids have both *R*- and *S*-enantiomeric form as they are chemically synthesized compounds.

8.2   *Diastereoisomers of 3-Hydroxy Fatty Acids in Saliva and Newborn Feces*

Figure 2 exemplifies the 3-OH FAs profile of the saliva samples revealed as trimethylsilyl methyl esters, the derivatives which do not distinguish between *R* and *S* isoforms. By using phenylethylamide derivatives only *R*-forms – that is of bacterial origin – of the respected hydoxy fatty acids were detected (Fig. 5).

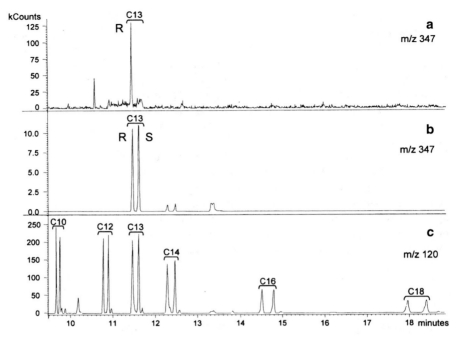

Fig. 3 *Pectinatus cerevisiiphilus* LPS, 3-OH FAs phenylethylamide derivative. GC-MSMS analysis, EI-SIM mode, ions 120 and 347 monitored (3-OH 13:0). (**a**) *P. cerevisiiphilus* LPS; (**b**) 3-OH 13:0 standard; (**c**) 3-hydroxy 10:0, 12:0, 13:0, 14:0, 16:0 and 18:0 standards.

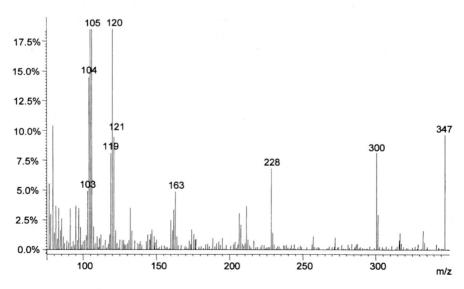

Fig 4 Spectrum of *R*-3-OH 13:0 phenylethylamide derivative, EI mode

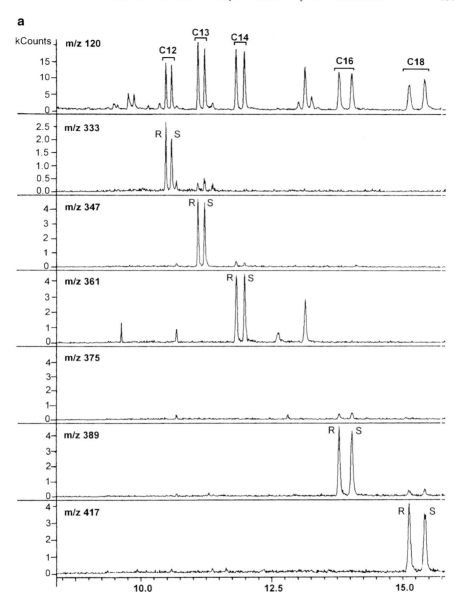

Fig. 5 Phenylethylamide derivatives 3-OH FAs of saliva sample. GC-tandem MS analysis, EI-SIM mode, ions 120 and molecular ions (M⁺), respectively: C12:0 (m/z 333), C13:0 (m/z 347), C14:0 (m/z 361), C15:0 (m/z 375), C16:0 (m/z 389), 3-OH C17:0 (m/z 403), and C18:0 (m/z 417). (**a**) 3-Hydroxy fatty acid standards

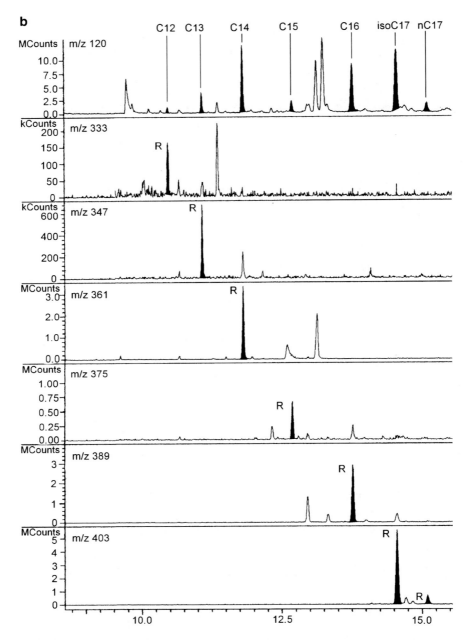

Fig. 5 (continued) (**b**) saliva sample

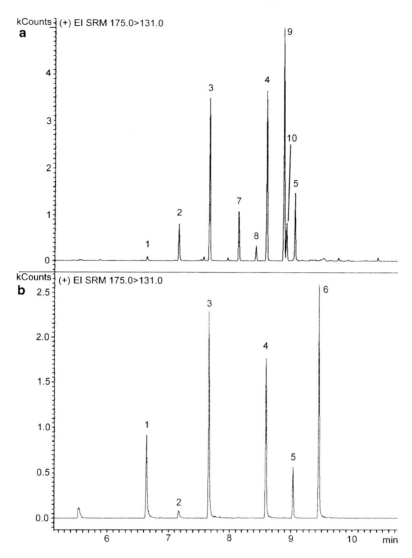

Fig. 6 Trimethylsilyl derivatives of 3-hydroxy fatty acid profiles of (**a**) newborn feces and (**b**) standards. 3-Hydroxy fatty acids are identified by chain length and retention time. 3-Hydroxy fatty acids: 1. *n*C12:0; 2. *n*C13:0; 3. *n*C14:0; 4. *n*C16:0; 5. *n*C17:0; 6. *n*C18:0; 7. *n*C15:0; 8. *iso*C16:0; 9. *iso*C17:0; 10. *anteiso*C17:0

Also in a sample from newborn feces the *R*-3-OH fatty acids were found (Fig. 6).

A sufficient amount of 3-OH is essential for the chiral analysis of FAs in the sample; this method is about 20-fold less sensitive than the trimethylsilyl derivatives method, mainly due to several extractions which decrease the yield of the final analyte that is injected on the GC-MS/MS instrument. Samples of saliva and newborn

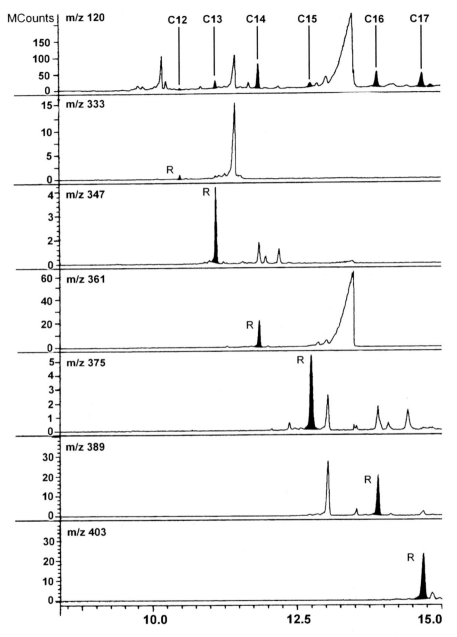

Fig. 7 Phenylethylamide derivatives of 3-OH FAs of a newborn feces sample. GC-tandem MS analysis, EI-SIM mode, ions m/z 120 and molecular ions (M⁺), respectively: C12:0 (m/z 333), C13:0 (m/z 347), C14:0 (m/z 361), C15:0 (m/z 375), C16:0 (m/z 389), and C17:0 (m/z 403)

feces were analysed for *R*- and *S*-configurations, and *R*-forms were the only ones present in the chromatograms. The derivatives of the 3-hydroxy fatty acids optical isomers were identified according to their molecular mass and retention time as obtained for the external standards (Figs. 5–7).

Both ways of ionization, electron impact (EI) and positive chemical ionization (positive CI) are equally good for quick and easy determination of absolute configuration of the 3-OH FAs, although mild conditions of chemical ionization in the quadrupole instrument better preserve the molecular ions and allow the optimal detection with the best sensitivity. Results obtained from TMS procedure for saliva and feces sample confirmed the general profile of 3-hydroxy acids, usually present as branched fatty acids, i.e. *R*-3-OH *iso*C17:0.

9 Conclusions

Microbial chemical markers provide many opportunities for further applications when complex sample matrix and/or microbial association need to be characterized, validated and applied in clinical studies. They can be used as an accomplished method to the diagnostic methods panel. As a valuable tool in indoor air – adverse health effects investigations, can serve for clinically related case- and epidemiological studies.

Acknowledgements The work was partially supported by the Ministry of Science and Higher Education of Poland, grant No. NN402 392235.

References

1. Bäckhed F, Ley RE, Sonnenburg JL, Peterson DA, Gordon JI (2005) Host-bacterial mutualism in the human intestine. Science 307:1915–1920
2. Björkström MV, Hall L, Söderlund S, Håkansson EG, Håkansson S, Domellöf M (2009) Intestinal flora in very low-birth weight infants. Acta Paediatr 98:1762–1767
3. Brandenburg K, Howe J, Gutsman T, Garidel P (2009) The expression of endotoxic activity in the Limulus test as compared to cytokine production in immune cells. Curr Med Chem 16:2653–2660
4. Brun-Buisson C, Doyon F, Carlet J (1996) Bacteremia and severe sepsis in adults: a multicenter prospective survey in ICUs and wards of 24 hospitals. French Bacteremia-Sepsis Study Group. Am J Respir Crit Care Med 154:617–624
5. Cohen J (2000) The detection and interpretation of endotoxaemia. Intensive Care Med 26(Suppl 1):S51–56
6. De La Cochetiere MF, Rouge C, Szponar B, Larsson L, Potel G (2007) 3-Hydroxy fatty acids in faeces of patients with Crohn's disease or ulcerative colitis as a non-invasive diagnostic approach. Microb Ecol Health Dis 19:1–6
7. Eckburg PB, Relman DA (2007) The role of microbes in Crohn's disease. Clin Infect Dis 44:256–262

8. Elson G, Dunn-Siegrist I, Daubeuf B, Pugin J (2007) Contribution of Toll-like receptors to the innate immune response to Gram-negative and Gram-positive bacteria. Blood 109:1574–1583

9. Ferrando R, Szponar B, Sánchez A, Larsson L, Valero-Guillén PL (2005) 3-Hydroxy fatty acids in saliva as diagnostic markers in chronic periodontitis. J Microbiol Meth 62:285–291

10. Gradowska W, Larsson L (1994) Determination of absolute configurations of 2-and 3-hydroxy fatty acids in organic dust by gas chromatography-mass spectrometry. J Microbiol Meth 20:55–67

11. Helander IM, Haikara A, Sadovskaya I, Vinogradov E, Salkinoja-Salonen MS (2004) Lipopolysaccharides of anaerobic beer spoilage bacteria of the genus *Pectinatus* -lipopolysaccharides of a Gram-positive genus. FEMS Microbiol Rev 28:543–552

12. Helander IM, Moran AP, Makela PH (1992) Separation of two lipopolysaccharide populations with different contents of O-antigen factor 122 in *Salmonella enterica* serovar typhimurium. Mol Microbiol 6:2857–2862

13. Hines CJ, Waters MA, Larsson L, Petersen MR, Saraf A, Milton DK (2003) Characterization of endotoxin and 3-hydroxy fatty acid levels in air and settled dust from commercial aircraft cabins. Indoor Air 13:166–173

14. Hyvärinen A, Sebastian A, Pekkanen J, Larsson L, Korppi M, Putus T, Nevalainen A (2006) Characterizing microbial exposure with ergosterol, 3-hydroxy fatty acids, and viable microbes in house dust: determinants and association with childhood asthma. Arch Environ Occup Health 61:149–157

15. Jones PM, Burlina AB, Bennett MJ (2000a) Quantitative measurement of total and free 3-hydroxy fatty acids in serum or plasma samples: short-chain 3-hydroxy fatty acids are not esterified. J Inherit Metab Dis 23:745–750

16. Jones PM, Quinn R, Fennessey PV, Tjoa S, Goodman SI, Fiore S, Burlina AB, Rinaldo P, Boriack RL, Bennett MJ (2000b) Improved stable isotope dilution-gas chromatography-mass spectrometry method for serum or plasma free 3-hydroxy-fatty acids and its utility for the study of disorders of mitochondrial fatty acid beta-oxidation. Clin Chem 46:149–155

17. Krahmer M, Fox K, Fox A, Saraf A, Larsson L (1998) Total and viable airborne bacterial load in two different agricultural environments using gas chromatography-tandem mass spectrometry and culture: a prototype study. Am Ind Hyg Assoc J 59:524–531

18. Krasnik L, Szponar B, Walczak M, Larsson L, Gamian A (2006) Routine clinical laboratory tests correspond to increased serum levels of 3-hydroxy fatty acids, markers of endotoxins, in cardiosurgery patients. Arch Immunol Ther Exp 54:55–60

19. Larsson L, Fredborn Larsson P (2001) Analysis of chemical markers as a means of characterising airborne microorganisms in indoor environments: a case study. Indoor Built Environ 10:232–237

20. Larsson L, Saraf A (1997) Use of gas chromatography-ion trap tandem mass spectrometry for the detection and characterization of microorganisms in complex samples. Mol Biotechnol 7:279–287

21. Lepage P, Seksik P, Sutren M, de la Cochetière M-F, Jian R, Marteau P, Doré J (2005) Biodiversity of the mucosa-associated microbiota is stable along the distal digestive tract in healthy individuals and patients with IBD. Inflamm Bowel Dis 11:473–480

22. Mackie RI, Sghir A, Gaskins HR (1999) Developmental microbial ecology of the neonatal gastrointestinal tract. Am J Clin Nutr 69:1035S–1045S

23. Marteau P, Lepage P, Mangin I, Suau A, Doré J, Pochart P, Seksik P (2004) Gut flora and inflammatory bowel disease. Aliment Pharmacol Therapeut 20(Suppl 4):18–23

24. Munford RS, Varley AW (2006) Shield as signal: lipopolysaccharides and the evolution of immunity to Gram-negative bacteria. PLoS Pathog 2:e67

25. Munford RS (2008) Sensing Gram-negative bacterial lipopolysaccharides: a human disease determinant? Infect Immun 76:454–465

26. Nilsson A, Kihlström E, Lagesson V, Wessén B, Szponar B, Larsson L, Tagesson C (2004) Microorganisms and volatile organic compounds in airborne dust from damp residences. Indoor Air 14:74–82

27. Papaioannou W, Gizani S, Haffajee AD, Quirynen M, Mamai-Homata E, Papagiannoulis L (2009) The microbiota on different oral surfaces in healthy children. Oral Microbiol Immunol 24:183–189
28. Park JH, Szponar B, Larsson L, Gold DR, Milton DK (2004) Characterization of lipopolysaccharides present in settled house dust. Appl Environ Microbiol 70:262–267
29. Rietschel ET (1976) Absolute configuration of 3-hydroxy fatty acids present in lipopolysaccharides from various bacterial groups. Eur J Biochem 64:423–428
30. Rybka J, Gamian A (2006) Determination of endotoxin by the measurement of the acetylated methyl glycoside derivative of Kdo with gas-liquid chromatography-mass spectrometry. J Microbiol Meth 64:171–184
31. Saraf A, Park JH, Milton DK, Larsson L (1999) Use of quadrupole GC-MS and ion trap GC-MS-MS for determining 3-hydroxy fatty acids in settled house dust: relation to endotoxin activity. J Environ Monit 1:163–168
32. Sebastian A, Larsson L (2003) Characterization of the microbial community in indoor environments: a chemical-analytical approach. Appl Environ Microbiol 69:3103–3109
33. Sebastian A, Szponar B, Larsson L (2005) Characterization of the microbial community in indoor environments by chemical marker analysis: an update and critical evaluation. Indoor Air 15(suppl 9):20–26
34. Seksik P, Rigottier-Gois L, Gramet G, Sutren M, Pochart P, Marteau P, Jian R, Doré J (2003) Alterations of the dominant faecal bacterial groups in patients with Crohn's disease of the colon. Gut 52:237–242
35. Szponar B, Larsson L (2001) Use of mass spectrometry for characterising microbial communities in bioaerosols. Ann Agric Environ Med 8:111–117
36. Szponar B, Krasnik L, Hryniewiecki T, Gamian A, Larsson L (2003) Distribution of 3-hydroxy fatty acids in tissues after intraperitoneal injection of endotoxin. Clin Chem 49:1149–1153
37. Szponar B, Norin E, Midtvedt T, Larsson L (2002) Limitations in the use of 3-hydroxy fatty acid analysis to determine endotoxin in mammalian samples. J Microbiol Meth 50:283–289
38. Tapiainen T, Ylitalo S, Eerola E, Uhari M (2006) Dynamics of gut colonization and source of intestinal flora in healthy newborn infants. APMIS 114:812–817
39. Vael C, Nelen V, Verhulst SL, Goossens H, Desager KN (2008) Early intestinal *Bacteroides fragilis* colonisation and development of asthma. BMC Pulm Med 26:19–24
40. van der Waaij LA, Harmsen HJ, Madjipour M, Kroese FG, Zwiers M, van Dullemen HM, de Boer NK, Welling GW, Jansen PL (2005) Bacterial population analysis of human colon and terminal ileum biopsies with 16S rRNA-based fluorescent probes: commensal bacteria live in suspension and have no direct contact with epithelial cells. Inflamm Bowel Dis 11:865–871
41. Wady L, Shehabi A, Szponar B, Pehrson C, Sheng Y, Larsson L (2004) Heterogeneity in microbial exposure in schools in Sweden, Poland and Jordan revealed by analysis of chemical markers. J Expo Anal Environ Epidemiol 14:293–299
42. Wollenweber H-W, Rietschel ET (1990) Analysis of lipopolysaccharide (lipid A) fatty acids. J Microbiol Meth 11:195–211
43. Zhang L, Henson BS, Camargo PM, Wong DT (2009) The clinical value of salivary biomarkers for periodontal disease. Periodontol 2000 51:25–37
44. Zhao Z, Sebastian A, Larsson L, Wang Z, Zhang Z, Norbäck D (2008) Asthmatic symptoms among pupils in relation to microbial dust exposure in schools in Taiyuan, China. Pediatr Allergy Immunol 19:455–465
45. Zoetendal EG, Vaughan EE, de Vos WM (2006) A microbial world within us. Mol Microbiol 59:1639–1650

Imaging Mass Spectrometry

Michelle L. Reyzer and Richard M. Caprioli

Abstract Imaging Mass Spectrometry (IMS) is a powerful analytical technology that provides both molecular and spatial information from a single sample. This chapter provides a brief history of Imaging Mass Spectrometry, including early work with secondary ion mass spectrometry (SIMS) and laser desorption/ionization (LDI) techniques. A more in-depth account of recent applications utilizing matrix-assisted laser desorption/ionization (MALDI) mass spectrometry for high-molecular weight imaging is presented.

1   Introduction

Modern mass spectrometry is one of the most sensitive, robust, and rapid analytical methodologies available today. Individual instrument configurations vary, and may include one of several different types of ionization sources, mass analyzers, and detectors in order to detect atomic or molecular species. In acquiring molecular information in a position-dependent manner, one can obtain an image of the molecular distribution of analytes as a two-dimensional ion density map. Most early work that employed MS for imaging involved the localization of elemental and fragmented molecular species. With the advent of MALDI, the field has expanded to include high molecular weight compounds, including proteins. This chapter is intended to provide a brief history of the use of mass spectrometry for imaging applications and a more in-depth account of recent imaging applications that utilize MALDI as the ionization source.

M.L. Reyzer (✉) and R.M. Caprioli
Mass Spectrometry Research Center, Vanderbilt University, Nashville, TN 37232
e-mail: r.caprioli@vanderbilt.edu

J. Banoub (ed.), *Detection of Biological Agents for the Prevention of Bioterrorism,*   267
NATO Science for Peace and Security Series A: Chemistry and Biology,
DOI 10.1007/978-90-481-9815-3_17, © Springer Science+Business Media B.V. 2011

2   History

2.1   Secondary Ion Mass Spectrometry (SIMS) Imaging

Mass spectrometry has a long history in the field of imaging, dating back at least 45 years to early studies with secondary ion mass spectrometry (SIMS) performed in microscope[1] and scanning microprobe mode.[2] Either mode is capable of generating images at submicron lateral resolution. SIMS utilizes a primary ion beam (for example, Ga^+, In^+, O_2^+, Au^+, or Cs^+) to sputter ions and neutrals from a surface. The sputtered (secondary) ions are directed into a mass spectrometer for analysis, typically a time-of-flight (TOF), quadrupole, or magnetic sector instrument.[3] Due to the energetic nature of the primary ion beam, the SIMS process predominantly generates elemental ions, atomic clusters, and organic fragments. As a result, many early SIMS imaging experiments involved elemental distribution, such as the distribution of metal or contaminant ions in alloys, ceramics, or semiconductors.[3]

One of the main advantages of SIMS imaging is the high lateral resolution attainable with the highly focused liquid metal ion guns typically used for the primary ion beams. The beam can be focused onto the target with a diameter in the submicron range (~10–100's nm), that is several orders of magnitude smaller than the diameter of an average human cell (~10–50 μm). Thus, there is considerable interest in using SIMS for subcellular imaging of biologically relevant species. Indeed, examples have been reported for the measurement of K^+ and Na^+ distributions in whole *Paramecia*,[4] Ca^+ and $C_5H_{15}PNO_4^+$ (lipid fragment) distributions in the single-cell foraminifer,[5] and K^+, Na^+, Ca^+ distributions in human glioblastoma cells.[6]

Subcellular localization coupled to quantitation has been demonstrated and is termed multi-isotope imaging mass spectrometry (MIMS).[7-10] Components of mammalian and bacterial cells have been analyzed using this technique. For example, the accumulation of oleic acid in cultured adipocytes was investigated by incubating the cells with ^{13}C-labeled oleic acid. SIMS images of $^{13}C^-$ and $^{12}C^-$ were obtained after 20 min incubation time and the ratio of $^{13}C^-/^{12}C^-$ was measured as an indication of the location of the oleic acid within the cells. Figure 1 shows the high level of excess $^{13}C^-$ found in intracellular lipid droplets, indicating that is where the oleic acid accumulates.[10] Nitrogen fixation by individual bacteria within eukaryotic host cells was also imaged by measuring the incorporation of gaseous ^{15}N into CN^- ions from the cells.[8]

The subcellular localization of pharmaceutical compounds containing an element not natively found in cells and tissues can be determined with SIMS by detecting ions originating from the unique element. For example, two drugs used for boron neutron capture therapy (BCNT) were synthesized with either ^{10}B or ^{11}B and used to treat T98G human glioblastoma cells. The subcellular distribution of the drugs was obtained by analyzing the boron elemental ions $^{10}B^+$ and $^{11}B^+$. The ^{10}B-containing drug, *p*-boronophenylalanine-fructose, was found to localize heterogeneously within the cells, with a lower concentration found in one perinuclear region. Conversely, the ^{11}B-containing drug, sodium borocaptate, was found

Fig. 1 MIMS image of $^{13}C^{14}N^-/^{12}C^{14}N^-$ ratio showing localization of ^{13}C-labeled oleic acid in intracellular lipid drops of a cultured adipocyte. Scale bar is 5 μm

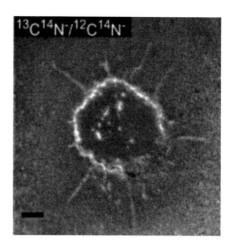

homogeneously distributed throughout the cells.[6] Another example showed the distribution of the chemotherapeutic agent 5-fluorouracil in human normal and neoplastic gastric cells via $^{19}F^-$ imaging.[11] In addition, 4'-iododioxyrubicin was localized to the nuclei of squamous cell carcinoma metastatic cells by analyzing $^{127}I^-$.[11]

The ability to image native organic species in cells and tissues is also of great interest, although the SIMS technique has a relatively low efficiency of molecular ionization.[12] Compounds such as lipids and cholesterol are thus frequently analyzed via fragment ions including m/z 184 ($C_5H_{15}NO_4P^+$, phosphocholine headgroup), m/z 166 ($C_5H_{13}NO_3P^+$, phosphocholine-H_2O), m/z 86 ($C_5H_{12}N^+$, choline), m/z 156 ($C_3H_{11}NO_4P^+$, [phospho-n-monomethyl ethanolamine headgroup + H]$^+$), and m/z 369 ($C_{27}H_{45}^+$, [cholesterol + H − H_2O] $^+$).[13-17] Recent advances to improve the formation and detection of molecular ions with SIMS include the use of cluster ion primary beams (such as C_{60}^+, Au_3^+, Ga_3^+, Bi_3^+, SF_5^+),[12,17-21] post-ionization methods to ionize desorbed neutrals,[3,5] and improvements in sample preparation that include the addition of a matrix.[22-25] These approaches have greatly improved the generation of intact molecular species. For example, many triacylglycerols, cholesterol, vitamin E, phosphatidylcholines, phosphatidylethanolamines, and free fatty acids were detected as their molecular species (either M$^+$, [M + H]$^+$, [M + Na$^+$]or [M − H]$^-$) in a study of nonalcoholic fatty liver tissue using a Bi_3^+ cluster beam.[21] Another report utilized both a C_{60}^+ ion beam as well as addition of matrix (sinapinic acid and α-cyano-4-hydroxycinnamic acid) to detect intact digitonin ([M + Na$^+$] at m/z 1251.56).[26] In addition to the sodium adduct, cluster SIMS also produced abundant lower mass fragment ions, corresponding to losses of successive glycosyl ring units. The addition of matrix greatly reduced the in-source fragmentation and increased the signal for intact digitonin.[26]

The increase in molecular ionization efficiency comes at the expense of the lateral resolution. Cluster ion beams cannot be as tightly focused as atomic ion beams, and in addition, the molecular ionization efficiency is related to the absolute amount of desorbed material, i.e., the larger the desorption site, the more material available to ionize. Thus the resulting molecular images have effective spatial resolutions on the micron scale (~1–30 μm). Addition of a matrix can affect ultimate resolution as well, depending on how the matrix was applied and what effect it may have on the spatial integrity of the analytes.

2.2 Laser Desorption/Ionization (LDI) Imaging

An alternative strategy for solid/surface ionization and imaging involves the use of lasers instead of ion beams. Laser desorption/ionization (LDI) has existed as an ionization source for mass spectrometry almost as long as there have been lasers (early 1960s).[27,28] In principle, any type of laser may be used, however each type (excimer, solid-state, IR, visible, UV, etc.) has different desorption, focusing, energy and pulse frequency characteristics, that make some more suitable than others for a given purpose. Overall, UV lasers are most commonly used for LDI imaging, although there have been reports of IR lasers being used to generate images of metabolites in plants[29] and visible lasers being used to generate images of rhodamine dye on a metal surface.[30] The first commercially available laser microprobe instrument, the LAMMA-500 (laser microprobe mass analyzer), coupled a Nd:YAG laser (256 nm) with a TOF mass analyzer.[31,32] That basic instrument configuration was modified to support automatic acquisition of ion images and subsequent image generation and processing.[33]

Compared to the SIMS process, LDI typically produces more intact, larger molecular ions, although the mass range is still generally limited to analytes <1,500 Da.[34] One approach to increasing the accessible mass range is to combine immunohistochemistry with LDI, utilizing antibodies with photocleavable tags.[35] In this method antibodies are applied to a tissue where they interact with the targeted protein. Upon exposure to laser irradiation, the tag is released, and the MS signature of the tag is recorded. For example, images from synaptophysin (38 kDa) were detected from human pancreas tissue by monitoring the cleaved tag at m/z 530. The spatial resolution in this case was ~50 μm.[35] This procedure may be multiplexed, with different mass tags attached to different antibodies, so that many proteins may be detected in one experiment.

The spatial resolution attainable from a laser source is primarily limited by the laser spot size on target, which is typically in the 1–5 μm range or greater.[27] As with SIMS, there is a tradeoff between spot size and sensitivity (ie, the ion signal decreases per pixel with increasing spatial resolution) which affects the ultimate practical resolution. Also, there are many neutrals desorbed from the surface that are not ionized, and postionization strategies have been employed to increase the chemical information obtainable by the LDI method.[36] One LDI imaging application used postionization to spatially resolve polycyclic aromatic hydrocarbons (PAHs) along a line from the fusion crust to the interior of a martian meteorite.[37] In this case, a CO_2 IR laser (10.6 μm) was used to desorb compounds intact, and a Nd:YAG laser was used to selectively ionize molecules that absorb energy at the wavelength of the laser (266 nm), ie, PAHs.[38] It was shown that there was no signal for phenanthrene (m/z 178) or pyrene (m/z 202) within the fusion crust or a zone ~500 μm into the interior of the meteor fragment. However, the signals increased further into the interior of the meteorite, providing evidence that organic PAH's were not introduced by terrestrial contamination through cracks in the meteorite while it was buried in the Antarctic ice sheet, and supporting the notion that the organic content found in the meteorite may have formed from biogenic processes.[37]

Postionization utilizing inductively coupled plasma (ICP) has resulted in a hyphe-nated technique known as laser ablation (LA)-ICP-MS that has been quite promising for elemental imaging. The simultaneous determination and quantitation of Zn, Cu, Th, and U in human brain tissue (hippocampus)[39] and the subsequent determination and quantitation of Cu, Zn, Pb, and U in human glioblastoma tissue has been reported.[40] The distribution of Zn and Cu in human hippocampus is shown in Fig. 2.

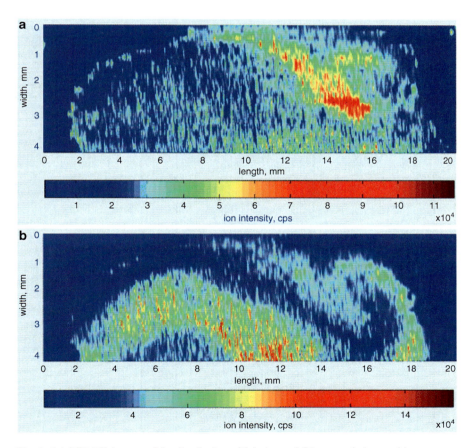

Fig. 2 LA-ICP-MS images of the distribution of (**a**) zinc and (**b**) copper in human hippocampus tissue

Similar studies were undertaken for the analysis of Mn, Fe, Cu, and Zn in lesioned mouse brain tissue, as elevated concentrations of these metals in the substantia nigra are implicated in the development of Parkinson's disease.[41] Indeed, elevated levels of Fe were detected bilaterally in the substantial nigra after unilateral lesioning of the brain with 6-hydroxydopamine.

3   Matrix-Assisted Laser Desorption/Ionization (MALDI) Imaging

In the late 1980s, matrix-assisted laser desorption/ionization (MALDI)[42–44] and electrospray ionization (ESI)[45,46] were developed, allowing intact high molecular weight gas-phase ions to be detected. MALDI is a laser desorption technique that requires the addition of an energy-absorbing matrix to the samples of interest. This allows the energy of the desorbing laser to be more gently transferred to the analytes, resulting in a much 'softer' desorption/ionization process with more intact compounds transferred the gas-phase and generally detected as pseudo-molecular ions $(M + H)^+$.

The use of MALDI for spatial localization of tissues was first reported in 1997 where the spatially resolved detection of protonated insulin (m/z 5,802) in a rat pancreas tissue section was shown.[47] Mass spectra were manually acquired every 25 μm across a 450 × 75 μm region, and the intensity of the m/z 5,802 ion was plotted as a bar graph as a function of location on the tissue section. A clear increase in signal was observed for the insulin signal from an islet cell. In the same report, ions up to m/z 22,549 were detected from MALDI MS of a rat pituitary section blotted onto a C-18 membrane, and ions up to m/z 11,504 were detected from an image of human buccal mucosa (cheek) cells. Further early work focused on improving the technology (automation[48], sample preparation[49–51]) and evaluating its applicability to clinical samples.[52]

3.1   Applications to Cancer Biology

One of the first applications of MALDI IMS was in the field of cancer biology.[52] Many cancers form solid tumors that can be biopsied and/or resected during surgery, providing a ready source of tissue for analysis. In addition, there are significant clinical needs in terms of the discovery of biomarkers that may be used for early detection of cancer, determination of metastatic potential, and response to therapy. Initial studies in our laboratory utilized mouse models of tumors to assess proteomic changes in cancer development. For example, a transgenic model of prostate cancer involving neuroendocrine cells was examined by IMS.[53] A group of five signals, putatively identified as histone proteins, was detected at high expression levels compared to normal in four different lobes of a transgenic mouse prostate that had developed prostate cancer, as well as in the lymph node and liver. These signals were not detected in analogous tissues from a nontransgenic control mouse, suggesting that they "may be indicative of active proliferation through high transcription of neuroendocrine cancer cells".[53] Similar studies were performed in mouse models of colon cancer, where several tumor specific protein markers were identified, including calgranulins A and B and calgizzarin.[54]

Human tissue has also been examined by IMS, both in protein profiling mode as well as the imaging mode. Protein profiles from lung cancer were determined using a collection of 79 lung tumors and 14 histologically normal lung tissues that were resected from patients with non-small-cell lung cancer (NSCLC) or metastases to lung.[55] Figure 3 shows an example of spectra taken from histologically normal lung and from two lung cancer tissues. The differences between the normal and tumor spectra are immediately obvious, with a complete statistical analysis defining 82 discriminatory peaks, some of which are marked with asterisks in Fig. 3. A hierarchical clustering of tumor and non–tumor lung tissue based on the 82 peaks is also shown in Fig. 3, resulting in complete differentiation of tumor from non-tumor tissue. Further evaluation of the data resulted in protein profiles that could distinguish non-tumor lung from primary tumor, primary tumor from other lung tumor (primarily lung metastases), as well as subtypes (for example, adenocarcinoma from squamous-cell or large cell). Protein profiles were also correlated with patient survival, where a set of 15 signals was able to distinguish a group of patients with poor prognosis (median survival 6 months) from a group with good prognosis (median survival 33 months).[55]

The work that resulted from the analysis of solid lung tumor tissue has been followed up by extending the technology to derive useful protein profiles from fine needle aspirates from lung cancer patients,[56] to discover proteomic signatures that classify survival of lung cancer patients after targeted chemotherapy,[57] and to determine its applicability for noninvasive screening of unfractionated serum.[58]

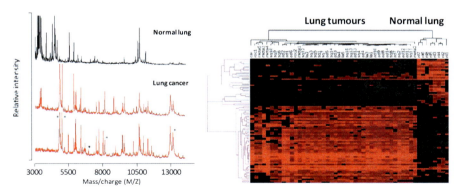

Fig. 3 Unique protein profiles are revealed for non-tumor lung versus lung tumor tissue by MALDI MS analysis

A similar study was undertaken with human glioblastoma samples.[59,60] In this case, a total of 162 tissue samples from 127 patients were analyzed, including patients undergoing surgery for nonneoplastic disease (non-tumor tissue), and patients with grades II, III, and IV glioma. Two independent supervised methods were used to analyze the protein profiles. Similar results were obtained from both statistical methods which were able to successfully classify tumor from non-tumor tissue and grade II and grade III from grade IV with ~90% classification accuracy. Protein profiles were also used to classify patients into a short-term survival (STS) group (average <15 months) and a long-term survival (LTS) group (average >90 months) independently of other known prognostic factors (age, tumor grade). A group of two unique features was even able to stratify grade IV glioma patients into STS (average 10.9 months) and LTS (average 16.8 months) groups.[60] While most tumors are accurately classified by grade and stage based on histology, prognostic indications are not directly observable by a pathologist. Classifications based on molecular features, whether for diagnostic or prognostic purposes, such as those derived from MALDI MS profiling, thus have tremendous clinical potential.

Due to the heterogeneity present in many different tumor types, different approaches have been explored to refine the proteomic profiling experiment to make it more specific and targeted. One approach, histology-directed protein profiling,[61] involves the automatic deposition of picoliter volumes of matrix onto discrete areas on a tissue that have been marked by a pathologist. The goal is to desorb proteins from discrete areas that are highly enriched in a certain cell type (ie, normal epithelial cells, tumor epithelial cells, stroma, etc.). This approach can be considered a targeted imaging experiment, because while an overall picture of proteomic distribution is not obtained, each protein profile is linked to a specific location on the tissue surface. A complimentary approach is to extract specific cells of interest from the section using laser capture microdissection (LCM), and subsequently obtain protein profiles from the extracted cells. Human breast cancer samples and tissue from mouse models of colorectal cancer have been analyzed in this way.[62-65]

3.2   Applications to Neurology

MALDI IMS profiling and imaging has also been applied to research investigations in neurology. For example, single neuron cells from *Aplysia californica* have been analyzed by MALDI MS.[66] These cells are larger than typical mammalian cells, ~50 μm up to 500 μm in diameter, and contain a number of signaling peptides. Vesicles from the exocrine atrial gland (~1.5 μm diameter) have also been analyzed by MALDI MS after manual microdissection.[67] An alternative approach to microdissection has also been reported, termed the stretched sample method.[68,69] In this case, thin tissue sections are adhered to a glass bead array attached to a stretchable membrane. When the membrane is stretched, the tissue fragments into many individual pieces, each attached to a glass bead. Application of matrix facilitates extraction of analytes, and the underlying hydrophobic membrane minimizes

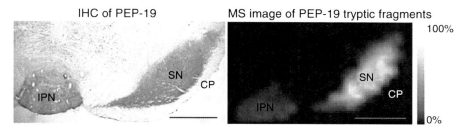

Fig. 4 Immunohistochemical analysis (*left*) and MS image analysis (*right*) of a section of rat brain at the same rostrocaudal level showing similar localizations of PEP-19. Scale bar is 1 mm

analyte delocalization. In this manner, the automated analysis of individual, single-cell-sized samples is possible. The original method has also been adapted for imaging by utilizing algorithms to reconstruct ion images from the stretched samples.[69]

Studies involving neurological disorders have been done using this technology. For example, studies of animal models of Parkinson's disease have been undertaken, with IMS analysis revealing localized decreased expression of PEP-19 (6.7 kDa) in the striatum of Parkinson's mouse brains (following MPTP treatment),[70] and increased expression of the neuroimmunophilin FKBP-12 (m/z 11,791) in the dorsal and middle part of lesioned rat brains.[71] In addition, three proteins also showed differential expression as a result of dopamine depletion, similar to controls after L-dopa administration.[72] Recently, the distribution of PEP-19 in rat brain has been obtained by on-tissue tryptic digestion and detection of PEP-19 tryptic fragment ions.[73] Figure 4 shows excellent agreement between the reconstructed MS ion image on the right with the immunohistochemically stained section on the left, with PEP-19 localizing to the substantia nigra (SN) and the interpeduncular nucleus (IPN).

3.3 Applications to Developmental Biology

MALDI IMS has been employed to study various aspects of development, including developmental neurology, with a study of proteomic changes in mouse cerebellum at different stages of development, from postnatal day 7, to day 14, and to adults.[74] Two sets of age-related proteomic changes were detected, illustrating the power of the technology to detect relatively small, perhaps gradual changes in proteomic expression as a function of normal development. Areas of reproductive biology have also been studied. The spatial and temporal distributions of proteins and lipids have been determined by IMS for embryo implantation and inter-implantation sites in mouse uterus.[75,76] A number of proteins were found to have differential expression depending on day post-pregnancy and the presence or absence of a blasyocyst. The expression of ubiquitin (m/z 8,565), calcyclin (m/z 9,962), calgizzarin (m/z 10,952),

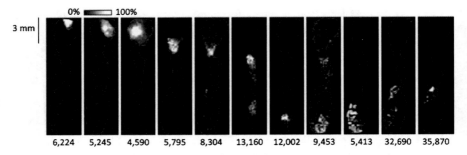

Fig. 5 MALDI IMS analysis of protein distribution in a mouse epididymis, showing selected proteins localized to distinct regions along the length of the organ

and transthyretin (m/z 13,641) determined by IMS were similar to their mRNA expression determined by *in situ* hybridization.[75] Mouse prostate development[77] and protein expression in mouse epididymis[78,79] have also been investigated. The epididymis is an organ where spermatozoa mature by interacting with proteins present in the epididymal fluid along the length of the organ. Figure 5 shows eleven proteins (m/z listed below image) with distinct localization patterns in the mouse epididymis, illustrating how the proteomic environment changes during the maturation process.

4 Technology Description

MALDI IMS technology is generally applied in two distinct but related modes, imaging and profiling. The mode that is used will depend primarily on the overall goal of the experiment, but also somewhat on the specific instruments and accessories at hand. Profiling is typically used for biomarker discovery experiments, where there is a sizeable amount of tissue samples to be analyzed, for example from human tissue databanks or ongoing accrual of clinical tissue. The goal is to acquire specific data from a large number of individuals and subject the results to biostatistical processing. Imaging is used when more fine detail of protein localization is desired. The two modes are also frequently used together, for example, when a few representative samples are imaged at high resolution to complement and validate a larger profiling experiment.

The sample preparation for the two modes is similar.[61,80,81] Fresh frozen tissue samples are sectioned on a cryostat to ~5–20 µm thick. For protein analyses the tissues are washed to remove lipids and salts (which interfere with matrix crystallization and may cause ionization suppression) and to dehydrate and fix the proteins, while maintaining tissue architecture. This is typically achieved by successive washes in graded ethanol (70%, 90%, 95% for 30 s each). Serial sections are often obtained, stained with hematoxylin and eosin (H&E), and used to either guide deposition of matrix (ie, in a histology-directed fashion) or to compare MS results to tissue histology.

Application of matrix is a critical step of the process. Typical MALDI matrices are applied to tissues in solution with the goals of optimal analyte extraction, minimal analyte migration, and sufficient crystal formation. Sinapinic acid (SA) is primarily used for protein analysis, while α-cyano-4-hydroxycinnamic acid (CHCA) is used for peptides and 2,5-dihydroxybenzoic acid(DHB) is used for lipids and small molecules. For tissues in general, manual application of ~250 nL–1 μL matrix solution produces the best MS signals but results in spots 1 mm or greater in diameter. This procedure is thus often used to test and validate instrument performance prior to data acquisition. For profiling experiments, smaller, accurately placed spots are desired and may be obtained through the use of automated robotic spotting. For example, an acoustic printer (Portrait 360, Labcyte), that is capable of depositing pL volumes at specified x,y coordinates on tissue is commercially available. This results in matrix spots of ~200 μm on tissue, with multiple passes required for adequate analyte extraction and crystal formation. A typical procedure on the Portrait 630 instrument involves six passes of 13 drops each of ~120 pL/drop for a total of ~9 nL deposited per matrix spot.

Imaging may be performed either from an ordered array of matrix spots or from a homogeneous coating of matrix on the tissue. Arrays of spots have the advantage of minimizing analyte migration to the diameter of the matrix spots, but suffer from having lower spatial resolution (~200–300 μm center-to-center spacing). Obtaining a homogeneous coating of matrix has typically been done by spraying the matrix solution over the tissue section (by electrospray,[47,48] airbrush,[19] or glass reagent sprayer[80]) in order to form a thin layer of small crystals. One commercial instrument that is available for this purpose is the ImagePrep (Bruker). The best results are obtained, either manually or robotically, when the tissue is sufficiently wetted to allow efficient extraction of analyte without inducing delocalization and small crystals are uniformly deposited over the surface of the tissue. At this point, the factor limiting the maximum achievable spatial resolution is the diameter of the laser, which is ~30–200 μm in diameter on commercially available instruments. Optimal matrix application often takes multiple cycles of spraying and drying to allow a matrix crystal layer to slowly build up on the tissue.

Data acquisition is performed similarly for profiling and imaging. Most mass spectrometer manufacturers provide software to facilitate image generation. Precision stepper motors under software control move the sample stage under the laser focus position, where multiple shots are fired. UV lasers are commonly used (N$_2$ at 337 nm or Nd:YAG at 355 nm), capable of firing at repetition rates of ~60 Hz–1 kHz. Typically ~50–400 laser shots are summed for a single spectrum at a given location. Because most array spots are larger than the diameter of the laser beam, these spots may be sub-sampled, with fresh crystals desorbed over the entire area of the spot. The spectral quality from array spots is typically better than from a spray coated tissue because of enhanced protein extraction, a larger analyte concentration per pixel, and the ability to sample fresh crystals. Protein analyses are typically performed on TOF or TOF/TOF instruments, although commercial ion mobility and FT-ICR instruments have also been used. Mass spectra are typically acquired in delayed extraction mode in a mass range from ~2,000 to 50,000 Da. Higher molecular weight proteins over 200 kDa have been measured from tissue,[82] however the analysis is not routine, requiring optimized instrumental parameters and typically additional acquisition time.

Mass spectra acquired in a MALDI IMS experiment typically contain protein signals that appear as peaks at unique m/z values. The mass accuracy in the protein mass range is typically 2–10 ppm for TOF instruments. After data acquisition, the data must be processed, typically by baseline subtraction, noise reduction, and normalization to total ion current (TIC). For profiling, features of interest, detected peak areas that meet certain threshold criteria (i.e., minimal S/N, prevalence in certain percentage of sampled spectra) are exported for biostatistical analysis. MatLab macros, ProTS Data and ProTS Marker (Biodesix), as well as collaboration with various biostatistical groups have been used to evaluate the data. In addition, one software package, ClinPro Tools (Bruker) enables preprocessing as well as statistical evaluation (average values, standard deviation, t-test) and classification (via hierarchical clustering, genetic algorithm, support vector machine, etc.) directly from the generated spectra.

Image processing may be performed on instrument-associated software, such as flexImaging (Bruker), or exported into the freely available BioMap program. The major goals of an imaging experiment are to identify unique localizations, to identify signals that may co-localize or exhibit correlated localizations, and to identify analytes that correlate with known histology.

5   Current Status and Future Directions

Over the last decade, MALDI imaging has undergone tremendous growth and development, with significant progress made in obtaining higher resolution images, improving sensitivity, better sample preparation and faster data acquisition and analysis. The current state of the art is capable of generating images directly from tissue sections at the ~30–50 μm scale with 2–10 ppm mass accuracy with commercial instruments. Most of the leading mass spectrometer manufacturers have implemented image-generating and image-processing software, facilitating high throughput analyses and data processing. High-frequency lasers and faster electronics have dramatically reduced analysis time from ~1 min/pixel in 1997[47] to ~100 ms/pixel, depending on desired resolution and sensitivity. Commercial availability of robotic devices capable of automatic matrix deposition has contributed to more robust and reproducible sample preparation, which along with a better understanding of sample pre-treatment, have allowed very targeted analyses to be undertaken with minimal variability.[55,83]

With these improvements to the technology, more ambitious projects may be undertaken. With more automation and data handling capabilities, extending molecular imaging to three dimensions has been sucessfully undertaken.[73,84,85]

An initial report from our laboratory utilized 264 coronal sections of mouse brain acquired across the corpus collosum to form a 3-D volume. MS images from ten of the brain sections were acquired and inserted into the volume. Subsequent applications involved the 3-D reconstruction of an entire mouse head and correlation of MS imaging data with MRI data of the same mouse brain,[85] and reconstruction of smaller brain structures including the substantia nigra and interpeduncular nucleus, in order to view the molecular images of small neuroproteins and peptides in rat brain.[73] In addition, MALDI IMS capability has been assessed with other MS platforms, including ion mobility[86] and high resolution Fourier transform ion cyclotron resonance (FT-ICR) MS.[76,87] Both ion mobility and FT-ICR MS allow separation of nominally isobaric ions in different ways, thus facilitating the acquisition of more specific ion images and potentially more images overall due to detection of otherwise unresolved masses.

One of the main challenges with direct protein analysis by IMS is the need for identification of the proteins. This is challenging because protein databases are based on sequences predicted from genomics and typically do not contain post-translational modifications that are present in active proteins. Current strategies for protein identification rely on homogenized samples for which all spatial information is lost. An alternative approach is to digest proteins directly on the tissue, thus maintaining spatial integrity while also obtaining peptide fingerprint and/or sequence data to identify the proteins of interest directly. This approach was initially applied to rat brain tissue, with one section subjected to standard IMS and a serial section subjected to on tissue tryptic digestion, followed by MS/MS analysis. Protein identifications were obtained from direct MS/MS sequencing of the tryptic peptides, and further validated by comparing the spatial distribution of the parent protein ion with putative tryptic peptide ions.[88] This approach not only allows for the identification of proteins directly from tissue, but also allows access to proteins not typically accessible to MALDI MS analysis, either due to very high molecular weight or solubility issues. For example, an image was generated for a tryptic peptide from the 71 kDa neuronal protein, synapsin I, that is not normally observed in typical IMS experiments. This application was further expanded to formalin-fixed paraffin-embedded clinical samples (where formalin crosslinking precludes direct MS analysis of proteins).[89,90] The ability to analyze FFPE samples opens up the possibility of using the vast collections of archival tissue deposited in tissue banks over many years, that are normally stored as FFPE specimens.

Although IMS has demonstrated impressive capabilities, significant challenges remain. Spatial resolution is still not sufficient to allow effective subcellular analyses. The protein complement typically measured is limited to proteins that are moderately abundant and of lower molecular weight (<50 kDa). In order to obtain information on membrane bound proteins or low abundance proteins, alternative sample preparation methodologies or *in situ* chemistries must be employed. As resolution improves, the amount of data generated for the same area will greatly increase, providing a need for advances in data storage and data processing as well as for increasing analysis speed. Higher frequency lasers and continuous acquisition of data (rather than step by step) will allow significantly shorter analysis times.

Given the tremendous improvements that have taken place in the past decade, it is likely these challenges will be successfully overcome as the technology continues to develop.

References

1. Castaing R, Slodzian G (1962) Microanalysis by secondary ionic emission. J Microsc 1:395–410
2. Liebl H (1967) Ion microprobe mass analyzer. J Appl Phys 38:5277–5283
3. Pacholski ML, Winograd N (1999) Imaging with mass spectrometry. Chem Rev 99:2977–3006
4. Colliver TL et al (1997) Atomic and molecular imaging at the single-cell level with TOF-SIMS. Anal Chem 69:2225–2231
5. Arlinghaus HF (2008) Possibilities and limitations of high-resolution mass spectrometry in life sciences. Appl Surf Sci 255:1058–1063
6. Chandra S, Lorey ID, Smith DR (2002) Quantitative subcellular secondary ion mass spectrometry (SIMS) imaging of boron-10 and boron-11 isotopes in the same cell delivered by two combined BNCT drugs: in vitro studies on human glioblastoma T98G cells. Radiat Res 157:700–710
7. Lechene C et al (2006) High-resolution quantitative imaging of mammalian and bacterial cells using stable isotope mass spectrometry. J Biol 5:20
8. Lechene CP, Luyten Y, McMahon G, Distel DL (2007) Quantitative imaging of nitrogen fixation by individual bacteria within animal cells. Science 317:1563–1566
9. Hallegot P, Peteranderl R, Lechene C (2004) In-situ imaging mass spectrometry analysis of melanin granules in the human hair shaft. J Invest Dermatol 122:381–386
10. Kleinfeld AM, Kampf JP, Lechene C (2004) Transport of ^{13}C-oleate in adipocytes measured using multi imaging mass spectrometry. J Am Soc Mass Spectrom 15:1572–1580
11. Fragu P, Kahn E (1997) Secondary ion mass spectrometry (SIMS) microscopy: a new tool for pharmacological studies in humans. Microsc Res Tech 36:296–300
12. Winograd N (2003) Prospects for imaging TOF-SIMS: from fundamentals to biotechnology. Appl Surface Sci 203–204:13–19
13. Magnusson Y et al (2008) Lipid imaging of human skeletal muscle using TOF-SIMS with bismuth cluster ion as a primary ion source. Clin Physiol Funct Imaging 28:202–209
14. Pacholski ML, Cannon DM Jr, Ewing AG, Winograd N (1998) Static time-of-flight secondary ion mass spectrometry imaging of freeze-fractured, Frozen-hydrated biological membranes. Rapid Commun Mass Spectrom 12:1232–1235
15. McCandlish CA, McMahon JM, Todd PJ (2000) Secondary ion images of the rodent brain. J Am Soc Mass Spectrom 11:191–199
16. Eijkel GB et al (2009) Correlating MALDI and SIMS imaging mass spectrometric datasets of biological tissue surfaces. Surf Interface Anal 41:675–685
17. Sjovall P, Lausmaa J, Johansson B (2004) Mass spectrometric imaging of lipids in brain tissue. Anal Chem 76:4271–4278
18. Touboul D et al (2004) Tissue molecular ion imaging by gold cluster ion bombardment. Anal Chem 76:1550–1559
19. Todd PJ, Schaaff TG, Chaurand P, Caprioli RM (2001) Organic ion imaging of biological tissue with secondary ion mass spectrometry and matrix-assisted laser desorption/ionization. J Mass Spectrom 36:355–369
20. Wong SCC et al (2003) Development of a C_{60}^+ ion gun for static SIMS and chemical imaging. Appl Surface Sci 203–204:219–222
21. Debois D, Bralet M-P, Le Naour F, Brunelle A, Laprevote O (2009) In situ lipidomic analysis of nonalcoholic fatty liver by cluster TOF-SIMS imaging. Anal Chem 81:2823–2831

22. Nicola AJ, Muddiman DC, Hercules DM (1996) Enhancement of ion intensity in time-of-flight secondary-ionization mass spectrometry. J Am Soc Mass Spectrom 7:467–472
23. McDonnell LA et al (2005) Subcellular imaging mass spectrometry of brain tissue. J Mass Spectrom 40:160–168
24. Nygren H, Malmberg P, Kriegeskotte C, Arlinghaus HF (2004) Bioimaging TOF-SIMS: localization of cholesterol in rat kidney sections. FEBS Lett 566:291–293
25. Wu KJ, Odom RW (1996) Matrix-enhanced secondary ion mass spectrometry: a method for molecular analysis of solid surfaces. Anal Chem 68:873–882
26. Carado A et al (2008) C60 secondary ion mass spectrometry with a hybrid-quadrupole orthogonal time-of-flight mass spectrometer. Anal Chem 80:7921–7929
27. Van Vaeck L, Struyf H, Van Wim R, Fred A (1994) Organic and inorganic analysis with laser microprobe mass spectrometry. Part I: Instrumentation and methodology. Mass Spectrom Rev 13:189–208
28. Denoyer E, Van Grieken R, Adams F, Natusch DFS (1982) Laser microprobe mass spectrometry. 1. Basic principles and performance characteristics. Anal Chem 54:26A–41A
29. Li Y, Shrestha B, Vertes A (2008) Atmospheric pressure infrared MALDI imaging mass spectrometry for plant metabolomics. Anal Chem 80:407–420
30. Bradshaw JA, Ovchinnikova OS, Meyer KA, Goeringer DE (2009) Combined chemical and topographic imaging at atmospheric pressure via microprobe laser desorption/ionization mass spectrometry-atomic force microscopy. Rapid Commun Mass Spectrom 23:3781–3786
31. Hercules DM, Day RJ, Balasanmugam K, Dang TA, Li CP (1982) Laser microprobe mass spectrometry. 2. Applications to structural analysis. Anal Chem 54:280A–305A
32. Wechsung R et al (1978) LAMMA – a new laser-microprobe-mass-analyzer. Microsc Acta Suppl 2:281–296
33. Wilk ZA, Hercules DM (1987) Organic and elemental ion mapping using laser mass spectrometry. Anal Chem 59:1819–1825
34. Karas M et al (1990) Principles and applications of matrix-assisted UV-laser desorption/ionization mass spectrometry. Anal Chim Acta 241:175–185
35. Thiery G et al (2007) Multiplex target protein imaging in tissue sections by mass spectrometry–TAMSIM. Rapid Commun Mass Spectrom 21:823–829
36. Savina MR, Lykke KR (1997) Chemical imaging of surfaces with laser desorption mass spectrometry. TrAC, Trends Anal Chem 16:242–252
37. McKay DS et al (1996) Search for past life on mars: possible relic biogenic activity in martian meteorite ALH84001. Science 273:924–930
38. Kovalenko LJ et al (1992) Microscopic organic analysis using two-step laser mass spectrometry: application to meteoritic acid residues. Anal Chem 64:682–690
39. Becker JS, Zoriy MV, Pickhardt C, Palomero-Gallagher N, Zilles K (2005) Imaging of copper, zinc, and other elements in thin section of human brain samples (hippocampus) by laser ablation inductively coupled plasma mass spectrometry. Anal Chem 77:3208–3216
40. Zoriy MV, Dehnhardt M, Reifenberger G, Zilles K, Becker JS (2006) Imaging of Cu, Zn, Pb and U in human brain tumor resections by laser ablation inductively coupled plasma mass spectrometry. Int J Mass Spectrom 257:27–33
41. Hare D et al (2009) Quantitative elemental bio-imaging of Mn, Fe, Cu and Zn in 6-hydroxy-dopamine induced Parkinsonism mouse models. Metallomics 1:53–58
42. Karas M, Bachmann D, Bahr U, Hillenkamp F (1987) Matrix-assisted ultraviolet laser desorption of non-volatile compounds. Int J Mass Spectrom Ion Processes 78:53–68
43. Karas M, Hillenkamp F (1988) Laser desorption ionization of proteins with molecular masses exceeding 10, 000 daltons. Anal Chem 60:2299–2301
44. Tanaka K et al (1988) Protein and polymer analyses up to m/z 100 000 by laser ionization time-of-flight mass spectrometry. Rapid Commun Mass Spectrom 2:151–153
45. Whitehouse CM, Dreyer RN, Yamashita M, Fenn JB (1985) Electrospray interface for liquid chromatographs and mass spectrometers. Anal Chem 57:675–679
46. Fenn JB, Mann M, Meng CK, Wong SF, Whitehouse CM (1989) Electrospray ionization for mass spectrometry of large biomolecules. Science 246:64–71

47. Caprioli RM, Farmer TB, Gile J (1997) Molecular imaging of biological samples: localization of peptides and proteins using MALDI-TOF MS. Anal Chem 69:4751–4760
48. Stoeckli M, Farmer TB, Caprioli RM (1999) Automated mass spectrometry imaging with a matrix-assisted laser desorption ionization time-of-flight instrument. J Am Soc Mass Spectrom 10:67–71
49. Chaurand P, Stoeckli M, Caprioli RM (1999) Direct profiling of proteins in biological tissue sections by MALDI mass spectrometry. Anal Chem 71:5263–5270
50. Chaurand P et al (2004) Integrating histology and imaging mass spectrometry. Anal Chem 76:1145–1155
51. Aerni H-R, Cornett DS, Caprioli RM (2006) Automated acoustic matrix deposition for MALDI sample preparation. Anal Chem 78:827–834
52. Stoeckli M, Chaurand P, Hallahan DE, Caprioli RM (2001) Imaging mass spectrometry: a new technology for the analysis of protein expression in mammalian tissues. Nat Med 7:493–496
53. Masumori N et al (2001) A probasin-large T antigen transgenic mouse line develops prostate adenocarcinoma and neuroendocrine carcinoma with metastatic potential. Cancer Res 61:2239–2249
54. Chaurand P, DaGue BB, Pearsall RS, Threadgill DW, Caprioli RM (2001) Profiling proteins from azoxymethane-induced colon tumors at the molecular level by matrix-assisted laser desorption/ionization mass spectrometry. Proteomics 1:1320–1326
55. Yanagisawa K et al (2003) Proteomic patterns of tumour subsets in non-small-cell lung cancer. Lancet 362:433–439
56. Amann JM et al (2006) Selective profiling of proteins in lung cancer cells from fine-needle aspirates by matrix-assisted laser desorption ionization time-of-flight mass spectrometry. Clin Cancer Res 12:5142–5150
57. Taguchi F et al (2007) Mass spectrometry to classify non-small-cell lung cancer patients for clinical outcome after treatment with epidermal growth factor receptor tyrosine kinase inhibitors: a multicohort cross-institutional study. J Natl Cancer Inst 99:838–846
58. Yildiz PB et al (2007) Diagnostic accuracy of MALDI mass spectrometric analysis of unfractionated serum in lung cancer. J Thorac Oncol 2:893–901
59. Schwartz SA, Weil RJ, Johnson MD, Toms SA, Caprioli RM (2004) Protein profiling in brain tumors using mass spectrometry: feasibility of a new technique for the analysis of protein expression. Clin Cancer Res 10:981–987
60. Schwartz SA et al (2005) Proteomic-based prognosis of brain tumor patients using direct-tissue matrix-assisted laser desorption ionization mass spectrometry. Cancer Res 65:7674–7681
61. Cornett DS et al (2006) A novel histology-directed strategy for MALDI-MS tissue profiling that improves throughput and cellular specificity in human breast cancer. Mol Cell Proteomics 5:1975–1983
62. Palmer-Toy DE, Sarracino DA, Sgroi D, LeVangie R, Leopold PE (2000) Direct acquisition of matrix-assisted laser desorption/ionization time-of-flight mass spectra from laser capture microdissected tissues. Clin Chem 46:1513–1516
63. Xu BJ, Caprioli RM, Sanders ME, Jensen RA (2002) Direct analysis of laser capture microdissected cells by MALDI mass spectrometry. J Am Soc Mass Spectrom 13:1292–1297
64. Sanders ME et al (2008) Differentiating proteomic biomarkers in breast cancer by laser capture microdissection and MALDI MS. J Proteome Res 7:1500–1507
65. Xu BJ et al (2009) Identification of early intestinal neoplasia protein biomarkers using laser capture microdissection and MALDI MS. Mol Cell Proteomics 8:936–945
66. Li L, Garden RW, Sweedler JV (2000) Single-cell MALDI: a new tool for direct peptide profiling. Trends Biotechnol 18:151–160
67. Rubakhin SS, Garden RW, Fuller RR, Sweedler JV (2000) Measuring the peptides in individual organelles with mass spectrometry. Nat Biotechnol 18:172–175
68. Monroe EB et al (2006) Massively parallel sample preparation for the MALDI MS analyses of tissues. Anal Chem 78:6826–6832
69. Zimmerman TA, Monroe EB, Sweedler JV (2008) Adapting the stretched sample method from tissue profiling to imaging. Proteomics 8:3809–3815

70. Skold K et al (2006) Decreased striatal levels of PEP-19 following MPTP lesion in the mouse. J Proteome Res 5:262–269

71. Nilsson A et al (2007) Increased striatal mRNA and protein levels of the immunophilin FKBP-12 in experimental Parkinson's disease and identification of FKBP-12-binding proteins. J Proteome Res 6:3952–3961

72. Pierson J et al (2004) Molecular profiling of experimental Parkinson's disease: direct analysis of peptides and proteins on brain tissue sections by MALDI mass spectrometry. J Proteome Res 3:289–295

73. Andersson M, Groseclose MR, Deutch AY, Caprioli RM (2008) Imaging mass spectrometry of proteins and peptides: 3D volume reconstruction. Nat Methods 5:101–108

74. Laurent C et al (2005) Direct profiling of the cerebellum by matrix-assisted laser desorption/ionization time-of-flight mass spectrometry: a methodological study in postnatal and adult mouse. J Neurosci Res 81:613–621

75. Burnum KE et al (2008) Imaging mass spectrometry reveals unique protein profiles during embryo implantation. Endocrinology 149:3274–3278

76. Burnum KE et al (2009) Spatial and temporal alterations of phospholipids determined by mass spectrometry during mouse embryo implantation. J Lipid Res 50:2290–2298

77. Chaurand P et al (2008) Monitoring mouse prostate development by profiling and imaging mass spectrometry. Mol Cell Proteomics 7:411–423

78. Chaurand P et al (2003) Profiling and imaging proteins in the mouse epididymis by imaging mass spectrometry. Proteomics 3:2221–2239

79. Cornett DS, Reyzer ML, Chaurand P, Caprioli RM (2007) MALDI imaging mass spectrometry: molecular snapshots of biochemical systems. Nat Meth 4:828–833

80. Schwartz SA, Reyzer ML, Caprioli RM (2003) Direct tissue analysis using matrix-assisted laser desorption/ionization mass spectrometry: practical aspects of sample preparation. J Mass Spectrom 38:699–708

81. Seeley EH, Oppenheimer SR, Mi D, Chaurand P, Caprioli RM (2008) Enhancement of protein sensitivity for MALDI imaging mass spectrometry after chemical treatment of tissue sections. J Am Soc Mass Spectrom 19:1069–1077

82. Chaurand P, Caprioli RM (2002) Direct profiling and imaging of peptides and proteins from mammalian cells and tissue sections by mass spectrometry. Electrophoresis 23:3125–3135

83. Khatib-Shahidi S, Andersson M, Herman JL, Gillespie TA, Caprioli RM (2006) Direct molecular analysis of whole-body animal tissue sections by imaging MALDI mass spectrometry. Anal Chem 78:6448–6456

84. Crecelius AC et al (2005) Three-dimensional visualization of protein expression in mouse brain structures using imaging mass spectrometry. J Am Soc Mass Spectrom 16:1093–1099

85. Sinha TK et al (2008) Integrating spatially resolved three-dimensional MALDI IMS with in vivo magnetic resonance imaging. Nat Methods 5:57–59

86. McLean JA, Ridenour WB, Caprioli RM (2007) Profiling and imaging of tissues by imaging ion mobility-mass spectrometry. J Mass Spectrom 42:1099–1105

87. Cornett DS, Frappier SL, Caprioli RM (2008) MALDI-FTICR imaging mass spectrometry of drugs and metabolites in tissue. Anal Chem 80:5648–5653

88. Groseclose MR, Andersson M, Hardesty WM, Caprioli RM (2007) Identification of proteins directly from tissue: in situ tryptic digestions coupled with imaging mass spectrometry. J Mass Spectrom 42:254–262

89. Groseclose MR, Massion PP, Chaurand P, Caprioli RM (2008) High-throughput proteomic analysis of formalin-fixed paraffin-embedded tissue microarrays using MALDI imaging mass spectrometry. Proteomics 8:3715–3724

90. Aerni HR, Cornett DS, Caprioli RM (2009) High-throughput profiling of formalin-fixed paraffin-embedded tissue using parallel electrophoresis and matrix-assisted laser desorption ionization mass spectrometry. Anal Chem 81:7490–7495

2 and 3D TOF-SIMS Imaging for Biological Analysis

John S. Fletcher

Abstract Secondary ion mass spectrometry (SIMS) is an established technique in the field of surface analysis but until recently has played only a very small role in the area of biological analysis. This chapter provides an overview of the application of secondary ion mass spectrometry to the analysis of biological samples including single cells, bacteria and tissue sections. The chapter will discuss how the challenges of biological analysis by SIMS have created an impetus for the development of new technology and methodology giving improved mass resolution, spatial resolution and sensitivity.

Keywords Secondary ion mass spectrometry • Imaging • Bio-analysis • Cells • Tissue • Bacteria

1   Introduction

Secondary ion mass spectrometry is a valuable technique for the characterization of solid samples by mass spectrometry. In SIMS a primary ion beam impacts the sample surface and initiates the ejection of secondary species a small percentage of which are ejected as ions and can therefore be analyzed by mass spectrometry. Traditionally SIMS analysis has been split into two distinct areas. The first method developed for analyzing samples with SIMS uses a high primary ion beam dose to erode the sample while analyzing changes in composition as a function of depth. Due to its application in the semi-conductor industry this method is still the most widely used. As the sample is constantly changing under the primary ion beam this mode of operation is termed *dynamic SIMS*. Dynamic SIMS instruments generally

J.S. Fletcher (✉)
Manchester Interdisciplinary Biocentre, University of Manchester, 131 Princess St.,
Manchester, UK

J. Banoub (ed.), *Detection of Biological Agents for the Prevention of Bioterrorism*,
NATO Science for Peace and Security Series A: Chemistry and Biology,
DOI 10.1007/978-90-481-9815-3_18, © Springer Science+Business Media B.V. 2011

use a quadrupole of magnetic sector mass analyzer. Using conventional monatomic primary ion beams, although including many 'small' cluster beams, the detection of intact molecular-type ions form the sample is limited to the upper most layers of the material. This is due to the accumulation of ion beam induced damage on the surface and in the sub-surface regions of the sample. Although the ejected material originates from the surface of the sample, the primary ion (or constituents of) actually penetrate and deposit their energy deep into the material. This results in disruption, including chemical bond breaking, of the sub-surface of the sample. If the same area of the sample is continuously analyzed the signals from the sample change. The higher mass species are lost and the character of the spectrum tends towards elemental and small fragment ions. A limit is imposed due to the loss of molecular signal on the primary ion beam dose density that can be used and therefore on the amount of useful molecular information can be extracted from the sample. This is generally referred to as the static limit and hence analysis within these conditions is termed *static SIMS*. For this method of analysis the time-of-flight (TOF) mass analyzer has become ubiquitous due to the simultaneous detection of all ejected species, compared to scanning mass spectrometers.

2   Cellular Analysis with SIMS

The ability of a range of ion beams particularly liquid metal ion sources such as Ga^+, In^+ and more recently Au_n^+ and Bi_n^+, to be finely focused to produce sub-micron spot sizes makes the technique of SIMS uniquely capable of producing mass spectrometric images of single cells. The surface sensitivity of the technique has meant that the much of the cell imaging has focused on the characterization of the cell membrane. Several studies have been reported on the analysis of model membrane systems to investigate lipid domain formation.[1-3]

In one example Ostrowski *et al.* describe the use of TOF-SIMS to image the changes in lipid composition during Tetrahymena mating using an In^+ primary ion beam.[4] During mating Tetrahymena, common fresh water protozoa, form a series of highly curved fusion pores at the conjugation site that allow the passage of micronuclei between the cells. Such radical changes in the shape/structure of the membrane would be expected to involve a change in the lipid composition in that area. SIMS imaging of the tetrahymena was performed following freeze fracture of the cells and inspection of the conjugation site, as hypothesized, showed a change in the lipid composition. A reduction in the signal from phosphocholine (m/z 184) was observed. Principal components analysis was performed on the image and a species was identified at m/z 126 that was associated with the conjugation junction. The peak at m/z 126 was assigned to the headgroup of a group of cone-shaped non-lamellar 2-aminoethylphosphonolipids (2-AEP). The localization of such lipids at the conjugation site is consistent with the formation of the high curvature structures (Fig. 1). The study shows that with careful experimental planning real biological problems can be tackled using TOF-SIMS however signal levels can be very low. As such a variety of

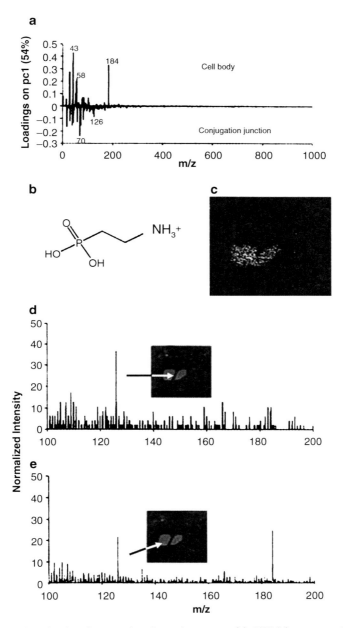

Fig. 1 The conjugation junction contains elevated amounts of 2-AEP. Mass spectra from pixels were generated by selecting the pixels of interest using software written in house. (**a**) Loadings plot from principal components analysis comparing the mass spectra of the cell bodies and the conjugation junction. (**b**) 2-AEP headgroup fragment corresponding to m/z 126. (**c**) SIMS image for m/z 126. (**d**) Mass spectrum from the pixels along the conjugation junction, as indicated in the inset. (**e**) Mass spectrum from the pixels in the cell bodies, as indicated in the inset (Reproduced with permission from reference[4])

methods have been suggested to increase the signal in SIMS including using metal addition and also addition of a MALDI matrix.[5, 6] Despite showing advantages in some cases the surface sensitivity and the desire for high resolution imaging can mean that these approaches result in more complications than benefits.

Arguably the greatest advance in the field of organic analysis with TOF-SIMS has been the implementation of polyatomic ion beams for routine analysis. Initial work from Appelhans and Delmore,[7] continued by Kotter and Benninghoven,[8] looking at methods for increasing secondary ion yields from polymers, reported that when the gas cluster ion SF_5^+ was used to analyse polymers there was not only the increase in secondary ion yield associated with the increase in mass, and the non-linear 'cluster effect' but also an increase in the amount of primary ion beam dose that could be used while maintaining characteristic molecular information. To paraphrase; the *static limit* could be relaxed. The group defined the efficiency of the ion beam as a ratio of the secondary ion yield to the disappearance cross section of a given ion. Hence the most efficient ion beam would be one that produced a high secondary ion yield with the slowest decline of molecular information as a function of primary ion beam dose. Gillen and co-workers had also observed a similar effect with small carbon clusters.[9] The Vickerman group, in collaboration with Ionoptika Ltd., developed a C_{60}^+ ion beam system for routine analysis in SIMS. The ion beam produced a further increase in sputter yield, again with a greater effect on the higher mass region of the spectrum.[10, 11]

A number of molecular dynamics simulations have been produced providing a comparison between the interaction of monatomic species, small clusters and C_{60}^+ with the sample surface during the sputtering event. The monatomic and small cluster ions such as Au_3^+ and/or Bi_3^+ penetrate the sample to a much greater depth than the C_{60}^+ ion, causing the disruption of many sub-surface layers with only a small proportion of the impact energy being transferred to the surface region of the sample and producing only a small yield of secondary particles. C_{60}^+, however, is expected to break up on impact and therefore is equivalent to the simultaneous impact of 60 carbon atoms each with 1/60th of the energy of the primary ion. The result is a massive ejection of material from the impact crater and, as each atom has only a relatively low energy, reduced disruption and damage of the sub-surface region.[12] The use of C_{60}^+ has allowed the static limit to be relaxed and in some cases ignored on a wide range of organic samples and has lead to the emergence of a whole new discipline in SIMS – *molecular* depth profiling.

In the area of cellular analysis the implementation of these ion beams has two implications. Firstly the signal per pixel in an image can be increased by allowing analysis beyond the static limit and thus sampling a volume and secondly when sufficient signal is available 3D molecular imaging can be performed.

The use of C_{60} to probe both the surface and sub-surface of biological samples is a unique capability of the SIMS technique. Vaidyanathan *et al.* applied TOF-SIMS to the imaging of *streptomyces coelicolor*.[13] The bacteria secrete different coloured anti bacterial agents as a defence mechanism and the secretion varies when cultured under normal and salt stressed conditions. Analysis by TOF-SIMS showed both agents on the surface of the bacteria but following C_{60} etching to remove the upper layer only the one of pigmented species remained suggesting a two stage defence mechanism.

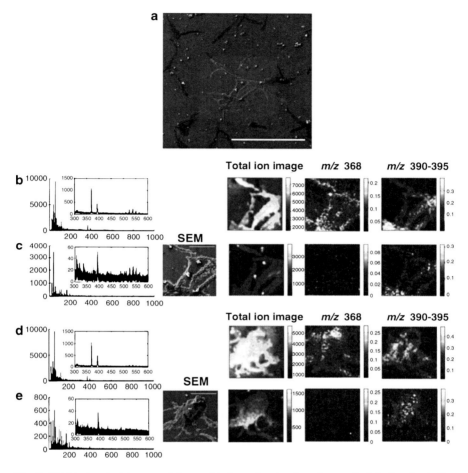

Fig. 2 3D biochemical images of freeze-dried oocyte, showing changes in (**a**) phosphocholine peaks m/z 58, 86, 166, and 184, (**b**) signal summed over the m/z range 540–650, (**c**) signal summed over the m/z range 815–960, and (**d**) cholesterol peak at m/z 369. Colour scale normalized for total counts per pixel for each variable (m/z range) (Reproduced with permission from reference [15])

Figure 2 shows the mass spectral imaging of the salt stressed bacteria before and after etching. The different coloured anti-bacterial agents are characterised by peaks at m/z 368 for one and several peaks in the m/z 390–395 region for the other.

2.1   3D Cellular Imaging SIMS

Chandra performed 3D single cell imaging using dynamic SIMS to generate a series of single mass images at increasing depths into a fractured freeze-dried T98G glioblastoma cell in the metaphase stage. The changing distribution of Ca^+ was observed as a function of depth.[14] However it is only with the advent of polyatomic ion beams that molecular information can be measured as a function of depth.

Fletcher *et al.*[15] demonstrated the initial application of TOF-SIMS using 40 keV C_{60}^+ ions to sputter etch and image *Xenopus laevis* oocyte, thus characterizing chemical changes in 3 dimensions of a biological cell with minimal sample preparation/intervention. *X. laevis* oocyte is a well established cell model that has been extensively used in many branches of experimental biology and pharmacological research. It is a large single cell about 0.8–1.3 mm in diameter, the cellular compartments of which are so large that they are accessible to cell biologists for manipulation and visualization. In addition, *X. laevis* oocyte is remarkably resistant to osmotic changes. This facilitated the preparation of the cells for mass spectral imaging as they are reasonably resistant to washing of exogenous substances using de-ionized water. Although the handling of the sample was simplified due to the size of the specimen the variation in topography over the analysis area resulted in reduced mass resolution and meant that over the course of the analysis only the outer section of the oocyte could be consumed. Despite this the sample provided clear proof of principle for molecular depth profiling of biological cells using C_{60}^+. Persistent secondary ion signals were observed through the analysis, where over 75 μm of material was eroded, for a range of species including cholesterol and diacylglycerides associated with previously identified, by extraction and MS analysis, lipid species known to be abundant in *X. laevis* oocyte membranes (Fig. 3).

The C_{60}^+ beam, arising from a gas phase electron impact source, is inherently difficult to finely focus while maintaining enough current to enable analysis on a useful time scale. A compromise can therefore be used where a liquid metal ion gun (LMIG) is used to perform the imaging cycles of the experiment while a polyatomic ion beam is used for interleaved etching periods to remove any damage generated by the LMIG beam. Brietenstein *et al.* employed this method of analysis for the 3D TOF-SIMS imaging of normal rat kidney (NRK) cells.[16] Bi_3^+ ions were used for the imaging while C_{60}^+ was used for the etching. The high spatial resolution imaging of the Bi_3^+ allows the sample to be imaged with sub-cellular resolution in 3D. Figure 4 shows the cells clearly outlined against the substrate while the combination of amino acid associated peaks also highlight the cells while the nuclear region can clearly be defined by the absence of pooled phospholipid signal. Although extremely impressive work it highlights a further compromise associated with high resolution TOF-SIMS imaging. The imaging was performed in a low mass resolution (*i.e.* using a longer than normal primary ion pulse) mode while a non-imaging high mass resolution mass spectrum must be acquired for accurate peak assignment. This is common with high resolution TOF-SIMS imaging where too rapid pulsing and/or bunching of the primary ion beam can prevent the realization of the full focus of the ion beam and the low ion currents from the focused and apertured ion guns means that the experiment time becomes much longer so it is desirable to increase the amount of primary ions in each shot thus increasing the duty cycle of the experiment in terms of ion fluence over time. This mode of operation can have a detrimental effect on the quality of the mass spectrum particularly the mass resolu-

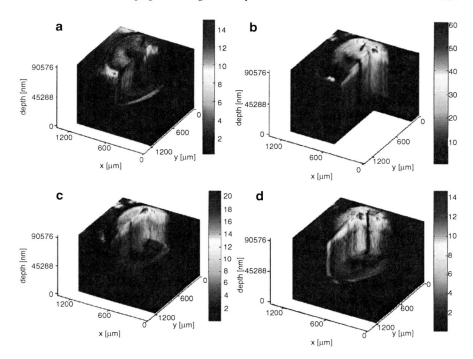

Fig. 3 Subsurface imaging of antibiotic distribution demonstrated on a salt-stressed bacterial population. An SEM image of one of the sample areas analyzed (bars – 500 μm) is shown with the etch crater visible (**a**). Two different areas are imaged before (**b** and **d**) and after (**c** and **e**) sputter etching the surface with a primary ion dose density of 6×10^{13} and 2×10^{14} ions/cm², respectively. The total ion spectrum for the four cases is shown on the *left* along with the corresponding total ion images (256×256 pixels) and the normalized images showing the distribution of the antibiotic peaks at m/z 368 and m/z 390–395 (128×128 pixels). An SEM image of the analyzed area (bars – 100 μm) is also shown for the etched surfaces for reference (Reproduced with permission from reference[13])

tion as this is directly coupled to the primary ion pulse length on a conventional TOF-SIMS instrument. Also, the ultimate sensitivity is reduced as more of the sample is discarded during the etching cycles than is sampled during the analysis period.

3 Developments in Instrumentation

There are however several drawbacks to the use of the TOF analyzer due to the pulsed, cyclic nature of the experiment. Particularly in terms of instrument duty cycle and the need to rapidly pulse the primary ion beam to maintain good mass resolution that become exacerbated when the goal is the interrogation of single cells.

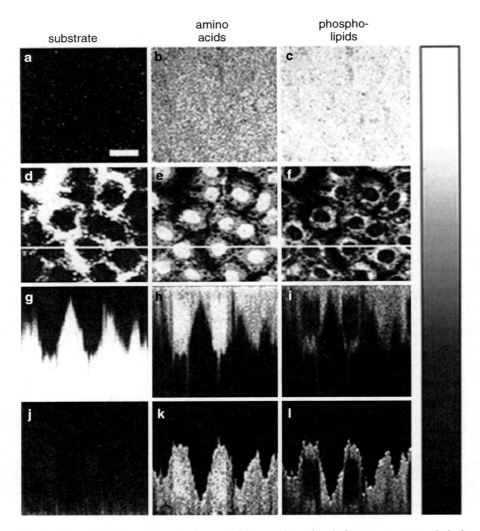

Fig. 4 Mass-resolved secondary ion images of the sample surface before any sputter cycle had been applied (**a–c**) and within the cell after the 45th sputter cycle (**d–f**). Image intensities are color coded on a black-and-white scale with white corresponding to high intensities. The color scale is normalized to the intensity in the brightest pixel. The scale bar in (**a**) corresponds to 20 μm. (**g–i**) xz Sections through the three-dimensional data stack along the *white line* in (**d–f**). Parts (**j–l**) show the data of parts (**g–i**) after a mathematical correction of the z axis. Mass-resolved images are based on the following secondary ions: (**a**), (**d**), (**g**), (**j**): Na+; (**b**), (**e**), (**h**), (**k**): amino acid fragment ions (pooled signal for masses 30 u, 44 u, 70 u, 101 u, 110 u, 136 u); (**c**), (**f**), (**i**), (**l**): phospholipid fragment ions (pooled signal for masses 58 u, 166 u, 184 u, 760 u) (Reproduced with permission from reference[16])

The desire to fully capitalize on the advantages delivered by polyatomic ion beams, coupled with the limitations in the application of conventional TOF instruments to these studies, have provided the impetus to develop new ways of applying SIMS to biological samples. Significant efforts have been directed towards this task by our laboratory and also by the Winograd group. The approaches may appear very different but have many fundamentally similar aspects. These similarities arise through both groups seeking to overcome the issues of duty cycle and spectral quality. Both of these can be overcome if the mass spectrometry can be decoupled from the sputtering aspect of the SIMS process.

The Winograd group have worked in collaboration with Applied Biosystems to modify the front end of a commercial MALDI platform (Applied Biosystems Q-Star XL) to allow the incorporation of a C_{60} ion beam system.[17] Here the C60 beam is used instead of the laser of the MALDI instrument front end and is run either continuously or using extremely long pulses. A continuous stream of secondary ions is extracted by gas flow into the mass spectrometer where it passes through a series of quadrupoles prior to electrostatic push-out into an orthogonally mounted TOF analyzer (Fig. 5).

This development has definitively overcome the issues raised above in regards to spectral quality. The mass spectrometry is completely decoupled from the secondary ion generation process removing the contribution from the ion beam

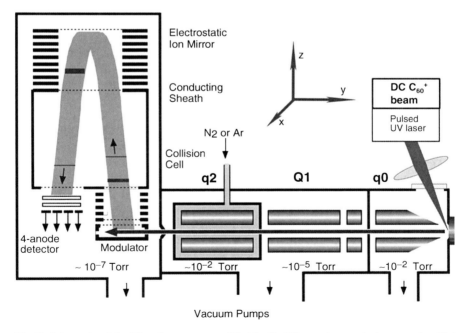

Fig. 5 Schematic of the QStar instrument modified by the Winograd group to incorporate a C_{60}^{+} primary ion beam system. The C_{60}^{+} can be used as an alternative to the MALDI laser (Reproduced with permission from reference [17])

pulse length and topographic effects. The result is a secondary ion mass spectrometer with excellent mass resolution (m/Δm~14000) in all modes of operation. The adaptation of the instrument allows complementary data to be obtained using the SIMS and MALDI modes of operation with the SIMS providing an improvement in the lateral resolution of the instrument. However at present the capabilities for the very high spatial resolution imaging normally associated with SIMS are limited. On the prototype instrument the primary ion beam cannot be rastered and the instrument is only fitted with a course sample stage making most single cell imaging impossible. If these technological drawbacks can be overcome the ultimate imaging resolution of the instrument is expected to be determined by the scattering of the primary ion beam by the relative high (~1 Torr) gas pressures in the sample region of the instrument. Secondary ion transit times through the quadrupole system will also limit the rate of image acquisition as the beam cannot be moved to the next pixel until all ions from the first pixel have reached the detector.

The approach in our laboratory has been somewhat different in that we have chosen to work closely with manufacturers of ion guns and mass spectrometers to develop an instrument purpose-built for 2D and 3D SIMS imaging using novel methodology (Fig. 6).[18]

The resulting instrument, the J105 3D Chemical Imager (Ionoptika Ltd, Southampton) exploits a unique linear buncher in conjunction with a harmonic reflectron TOF analyzer. As with the Q-Star platform, the instrument allows a continuous primary ion beam to be used for the analysis. The buncher, which is approximately 30 cm long, is filled with a portion of secondary ions. Next the buncher fires by suddenly applying an accelerating field that varies from 7 kV at the entrance of the buncher to 1 kV at the exit. This creates a time focus at the entrance of the TOF analyzer as the ions from the back of the buncher catch up with those from the front.

The ultimate mass resolution is, like the Q-Star, decoupled from the sputtering event and is now dependent on the quality of this focus. Due to the acceleration in the buncher the ions now have a 6 keV energy spread. A harmonic field TOF reflectron is required and employed such that the path of the ions is dependent only on the mass and charge, not the energy, of the secondary ions. The secondary ions undergo half a period of simple harmonic motion in the analyzer before impacting the detector with the same time spread as the focus from the buncher. The system is currently delivering m/Δm ~ 6,000 at mass 500 although an upgraded buncher is to be installed shortly that is expected to produce in excess of m/Δm > 10,000.

The sample handling and secondary ion generation and extraction of this instrument are more comparable to a conventional TOF-SIMS instrument than on the hybrid Q-Star instrument in that the design is based around an ultra high vacuum system and the secondary ions are extracted electrostatically. The sample handling is also optimized for manipulation of frozen biological specimens with insertion and analysis sample stages that can be cooled to liquid nitrogen temperatures and a glove box to prevent atmospheric water deposition during sample insertion. The instrument is thus able to fully exploit the high resolution imaging available using focused ion beams while increasing the duty cycle and improving the quality of the mass spectra.

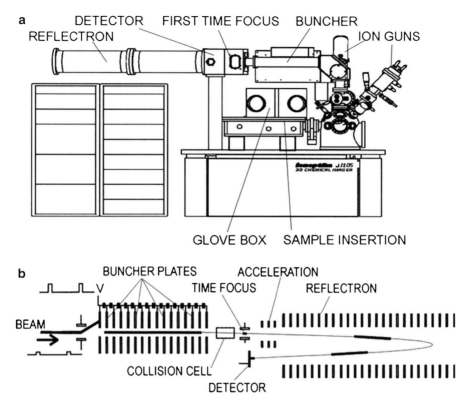

Fig. 6 Illustration of the Ionoptika J105 *3D Chemical Imager*. Sample insertion can be performed under an inert atmosphere using the glove box to prevent frosting of cryogenically preserved samples. Continuous primary ion beams generate a stream of secondary ions. A section (ca. 0.3 m) of the continuous secondary ion stream is bunched to a time focus and accelerated into the reflectron (Reproduced with permission from reference[18])

The duty cycle increase on the instruments described above allows depth profiling experiments to be performed without the interleaved analyze/etch protocol thus maximizing sensitivity and combining the speed of the dynamic SIMS instrumentation with the multiplexing advantage of the TOF analyzer. In the case of the J105 high lateral resolution 3D imaging is possible in a manageable time frame without the necessity for dual beam analysis and without discarding valuable material. The increase in acquisition rate associated with both instruments facilitates larger area imaging, for example sectioned tissue samples.

Figure 7 Illustrates the possibilities available for cellular imaging using continuous (as opposed to pulsed) polyatomic ion beams. A reconstructed image of benign prostatic hyperplasia cells comprising 16 stacked images allows the biochemistry to be explored in 3D. Phospholipid signal is observed around a central area in which there is an abundance of signal from adenine expected to be particularly

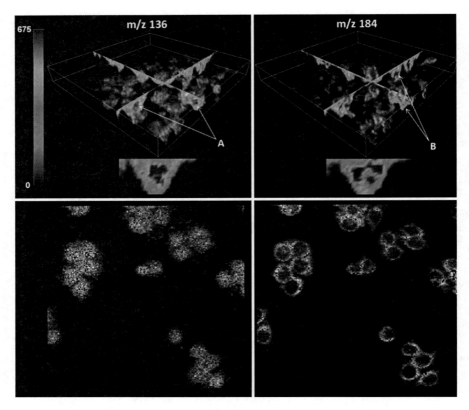

Fig. 7 Cell imaging using the J105. *Top half* of figure shows a 3D reconstruction of image data from BPH cells. Isosurface rendering shows the distribution of signal in 3D through the entire sample while orthogonal slices through the data set facilitate visualization of the chemical distribution within selected cells. Data shown for the m/z 136 ion (*left*), the protonated molecular ion from adenine, and m/z 184 (*right*) originating from phosphocholine-containing lipids. The adenine signal is localized to the centre of the cells (**a**) as it arises from the nuclear DNA while the m/z 184 signal is observed almost as rings around the edge of the cell (**b**) as it arises from the lipid membrane. A larger view of the orthogonal slice through a single cell is also shown for clarity. Field of view for the analysis was approximately $180 \times 180 \ \mu m^2$, 128×128 pixels; 16 images were acquired as the analysis progressed through the cells (Reproduced with permission from reference 52. *Lower images* are the same ions from a $256 \times 256 \ \mu m^2$, 256×256 pixel image of freeze dried HeLa cells. In this analysis the entire sample was consumed in the generation of one image. Result is comparable to a bright field microscope image (Reproduced with permission from reference[18])

high in the nucleus. The data was acquired in a morning where it would have taken months to years to perform, without the aforementioned compromises, on a conventional TOF-SIMS instrument. Also shown in the figure are the distributions of the same ions in HeLa cells. These cell images, at higher lateral resolution, were acquired using the alternative approach of consuming all of the available material in a single image. The result is more comparable to what one would observe in a bright field microscope image or the type of images associated with dynamic SIMS although for molecular species. The adenine signal is highest in the nucleus although is seen extending out through the cell. The phospholipid signal highlights the cell membrane with particular intensity around the nucleus most likely arising for the endoplasmic reticulum. Even though the pixel area is now only 1 μm, by analyzing beyond the static limit high signal levels are recorded; maximum counts per pixel is 1,886 for the adenine peak and 3,789 for the phosphocholine signal (from a 0.1 amu mass channel).

The development of novel instrumentation also has implications for large area analysis such as the imaging of tissue sections. Tissue imaging with SIMS is not uncommon but is complicated by the size of the tissue sections. The majority of sample stages are optimized for high spatial resolution imaging over a few hundred micrometers and not for analyzing tissue slices several centimeters long. Several groups have pursued this avenue of analysis however to good effect to image lipid distribution in a range of samples including rodent and bird brain samples.[19, 20] The higher resolution SIMS images are sometimes combined with MALDI images with higher available mass range highlighting the potential complimentarity of the techniques.[21] It is usual to acquire a series of adjacent images or tiles that are either stitched together during analysis or retrospectively using in house software. The use of polyatomic ion beam scan again offer increased signal per pixel by allowing continued analysis at each pixel while accumulating (as opposed to loosing) molecular information. Again however, the duty cycle can mean that this approach is not used on conventional TOF-SIMS instruments due to the time frame required for acquiring so many pixels at high dose.

Figure 8a illustrates the potential for both large and small area tissue imaging using continuous polyatomic ion beams. The images were acquired using 40 keV C_{60}^+ primary ions. The large area image (a) consists of 9×16, 600×600 μm^2 'tiles' each containing 64×64 pixels while the higher resolution images (b) are single 600×600 μm^2 images containing 256×256 pixels. Even in the large area overview of the rat brain fine structure is visible towards the centre of the brain while large striations of white matter can be seen in the surrounding areas. Figure 8b shows images corresponding to the total ion signal and characteristic peaks commonly observed in tissue imaging using TOF-SIMS. Cholesterol, $[M + H - H_2O]^+$, dipalmitoyl phosphatidylcholine (DPPC), $[M + H]^+$ and the ubiquitous phosphocholine head group peak (m/z 184). Three SIMS images were acquired, each to an ion beam fluence of 1×10^{13} ions/cm^2, displayed from left to right. The *static limit* on such samples would be expected to be in the range of $1-5 \times 10^{12}$ ions/cm^2 so the 1st image in Fig. 8b can be used to illustrate the benefits of polyatomic ion beam analysis to increased ion beam fluence. Even on a simple grey scale the contrast in the individual ion images is clear and the maximum counts per pixel are 1,842, 403 and 1,250 for cholesterol, DPPC and PC respectively. The benefits of increased ion beam fluence are clear, the dynamic range of the signal has been extended and contrast has been improved.

Again the use of the polyatomic ion beam allows chemical changes to be mapped as a function of depth and clear transition between the surface and subsurface can

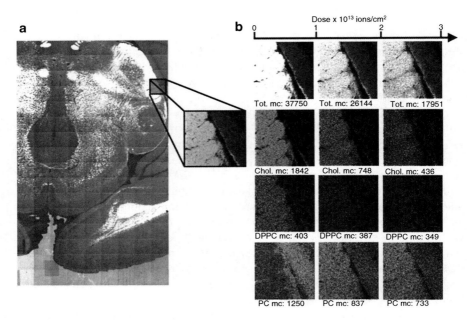

Fig. 8 Imaging of rat brain using 40 keV $C_{60}{}^{+}$ on the J105 3D Chemical Imager. Large area (5.4×9.6 mm², $576 \times 1,024$ pixels) image of total ion signal clearly illustrates structure of the brain with particularly high signal levels originating from the white matter rich regions, (**a**) Variation of signal from characteristic peaks at the white/grey matter boundary are imaged as a function of increase in ion beam fluence (600×600 μm², 256×256 pixels), (**b**) Total ion counts (Tot.), cholesterol (Chol.), dipalmitoyl phosphatidylcholine (DPPC) and the phosphocholine head group (PC). Maximum counts per pixel (mc) are also displayed

be seen particularly in the m/z 184 species that is initially most intense on the right, grey matter, side of the image but upon excavation of the sample becomes predominantly localized to the left, white matter, side of the image.

4   Conclusions

TOF-SIMS is becoming increasingly applicable to biological analysis.

Molecular SIMS has struggled in the past due to limitations arising from ion beam induced damage. However, with the introduction of cluster ion beams such as C_{60} the obtainable signal has increased and 3D imaging has become a new avenue for analysis. Many small molecules including pharmaceuticals are stable to bombardment with C_{60} and thus present an exciting target for TOF-SIMS analysis in cells and also tissue sections. Recent developments in instrumentation to fully exploit the properties of these ion beams may allow the potential to be realized.

References

1. Sostarecz AG, McQuaw CM, Ewing AG, Winograd N (2004) Phosphatidylethanolamine-induced cholesterol domains chemically identified with mass spectrometric imaging. J Am Chem Soc 126(43):13882
2. Biesinger MC, Miller DJ, Harbottle RR, Possmayer F, McIntrye SN, Peterson NO (2006) maging lipid distributions in model monolayers by ToF-SIMS with selectively deuterated components and principal components analysis. Appl Surf Sci 252(19):6957
3. Baker MJ, Zheng L, Winograd N, Lockyer NP, Vickerman JC (2008) Mass spectral imaging of glycophospholipids, cholesterol, and glycophorin A in model cell membranes. Langmuir 24(20):11803
4. Ostrowski SG, Van Bell CT, Winograd N, Ewing AG (2004) Mass spectrometric imaging of highly curved membranes during Tetrahymena mating. Science 305:71
5. Nygren H, Malmberg P (2004) Silver deposition on freeze-dried cells allows subcellular localization of cholesterol with imaging TOF-SIMS. J Microsc 215:156
6. McDonnell LA, Piersma SR, Alterlaar AFM, Mize TH, Luxembourg SL, Verhaert PDEM, Van Minnen J, Heeren RMA (2005) Subcellular imaging mass spectrometry of brain tissue. J Mass Spectrom 40:160
7. Appelhans AD, Delmore JE (1989) Comparison of polyatomic and atomic primary beams for secondary ion mass-spectrometry of organics. Anal Chem 61(10):1087
8. Kotter F, Benninghoven A (1998) Secondary ion emission from polymer surfaces under Ar^+, Xe^+ and SF_5^+ ion bombardment. Appl Surf Sci 133:47
9. Gillen G (2000) Microbeam analysis. Inst Phys Conf Ser 165:339
10. Wong SCC, Hill R, Blenkinsopp P, Lockyer NP, Weibel DE, Vickerman JC (2003) Development of a C_{60}^+ ion gun for static SIMS and chemical imaging. Appl Surf Sci 203–204:219
11. Weibel DE, Wong SCC, Lockyer NP, Hill R, Blenkinsopp P, Vickerman JC (2003) A C60 primary ion beam system for time of flight secondary ion mass spectrometry: Its development and secondary ion yield characteristics. Anal Chem 75:1754
12. Garrison BJ, Postawa Z (2008) Computational view of surface based organic mass spectrometry. Mass Spectrom Rev 27:289
13. Vaidyanathan S, Fletcher JS, Goodacre R, Lockyer NP, Micklefield J, Vickerman JC (1942) Subsurface biomolecular imaging of streptomycescoelicolor using secondary ion mass spectrometry. Anal Chem 2008:80
14. Chandra S (2004) 3D subcellular SIMS imaging in cryogenically prepared single cells. Appl Surf Sci 231–232:467
15. Fletcher JS, Lockyer NP, Vaidyanathan S, Vickerman JC (2007) TOF-SIMS 3D biomolecular imaging of Xenopus laevis oocytes using buckminsterfullerene (C_{60}) primary ions. Anal Chem 79:2199
16. Breitenstein D, Rommel CE, Mollers R, Wegener J, Hagenhoff B (2007) The chemical composition of animal cells and their intracellular compartments reconstructed from 3D mass spectrometry. Angew Chem Int Ed 46:5332
17. Carado A, Passarelli MK, Kozole J, Wingate JE, Winograd N, Loboda AV (2008) C-60 secondary ion mass spectrometry with a hybrid-quadrupole orthogonal time-of-flight mass spectrometer. Anal Chem 80(21):7921
18. Fletcher JS, Rabbani S, Henderson A, Blenkinsopp P, Thompson SP, Lockyer NP, Vickerman JC (2008) A new dynamic in mass spectral imaging of single biological cells. Anal Chem 80:9058
19. Sjovall P, Lausmaa J, Johansson B (2004) Mass spectrometric imaging of lipids in brain tissue. Anal Chem 76:4271
20. Amaya AR, Monroe EB, Sweedler JV, Clayton DF (2007) Lipid imaging in the zebra finch brain with secondary ion mass spectrometry. Int J Mass Spectrom 260:121
21. Brunelle A, Touboul D, Laprevote O (2005) Biological tissue imaging with time-of-flight secondary ion mass spectrometry and cluster ion sources. J Mass Spectrom 40:985

Application of Mass Spectrometry for the Analysis of Vitellogenin, a Unique Biomarker for Xenobiotic Compounds

Alejandro M. Cohen and Joseph H. Banoub

Abstract Vitellogenin is a complex phosphoglycolipoprotein that is secreted into the bloodstream of sexually mature, female, oviparous animals in response to circulating estrogens. It is then incorporated into the ovaries by receptor mediated endocytosis, where it is further cleaved to form the major constituents of the egg yolk proteins. It is generally accepted that these protein and peptide products serve as the main nutritional reserve for the developing embryo. Quantification of vitellogenin in blood is useful for different purposes. The reproductive status and degree of sexual maturation of oviparous animals can be assessed according to the levels of vitellogenin in plasma. The expression of this protein can also be induced in males under the effect of estrogenic compounds. Relying on this observation, vitellogenin has been used as a unique biomarker of environmental endocrine disruption in many species. In this respect, vitellogenin levels could potentially be used to assess the use of chemical warefare compounds with estrogenic activity. In this paper we review a technique developed for measuring vitellogenin plasma levels of different fish species using high performance liquid chromatography coupled to tandem mass spectrometry.

Keywords Vitellogenin • Biomarker • MALDI • ESI • Tandem mass spectrometry

A.M. Cohen
Environmental Sciences Division, Special Projects, Fisheries and Oceans Canada,
Science Branch, St John's, Newfoundland, A1C 5X1, Canada

J.H. Banoub (✉)
Department of Chemistry, Memorial University of Newfoundland, St. John's, Newfoundland
A1B 3V6, Canada
e-mail: banoubjo@dfo-mpo.gc.ca

J. Banoub (ed.), *Detection of Biological Agents for the Prevention of Bioterrorism*,
NATO Science for Peace and Security Series A: Chemistry and Biology,
DOI 10.1007/978-90-481-9815-3_19, © Springer Science+Business Media B.V. 2011

1   Introduction

The term vitellogenin (Vtg) was first coined in Pan *et al.* in 1969,[1] who refered to Vtg as the female specific proteins found in the blood of mature female insects. In this work, the female fat body of two insect species incorporated radio-labeled amino acids *in vitro* into substances which were precipitated by antibodies formed in response to the vitellogenic blood proteins. This was not observed in the fat bodies of males or even females before the appearance of the Vtgs in blood.[1] According to literature, vitellogenesis is defined as the estradiol-induced hepatic synthesis of Vtg, its secretion and transport in blood to the ovary, and its uptake into maturing oocytes.[2] This definition implies a series of sequential complex biochemical processes which are summarized and illustrated in Fig. 1.

In fish, the principal external factors that affect the endocrine regulation of the oocyte maturation are probably water temperature and the photoperiod[3] cycles, which trigger in the hypothalamus the release of the gonadotropin-releasing hormone. In response to these the pituitary gland secretes the gonadotropin hormones (GtH I and GtH II), which are structurally related to the follicle-stimulating hormone (FSH) and the luteinizing hormone (LH) found in humans. GtH I (FSH) is involved in vitellogenesis and zonagenesis while GtH II (LH) is related to oocyte maturation and ovulation.

GtH I ultimately induces in the follicular cells the secretion of 17β-estradiol into the bloodstream. Testosterone is initially synthesized by the theca cells and is further transformed to 17β-estradiol in the granulose cells by the action of cytochrome P450 aromatase (CYP19). Once exported to the blood, estradiol is transported to other organs by sex-hormone binding proteins, and is incorporated into the liver. Once inside the hepatocyte, estradiol initiates a long chain of biochemical events which finally results in the activation or enhanced transcription of Vtg. Once exported from the liver, Vtg circulates in the bloodstream and reaches the ovaries, where it is incorporated into the growing oocyte through receptor mediated endocytosis. An extensive list of comprehensive reviews on Vtg and vitellogenesis can be found in the literature.[4–6]

1.1   Vitellogenin as an Indicator of Sexual and Reproductive Maturation

Estimation of the sexual maturation in different oviparous animals is very important to assess their reproductive status.[7–10] Observation of increased gonad size is an obvious way to determine sexual status, although macroscopic changes are usually the last indicators. Microscopic histopathology is an earlier indicator, although both imply destructive techniques. More recently, invasive but non-destructive techniques have been proposed. The first biochemical biomarkers used for such purpose were serum calcium, total phosphoprotein phosphorus and alkali-labile phosphoprotein

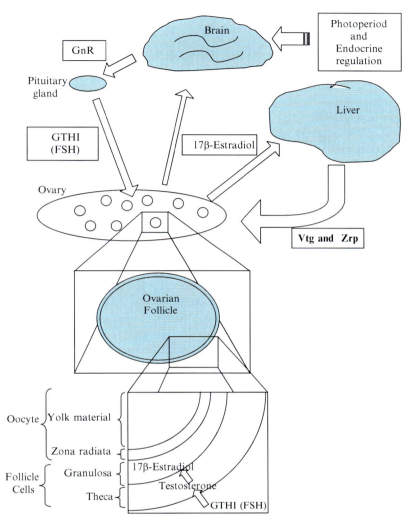

Fig. 1 Scheme of the hypothalamic-pituitary-gonadal-liver axis

phosphosphorus. All these have been shown to increase during vitellogenesis.[11] Other studies additionally propose serum Mg levels to increase during this stage.[12] All these parameters can be used as indicators of plasma Vtg levels.[6] Alternatively, the hormones of the hypothalamic-pituitary-gonadal-liver axis can be measured. However, direct measurement of plasma Vtg levels seems to be the most appropriate indicator since it is the last in a series of biochemical pathways leading to sexual maturation. Currently we are studying the relation between Vtg levels in Greenland halibut, a very poorly studied species, and other indicators of reproductive maturation. Preliminary results have shown a significant correlation between these parameters.[13]

1.2   Vitellogenin as a Biomarker

Vitellogenin synthesis is induced by 17β-estradiol, the physiological ligand of the estradiol receptor in the liver of vertebrate oviparous animals. In invertebrates, this synthesis can occur in analogue tissues or organs such as the fat body of insects or the hepatopancreas of crustaceans. There are, however, other molecules that can interact with the estradiol receptor too, producing an agonistic effect on Vtg synthesis. These compounds are given the generic name of "xenoestrogens".[14] The basic meaning of this word refers to any chemical foreign to that particular organism, which will have an estrogenic effect. Xenoestrogens are one of the many types of endocrine disruptors or modulators.[15, 16] These compounds affect not only wildlife, but also humans, as discussed in some publications.[17, 18] The xenoestrogens originate from diverse sources, and can be classified on different basis. Some occur naturally in the environment, such as the phytoestrogens.[19–21] Others are of anthropogenic nature, chemically synthesized by humans and released into the environment by agricultural and industrial activities. It is common to find in literature other names that refer to this type of compounds, namely, environmental estrogens, eco-estrogens, xenobiotic estrogens, environmental hormones, or hormone related toxicants, among others. These compounds are agents which could potentially be used in chemical warfare. Developing analytical assays towards all of these chemicals is challenging, however, a specific and sensitive method for detecting Vtg levels in target organisms could be used as an indirect method for monitoring the presence of any of these compounds in the environment.

2   Quantification of Vitellogenin

Over the last 2 decades, a wide spectrum of analytical methods has been developed to meet this purpose. The first indirect estimates of Vtg were performed colorimetrically by determining the alkaline-labile phosphorous content of fish plasma.[22, 23] Other techniques have also been used, such as immunoagglutination, densitometry following electrophoresis,[24] immunobloting (e.g. western blot),[25] radial immunodiffusion[26] and by liver tissue slice immunoassays.[27]

Synthesis of Vtg in the liver can also be estimated by quantification of Vtg mRNA levels in liver tissues and different approaches are possible for this task. mRNA can be qualitatively assayed by direct northern blotting while more sophisticated methods such as reverse transcriptase PCR and quantitative PCR provide more accurate results.[28–30] It was not until the development of radioimmunoassays[31–35] and enzyme-linked immunosorbent assays (ELISA)[36–40] that plasma or serum Vtg levels were first accurately and rapidly measured. Both techniques have had a wide acceptance in the last decades, mainly due to the very low limits of detection (LLOD) they have attained, in the order of nanograms per milliliter, and their high reproducibility. In spite of the many advantages offered by immunoassays, certain issues need to be considered before starting to develop a new technique. The first step in the process is obtaining a highly purified homogenous standard. In the case of Vtg, extra precautions should be taken since this is a thermo-labile and susceptible to the action of proteases. Purification of this protein usually yields degraded or aggregated species.[41–44] Some researchers have avoided these problems by raising the antibodies against purified yolk proteins.[5, 33, 45] The purification of this high quality standard is critical for the following step, the production of polyclonal or monoclonal antibodies. This is probably the most labor intensive and important step. Increased non-specific binding and background noise are the consequence of poor protein purification.

In this current work, we review a different approach for quantification of Vtg of different fish species using mass spectrometry. For this purpose, a diagnostic proteolytic peptide, the signature peptide, was chosen as an analytical surrogate of the precursor protein and monitored by HPLC-ESI-MS/MS for quantification.[13, 46–49] The approach explored in this work offers an alternative to developing immunoassays, especially when no specific antibodies or commercial kits are available for a particular species. The 'signature peptide' approach circumvents the need of protein purification and antibody production, two time consuming and costly steps required for immunoassay development.[50–52]

2.1   Characterization of Vitellogenin Tryptic Derived Peptides Using MALDI-MS

Trypsin in-gel digestion of Vtg was performed after running control (male or non-treated) and experimental (17B-estradiol induced) serum samples of fish plasma on SDS-PAGE. We have performed these analyses on a variety of species such as rainbow trout (*Oncorhynchus mykiss*), Atlantic salmon (*Salmo salar*), Atlantic cod (*Gadus morhua*) and Greenland halibut (*Reinhardtius hippoglossoides*).[53–55] A typical coomassie blue stained gel of such run for Atlantic cod is shown on Fig. 2.

The extracted tryptic peptides were analyzed by MALDI-MS, to obtain the characteristic peptide mass fingerprint for each species. A characteristic spectrum can be observed in Fig. 3.

The peak lists of all these experiment confirmed the nature of the protein studied, and often showed common tryptic peptides to Vtg of other species. Interestingly, in the case of Greenland halibut, the Vtg sequence was unknown at the time of analysis, however the peptide mass fingerprint confirmed the identity of the protein by comparison of the tryptic peptides of related fish species, such as the Barfin flounder (*Verasper moseri*), the European plaice (*pleuronectes platessa*) and the Atlantic halibut (*Hippoglossus hipoglossus)* as shown in Fig. 4 by the results of a database search using Mascot search engine.[56]

Fig. 2 Plasma from control
(C) and experimental (E) fish
were subjected to SDS-PAGE
and Coomassie blue staining.
Five μL of plasma
were loaded into each
lane and electrophorosed
in the conditions described in
Section 3.1.2. The banding
profile indicates the increased
synthesis of Vtg in the fish
treated with β-estradiol

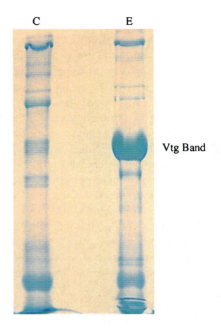

Vtg Band

2.2 Characterization of Vitellogenin Tryptic Derived Peptides Using ESI-MS/MS

Vtg is a relatively large protein, ranging from 170 to 400 KDa, depending on the species. The proteolytic digestion of such protein generates a large number of peptides, some of which result isobaric according to the mass resolution of certain mass spectrometers. For example, Table 1 shows for eight pairs of isobaric tryptic peptides resulting from the digestion of rainbow trout Vtg. For this reason, the tryptic peptides were all analyzed by HPLC-ESI-MS/MS using reverse phase chromatographic separations (30 × 2 mm Gemini 5μm C18 110Å column, Phenomenex, Torrance CA, USA). The column was interfaced in-line to a QqToF (QStar XL, Applied Biosystems, Foster City CA, USA) tandem mass spectrometer equipped with an ESI ion source (TurboIonSpray).

Acquisition of product ion spectra was achieved running an Information Dependent Acquisition (IDA) mode selecting for double charged molecular ions $[M + 2H]^{2+}$ between m/z 400 and 1,000. The product ion spectra were initially used for MS/MS Ion Search on the Mascot search engine. Again, the results of these searches confirmed the identity of Vtg in the studies gel piece. The highest quality spectra were then manually inspected to confirm the sequence by the presence of the diagnostic B and Y ions. 'De-novo' sequencing was also performed for those species where no sequence was available in the protein data bases. An example is portrayed in Fig. 5a, where the product ion spectra together with its fragmentation scheme are shown for a rainbow trout tryptic peptide. Similarly, a list of candidate signature peptides was built for each one of our studied fish species.

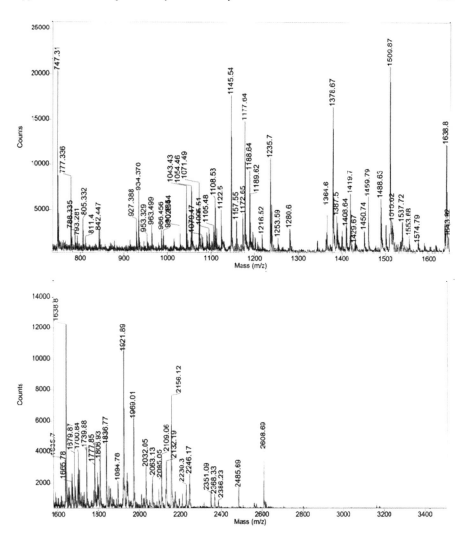

Fig. 3 MALDI-MS spectra of the trypsin digestion of purified RT Vtg. The abscissa of the spectrum was split into two m/z ranges to facilitate the visualization of the peaks. The *top* and *bottom* spectra show the m/z ranges from 700 to 1,650 and 1,600 to 3,500, respectively. The MALDI-MS was operated in reflectron mode using DHB as matrix

2.3   Selection of Signature Peptides

The first step necessary to develop a quantitative technique using the signature peptide approach is to select suitable candidate peptide.[46] This selection depends on theoretical, empirical and technical issues. From the theoretical point of view, sequences containing amino acid residues prone to oxidation during sample

{MATRIX} *{SCIENCE}* **Mascot Search Results**

```
User          : Alejandro Cohen
Email         : ale.cohen1@gmail.com
Search title  :
Database      : NCBInr 20091024 (9937670 sequences; 3390302707 residues)
Taxonomy      : Chordata (vertebrates and relatives) (1098836 sequences)
Timestamp     : 26 Oct 2009 at 15:51:23 GMT
Top Score     : 132 for gi|62241080, vitellogenin [Verasper moseri]
```

Probability Based Mowse Score

Protein score is -10*Log(P), where P is the probability that the observed match is a random event.
Protein scores greater than 73 are significant (p<0.05).

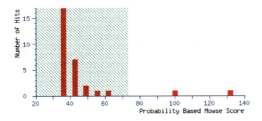

Concise Protein Summary Report

Format As   [Concise Protein Summary]   Help

Significance threshold p< 0.05   Max. number of hits AUTO

[Re-Search All]   [Search Unmatched]

1.   gi|62241080   Mass: 182217   Score: 132   Expect: 6.9e-08   Queries matched: 20
      vitellogenin [Verasper moseri]
      gi|16151381   Mass: 54415   Score: 36   Expect: 2.6e+02   Queries matched: 6
      vitellogenin [Pleuronectes platessa]

2.   gi|148299214   Mass: 181329   Score: 101   Expect: 8.7e-05   Queries matched: 17
      vitellogenin [Hippoglossus hippoglossus]

Fig. 4 Peptide mass fingerprint results for the 'in-gel' tryptic digest of a serum sample of Greenland halibut

(methionine or cysteine) were avoided. Also, all candidate sequences were checked to confirm the absence of any consensus sites for known post translational modifications. From an empirical point of view, those peptides providing the strongest signals on the ESI-MS/MS experiment were preferred over the others, since from a quantitative point of view, stronger signal is translated into higher sensitivity and low limits of detection. Taking this into consideration, the peptides most susceptible to CID fragmentation which provided the most intense high-mass fragment ions were selected as most suitable candidates. From a technical point of view, the candidate peptide should meet many of the design considerations for peptide synthesis, especially with respect to the very hydrophobic peptides. Furthermore, an isotopically labeled peptide should be also synthesized. In this respect, the residues and positions

Table 1 Isobaric peptides found in the tryptic digestion of RT Vtg. The following peaks identified by MALDI-MS matched isobaric tryptic peptides predicted by the simulated digestion of RT Vtg

Sequence	Formula	Theoretical $[M + H]^+$ m/z	Calculated $[M + H]^+$ m/z	$\Delta m/z$
TLDVILK	$C_{37}H_{68}N_8O_{11}$	801.51	801.36	−0.15
AGVKVISK	$C_{36}H_{68}N_{10}O_{10}$	801.52		−0.16
QVIVDDR	$C_{35}H_{61}N_{11}O_{13}$	844.45	844.29	−0.16
DPFVPAAK	$C_{40}H_{61}N_9O_{11}$	844.46		−0.17
DSQSTSNVISRSK	$C_{55}H_{97}N_{19}O_{24}$	1,408.70	1,408.43	−0.27
FAAQLDIANGNFK	$C_{64}H_{97}N_{17}O_{19}$	1,408.72		0.01
IHYLFSEVNAVK	$C_{67}H_{102}N_{16}O_{18}$	1,419.76	1,419.49	−0.27
DLNNCQQRIMK	$C_{56}H_{98}N_{20}O_{19}S_2$	1,419.68		−0.19
DASECLMKLESVK	$C_{62}H_{108}N_{16}O_{23}S_2$	1,509.73	1,509.61	−0.12
LLPVFGTAAAALPLR	$C_{72}H_{120}N_{18}O_{17}$	1,509.92		−0.31
HSVLISVKPSASEPAIER	$C_{84}H_{142}N_{24}O_{27}$	1,920.06	1,919.69	−0.37
EPRMVQEVAVQLFMDK	$C_{84}H_{138}N_{22}O_{25}S_2$	1,919.97		−0.28
ILNHLVTYNTAPVHEDAPLK	$C_{102}H_{161}N_{27}O_{30}$	2,245.20	2,244.77	−0.43
NVEDVPAERITPLIPAQGVAR	$C_{98}H_{165}N_{29}O_{31}$	2,245.23		−0.46

to be labeled should be considered when selecting the signature peptide. A mass difference of at least 8 Da is necessary to avoid overlapped signals during the SRM experiments in the final quantitative procedure. Figure 5b shows the product ion spectrum for the isotopically labeled internal standard selected for rainbow trout Vtg. Similarly, we selected signature peptides for each of the studied species.

2.4 Quantification of Vitellogenin by SRM-MS/MS

Recent advances in mass spectrometry have provided researchers with powerful relative and absolute quantitative techniques. The rationale behind the quantification of proteins using signature peptides is that peptides generated by the digestion of complex mixtures may be used as analytical surrogates for the protein from which they are derived.[51] This approach is based on the fact that single peptides are in many cases easier to separate and identify than the intact proteins. Furthermore, the synthesis of custom made peptide standards and internal standards is easier than the isolation and purification of the intact proteins for calibration purposes. The internal standards usually consist of homologues of the analyzed peptides in which one or more atoms have been replaced by stable isotopes (e.g. ^{13}C, ^{15}N, ^2H etc.). These peptides are spiked into each one of the studies blood samples prior to analysis. A diagram of the work scheme is summarized in Fig. 6.

In our studies, the characteristic 'signature' proteolytic peptides were monitored by high performance liquid chromatography (HPLC) coupled to electrospray ionization tandem mass spectrometers. Quantification by mass spectrometry has commonly been performed on triple quadrupole (QqQ) tandem mass spectrometers.

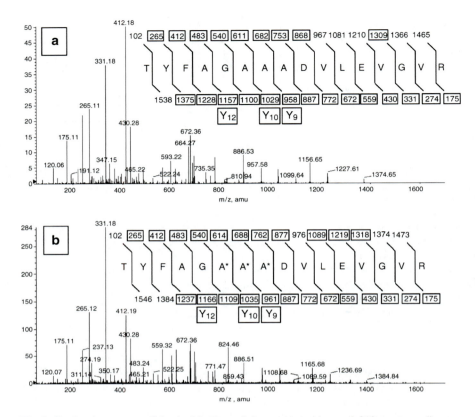

Fig. 5 Product ion spectra of the doubly charged signature peptide at *m/z* 819 corresponding to a rainbow trout Vtg tryptic peptide (Fig. 6a). Inset show the expected b and y-type peptide ions. The product ions observed in the spectra are *highlighted with boxes*. The Y_9, Y_{10} and Y_{12} diagnostic ions used for identification and quantification purposes during the SRM-MS/MS experiments. Figure 6b shows the product ion spectra of the corresponding deuterated isotopic homologue at *m/z* 824.4. A* stands for L-Alanine-3,3,3-D$_3$

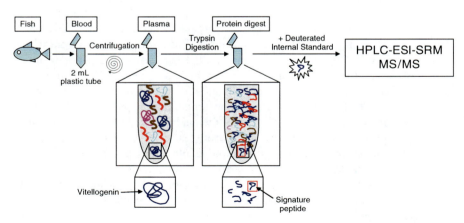

Fig. 6 Scheme of sample processing and analysis for the quantification of vtg using the signature peptides approach couples to LC-ESI-MSMS

These instruments have shown exceptional results when operated in Selected Reaction Monitoring (SRM) mode. Alternatively, we have also explored quantification using the hybrid QqToF-MS/MS instrument. This method, developed by adjusting and optimizing the tuning parameters of the QqToF analyzer was termed as 'pseudo' SRM. This CID-MS/MS method was programmed to operate in a mode that monitors the products of both the 'signature peptide' and its deuterated isotopic homologue, used as an internal standard. Three transitions were selected for each of the chosen signature peptides, and a SRM method was designed and optimized for each one. Parameters such as declustering potentials, collision energies, collision gas etc were set to values that would maximize signal to noise ratios. Figure 7 shows the exceptional specificity and sensitivity of this technique, illustrated by the low signal to noise chromatograms generated by the SRM acquisition of a real plasma digest of a rainbow trout sample.

2.5 Analyses of Serum Samples

Once all of the experimental parameters were adjusted, calibration curves were constructed for each of the product ions monitored. Limits of quantification (LOQ) were calculated, and these often translated into values in the pictogram per milliliter order. The digested plasma samples were then run under the same conditions as the standards. Figure 8 shows of a Rainbow trout sample prior to (Fig. 8a) and after β-estradiol induction (Fig. 8b). These figures show the overlaid extracted ion chromatograms of the Y_9, Y_{10} and Y_{12} product ions of both signature peptide and labeled internal standard at m/z 957.5, 1,028.57, 1,156.6 and m/z 960.5, 1,034.61, 1,165.7, respectively. The traces obtained from the 'signature peptide' before β-estradiol induction were hardly distinguishable from the baseline. The traces corresponding to the internal standard formed a clear peak in both samples. The inset shows the expanded scale of these traces, confirming the absence of Vtg before treatment (Fig. 8a).

This technique was then applied to a set of samples obtained from a group of rainbow trouts and Atlantic salmon blood specimens obtained before and after the induction with β-estradiol (Fig. 9). The presence of the signature peptide was confirmed by the presence of all three product-ions (Y_9, Y_{10} and Y_{12}) in the SRM analyses at the corresponding elution time. As expected, most of the juvenile Rainbow trout had non-detectable plasma levels of Vtg before β-estradiol induction, except for a couple of larger female trout which were probably at stages of early sexual maturation. As for the Atlantic salmon, two samples were confirmed as females, after the animals were killed and their ovaries found filled with eggs. This would explain the high levels of Vtg found in some of the fish before β-estradiol induction, characteristic of the vitellogenic reproductive phase. A t-test for paired samples was applied to each one of the groups (control Rainbow Trout, experimental Rainbow trout, control Atlantic salmon and experimental Atlantic salmon). As expected, a statistically significant ($P \ll 0.001$) increase in Vtg plasma concentration was established

a

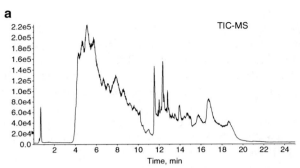

b

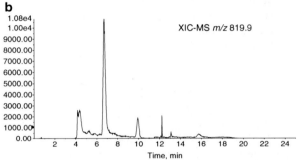

c

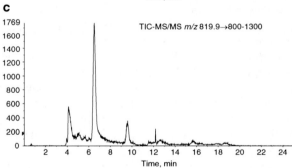

d

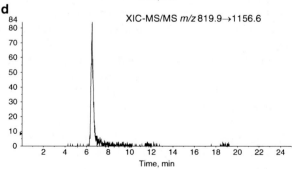

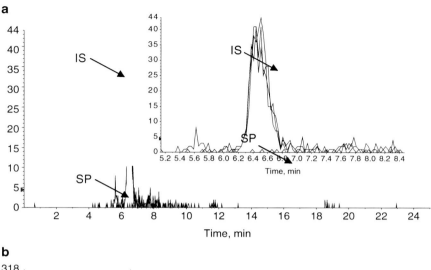

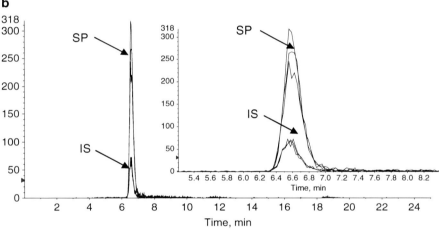

Fig. 8 Detection of the 'signature peptide' by HPLC-MS/MS The overlaid EICs obtained from the 'pseudo' SRM-MS/MS analysis for the same fish prior to (**a**) and following β-estradiol injection (**b**) are shown. The absence of a signal for the signature peptide (SP) is observed in the control sample (**a**). The trace of the internal standard (IS) is similar in both samples. The *inset* figures show an expanded scale of the peaks eluting at 6.5 min. In these insets, the overlaid signals produced by the Y_9, Y_{10} and Y_{12} product ions appear as overlapped traces

---

Fig. 7 A plasma digested sample from Rainbow trout was analyzed in both HPLC-QqToF-MS (Fig.8a and b) and MS/MS (Fig 8c and d) modes for comparison purposes. Figure 7a shows the complex TIC chromatogram acquired in full scan mode. Figure 3b shows the extracted ion chromatogram (XIC) from Fig. 8a for the ions at *m/z* 819.9. Figure 7c shows the TIC chromatogram obtained in the 'pseudo' SRM scan of the *m/z* 819.9 precursor. Figure 3D shows the superimposed XICs obtained from the Y_9, Y_{10} and Y_{12} diagnostic ions. The *inset* shows an expanded scale of the peak showing the traces obtained from each product

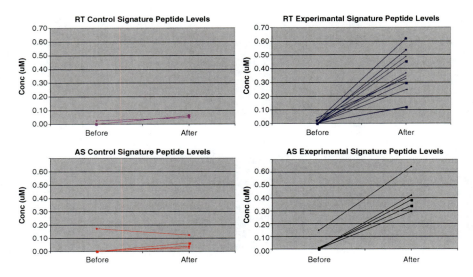

Fig. 9 Result obtained from analyzing Vtg of Atlantic salmon (AS) and rainbow trout (TR) blood samples using the signature peptide approach (HPLC-LC-SRM MS/MS) both before and after induction with β-estradiol

for all experimental fish, whereas no significant increase was observed for the control fish. The trends of Vtg concentration both before and after β-estradiol induction can be observed in Fig. 9.

3 Conclusions

Vitellogenin, the egg yolk precursor protein, has become a popular biomarker for measuring exposure of oviparous animals to estrogen or estrogen mimics. Vitellogenin is normally produced by females in response to normal cycles of estradiol during oogenesis. The gene for Vtg is also present in the liver of males but it is normally silent. Upon exposure to xenobiotic chemical possessing estrogen mimic potential, transcription of the Vtg gene is upregulated, leading to Vtg synthesis. After synthesis, it is exported into the blood where, in males, it remains until it is degraded or cleared out by the kidneys. In females, Vtg is taken up by the developing oocyte through receptor mediated endocytosis.

There is increasing concern about man-made synthetic chemicals or xenobiotic compounds; explosives such as nitroaromatics, nitroesters, and sulfonated aromatic compounds that can mimic oestrogens have been found in aquatic environments. This concern is due to the tendency of the xenobiotic chemicals to disrupt the fish reproductive function. Vitellogenin, a precursor of egg-yolk in fish and other oviparous animals, may be used as a biomarker for xenobiotic chemical which can mimic oestrogens exposure.

There are several assays in the literature for measuring Vtg levels in plasma. The easiest method is through antibody based assays including ELISA or by western blot. Competition or sandwich ELISAs are the most sensitive assays and they can detect Vtg in plasma in the nanogram to milligram per milliliter range.

In this rationale, we have presented the work developed in our laboratories for measuring the serum levels of Rainbow trout, Atlantic salmon, Atlantic cod, Atlantic haddock and Greenland halibut Vtgs by the signature peptide approach, using liquid chromatography electrospray ionization tandem mass spectrometric technique.

Currently, this technique is being used for larger scale field surveys and studies on Greenland halibut from the Gulf of St. Lawrence kept in experimental tanks to test the accuracy of macroscopically assigning female maturity status (i.e. immature vs mature) and to follow the annual female reproductive cycle. Together, these results should help better understand the reproductive biology of this deepwater species. This work further proves mass spectrometry to be an emerging tool not only in basic research, but also in applied biological studies. The reproducibility (coefficient of variation *ca.* 5%) and sensitivity (limit of quantification (LOQ) of 0.009 mg/ml) achieved by this simple assay allow it to be considered as an alternative more accurate and sensible method to immunological assays such as ELISA which uses monoclonal or polyclonal antibodies developed for different Vtg antigens extracted from fish species than are usually dissimilar from the fish species that is usually measured.

References

1. Pan ML, Bell WJ, Telfer WH (1969) Vitellogenic blood protein synthesis by insect fat body. Science 165:393–394
2. Tyler CR, Lubberink K (1996) Identification of four ovarian receptor proteins that bind vitellogenin but not other homologous plasma lipoproteins in the rainbow trout, Oncorhynchus mykiss. J Comp Physiol 166(1):11–20
3. Dey R, Bhattacharya S, Maitra SK (2005) Importance of photoperiods in the regulation of ovarian activities in Indian major carp Catla catla in an annual cycle. J Biol Rhythms 20(2):145–158
4. Tata JR, Smith DF (1979) Vitellogenesis: a versatile model for hormonal regulation of gene expression. Recent Prog Horm Res 35:47–95
5. Kunkel JG, Nordin JH, Gilbert LI, Miller TA (1986) "Yolk proteins", comprehensive insect physiology biochemistry pharmacology. Pergamon Press, London, pp 83–111
6. Arukwe A, Goksoyr A (2003) Eggshell and egg yolk proteins in fish: hepatic proteins for the next generation: oogenetic, population, and evolutionary implications of endocrine disruption. Comp Hepatol 2(1):4
7. Copeland PA, Sumpter JP, Walker TK, Croft M (1986) Vitellogenin levels in male and female rainbow trout (Salmo gairdneri richardson) at various stages of the reproductive cycle. Comp Biochem Physiol B 83(2):487–493
8. Kambysellis MP, Hatzopoulos P, Craddock EM (1989) The temporal pattern of vitellogenin synthesis in Drosophila grimshawi. J Exp Zool 251(3):339–348
9. Wallaert C, Babin PJ (1994) Age-related, sex-related, and seasonal changes of plasma lipoprotein concentrations in trout. J Lipid Res 35(9):1619–1633

10. Pavlidis M, Greenwood L, Mourot B, Kokkari C, Le Menn F, Divanach P, Scott AP (2000) Seasonal variations and maturity stages in relation to differences in serum levels of gonadal steroids, vitellogenin, and thyroid hormones in the common dentex (dentex dentex). Gen Comp Endocrinol 118(1):14–25

11. Nagler JJ, Ruby SM, Idler DR, So YP (1987) Serum phosphoprotein phosphorus and calcium levels as reproductive indicators of vitellogenin in highly vitellogenic mature female and estradiol-injected immature rainbow trout Salmo-gairdneri. Can J Zool 65(10): 2421–2425

12. Bjornsson BT, Haux C, Forlin L, Deftos LJ (1986) The involvement of calcitonin in the reproductive physiology of the rainbow trout. J Endocrinol 108(1):17–23

13. Cohen AM, Jahouh F, Sioud S, Rideout RM, Morgan MJ, Banoub JH (2009) Quantification of Greenland halibut serum vitellogenin: a trip from the deep sea to the mass spectrometer. Rapid Commun Mass Spectrom 23(7):1049–1060

14. Arukwe A, Celius T, Walther BT, Goksoyr A (2000) Effects of xenoestrogen treatment on zona radiata protein and vitellogenin expression in Atlantic salmon (Salmo salar). Aquat Toxicol (Amsterdam, Netherlands) 49(3):159–170

15. Bjerregaard P, Korsgaard B, Christiansen LB, Pedersen KL, Christensen LJ, Pedersen SN, Horn P (1998) Monitoring and risk assessment for endocrine disruptors in the aquatic environment: a biomarker approach. Arch Toxicol Suppl 20:97–107

16. Tyler CR, Van Aerle R, Hutchinson TH, Maddix S, Trip H (1999) An in vivo testing system for endocrine disruptors in fish early life stages using induction of vitellogenin. Environ Toxicol Chem 18(2):337–347

17. Toppari J (1996) Is semen quality declining? Andrologia 28(6):307–308

18. Damstra T (2002) Potential effects of certain persistent organic pollutants and endocrine disrupting chemicals on the health of children. J Toxicol 40(4):457–465

19. Pelissero C, Bennetau B, Babin P, Le Menn F, Dunogues J (1991) The estrogenic activity of certain phytoestrogens in the siberian sturgeon Acipenser baeri. J Steroid Biochem Mol Biol 38(3):293–299

20. Ishibashi H, Tachibana K, Tsuchimoto M, Soyano K, Tatarazako N, Matsumura N, Tomiyasu Y, Tominaga N, Arizono K (2004) Effects of nonylphenol and phytoestrogen-enriched diet on plasma vitellogenin, steroid hormone, hepatic cytochrome p450 1a, and glutathione-s-transferase values in goldfish (Carassius auratus). Comp Med 54(1):54–62

21. Mitsui N, Tooi O, Kawahara A (2007) Vitellogenin-inducing activities of natural, synthetic, and environmental estrogens in primary cultured Xenopus laevis hepatocytes. Comp Biochem Physiol C Toxicol Pharmacol 146(4):581–587

22. Wallace RA, Jared DW (1968) Studies on amphibian yolk. Vii. Serum phosphoprotein synthesis by vitellogenic females and estrogen-treated males of Xenopus laevis. Can J Biochem 46(8):953–959

23. Verslycke T, Vandenbergh GF, Versonnen B, Arijs K, Janssen CR (2002) Induction of vitellogenesis in 17alpha-ethinylestradiol-exposed rainbow trout (oncorhynchus mykiss): a method comparison. Comp Biochem Physiol C Toxicol Pharmacol 132(4):483–492

24. van Bohemen CG, Lambert JG (1981) Estrogen synthesis in relation to estrone, estradiol, and vitellogenin plasma levels during the reproductive cycle of the female rainbow trout, Salmo gairdneri. Gen Comp Endocrinol 45(1):105–114

25. Kwon HC, Hayashi S, Mugiya Y (1993) Vitellogenin induction by estradiol-17-beta in primary hepatocyte culture in the rainbow trout, Oncorhynchus mykiss. Comp Biochem Physiol B Comp Biochem 104(2):381–386

26. Hara A, Hirai H (1978) Comparative studies on immunochemical properties of female-specific serum protein and egg yolk proteins in rainbow trout (Salmo gairdneri). Comp Biochem Physiol B 59(4):339–343

27. Schmieder P, Tapper M, Linnum A, Denny J, Kolanczyk R, Johnson R (2000) Optimization of a precision-cut trout liver tissue slice assay as a screen for vitellogenin induction: comparison of slice incubation techniques. Aquat Toxicol 49(4):251–268

28. Celius T, Matthews JB, Giesy JP, Zacharewski TR (2000) Quantification of rainbow trout (Oncorhynchus mykiss) zona radiata and vitellogenin mRNA levels using real-time PCR after in vivo treatment with estradiol-17 beta or alpha-zearalenol. J Steroid Biochem Mol Biol 75(2–3):109–119

29. Lorenzen A, Casley WL, Moon TW (2001) A reverse transcription-polymerase chain reaction bioassay for avian vitellogenin mRNA. Toxicol Appl Pharmacol 176(3):169–180

30. Islinger M, Yuan H, Voelkl A, Braunbeck T (2002) Measurement of vitellogenin gene expression by rt-PCR as a tool to identify endocrine disruption in Japanese medaka (Oryzias latipes). Biomarkers 7(1):80–93

31. Redshaw MR, Follett BK (1976) Physiology of egg yolk production by the fowl: the measurement of circulating levels of vitellogenin employing a specific radioimmunoassay. Comp Biochem Physiol A Comp Physiol 55(4A):399–405

32. Gapp DA, Ho SM, Callard IP (1979) Plasma levels of vitellogenin in Chrysemys picta during the annual gonadal cycle: measurement by specific radioimmunoassay. Endocrinology 104(3):784–790

33. Campbell CM, Idler DR (1980) Characterization of an estradiol-induced protein from rainbow trout serum as vitellogenin by the composition and radioimmunological cross reactivity to ovarian yolk fractions. Biol Reprod 22(3):605–617

34. Asher C, Ramachandran J, Applebaum SW (1983) Determination of locust vitellogenin by radioimmunoassay with [3h]propionyl-vitellogenin. Gen Comp Endocrinol 52(2):207–213

35. Benfey TJ, Donaldson EM, Owen TG (1989) An homologous radioimmunoassay for coho salmon (Oncorhynchus kisutch) vitellogenin, with general applicability to other pacific salmonids. Gen Comp Endocrinol 75(1):78–82

36. Nunez Rodriguez J, Kah O, Geffard M, Le Menn F (1989) Enzyme-linked immunosorbent assay (ELISA) for sole (Solea vulgaris) vitellogenin. Comp Biochem Physiol – Part B: Biochem Physiol 92(4):741–746

37. Mourot B, Le Bail PY (1995) Enzyme-linked immunosorbent assay (ELISA) for rainbow trout (Oncorhynchus mykiss) vitellogenin. J Immunoassay 16(4):365–377

38. Lomax DP, Roubal WT, Moore JD, Johnson LL (1998) An enzyme-linked immunosorbent assay (ELISA) for measuring vitellogenin in English sole (Pleuronectes vetulus): development, validation and cross-reactivity with other pleuronectids. Comp Biochem Physiol B Biochem Mol Biol 121(4):425–436

39. Sherry J, Gamble A, Fielden M, Hodson P, Burnison B, Solomon K (1999) An ELISA for brown trout (Salmo trutta) vitellogenin and its use in bioassays for environmental estrogens. Sci Total Environ 225(1–2):13–31

40. Palumbo AJ, Koivunen M, Tjeerdema RS (2009) Optimization and validation of a California halibut environmental estrogen bioassay using a heterologous ELISA. Sci Total Environ 407(2):953–961

41. Kanungo J, Petrino TR, Wallace RA (1990) Oogenesis in Fundulus heteroclitus. VI. Establishment and verification of conditions for vitellogenin incorporation by oocytes in vitro. J Exp Zool 254(3):313–321

42. Tao Y, Hara A, Hodson RG, Woods LC III, Sullivan CV (1993) Purification, characterization and immunoassay of striped bass (Morone saxatilis) vitellogenin. Fish Physiol Biochem 12(1):31–46

43. Silversand C, Hyllner SJ, Haux C (1993) Isolation, immunochemical detection, and observations of the instability of vitellogenin from four teleosts. J Exp Zool 267(6):587–597

44. Specker JL, Sullivan CV, Davey KG, Peter RG, Tobe SS (1994) Vitellogenesis in fish: status and perspectives. In: Perspectives in comparative endocrinology. National Research Council of Canada, Ottawa, pp 304–315

45. Brown MA, Carne A, Chambers GK (1997) Purification, partial characterization and peptide sequences of vitellogenin from a reptile, the tuatara (Sphenodon punctatus). Comp Biochem Physiol B Biochem Mol Biol 117(2):159–168

46. Cohen AM, Mansour AA, Banoub JH (2006) Absolute quantification of Atlantic salmon and rainbow trout vitellogenin by the 'signature peptide' approach using electrospray ionization qqt of tandem mass spectrometry. J Mass Spectrom 41(5):646–658

47. Zhang F, Bartels MJ, Brodeur JC, Woodburn KB (2004) Quantitative measurement of fathead minnow vitellogenin by liquid chromatography combined with tandem mass spectrometry using a signature peptide of vitellogenin. Environ Toxicol Chem / SETAC 23(6):1408–1415

48. Wunschel D, Schultz I, Skillman A, Wahl K (2005) Method for detection and quantitation of fathead minnow vitellogenin (Vtg) by liquid chromatography and matrix-assisted laser desorption/ionization mass spectrometry. Aquat Toxicol (Amsterdam, Netherlands) 73(3):256–267

49. Palumbo AJ, Linares-Casenave J, Jewell W, Doroshov SI, Tjeerdema RS (2007) Induction and partial characterization of California halibut (Paralichthys californicus) vitellogenin. Comp Biochem Physiol 146(2):200–207

50. Geng M, Ji J, Regnier FE (2000) Signature-peptide approach to detecting proteins in complex mixtures. J Chromatogr A 870(1–2):295–313

51. Ji J, Chakraborty A, Geng M, Zhang X, Amini A, Bina M, Regnier F (2000) Strategy for qualitative and quantitative analysis in proteomics based on signature peptides. J Chromatogr B Biomed Sci Appl 745(1):197–210

52. Riggs L, Sioma C, Regnier FE (2001) Automated signature peptide approach for proteomics. J Chromatogr A 924(1–2):359–368

53. Banoub J, Thibault P, Mansour A, Cohen A, Heeley DH, Jackman D (2003) Characterisation of the intact rainbow trout vitellogenin protein and analysis of its derived tryptic and cyanogen bromide peptides by matrix-assisted laser desorption/ionisation time-of-flight-mass spectrometry and electrospray ionisation quadrupole/time-of-flight mass spectrometry. Eur J Mass Spectrom 9(5):509–524

54. Banoub J, Cohen A, Mansour A, Thibault P (2004) Characterization and de novo sequencing of Atlantic salmon vitellogenin protein by electrospray tandem and matrix-assisted laser desorption/ionization mass spectrometry. Eur J Mass Spectrom 10(1):121–134

55. Cohen AM, Mansour AA, Banoub JH (2005) 'De novo' sequencing of Atlantic cod vitellogenin tryptic peptides by matrix-assisted laser desorption/ionization quadrupole time-of-flight tandem mass spectrometry: similarities with haddock vitellogenin. Rapid Commun Mass Spectrom 19(17):2454–2460

56. Perkins DN, Pappin DJ, Creasy DM, Cottrell JS (1999) Probability-based protein identification by searching sequence databases using mass spectrometry data. Electrophoresis 20(18):3551–3567

Mass Spectrometry as a Powerful Analytical Technique for the Structural Characterization of Synthesized and Natural Products

Nour-Eddine Es-Safi, El Mokhtar Essassi, Mohamed Massoui, and Joseph Banoub

Abstract Mass spectrometry is an important tool for the identification and structural elucidation of natural and synthesized compounds. Its high sensitivity and the possibility of coupling liquid chromatography with mass spectrometry detection make it a technique of choice for the investigation of complex mixtures like raw natural extracts. The mass spectrometer is a universal detector that can achieve very high sensitivity and provide information on the molecular mass. More detailed information can be subsequently obtained by resorting to collision-induced dissociation tandem mass spectrometry (CID-MS/MS). In this review, the application of mass spectrometric techniques for the identification of natural and synthetic compounds is presented. The gas-phase fragmentation patterns of a series of four natural flavonoid glycosides, three synthesized benzodiazepines and two synthesized quinoxalinone derivatives were investigated using electrospray ionization mass spectrometry (ESI-MS) and tandem mass spectrometry techniques. Exact accurate masses were measured using a modorate resolution quadrupole orthogonal time-of-flight QqTOF-MS/MS hybrid mass spectrometer instrument. Confirmation of the molecular masses and the chemical structures of the studied compounds were achieved by exploring the gas-phase breakdown routes of the ionized molecules.

N.-E. Es-Safi
Laboratoire de Chimie Organique et d'Etudes Physico-Chimiques, Ecole Normale Supérieure, B.P 5118 Rabat, Morocco
e-mail: nouressafi@yahoo.fr

E.M. Essassi (✉)
Laboratoire de Chimie Organique Hétérocyclique, Université Mohammed V-Agdal, Rabat, Morocco
e-mail: emessassi@yahoo.fr

M. Massoui
Pôle de Compétences Pharmacochimie, BP 1014 Rabat, Morocco
e-mail: massoui@yahoo.fr

J. Banoub
Fisheries and Oceans Canada, Science Branch, Special Projects and Department of Chemistry, Memorial University of Newfoundland, St. John's Newfoundland, 1B 3V6, Canada
e-mail: banoubjo@dfo-mpo.gc.ca

J. Banoub (ed.), *Detection of Biological Agents for the Prevention of Bioterrorism*,
NATO Science for Peace and Security Series A: Chemistry and Biology,
DOI 10.1007/978-90-481-9815-3_20, © Springer Science+Business Media B.V. 2011

This was rationalized by conducting low-energy collision CID-MS/MS analyses (product ion- and precursor ion scans) using a conventional quadrupole hexapole-quadrupole (QhQ) tandem mass spectrometer.

Keywords ESI-MS • CID-MS/MS • Structural characterization • Flavonoids • Benzodiazepines • Quinoxalinones

1 Introduction

Structural elucidation of natural and synthesized compounds plays a major role in organic chemistry and is of a crucial importance in many fields of science. It is thus essential to identify relevant substances for the development of new drugs and for structure-activity relationship investigation and food quality control. Such identification should be achieved with the highest confidence possible because of the potential biological effect of the used substances. Research development in these areas is obviously dependent of those of structural elucidation techniques. This explains the importance of having access to rapid and reliable methods for the analysis and identification of compounds.

Nowadays, the elucidation of chemical structures involves the use of a combination of different techniques including NMR spectroscopy, mass spectrometry, ultraviolet and infrared spectrometry. Among all these techniques mass spectrometry(MS), due to its sensitivity, rapidity, and low levels of sample consumption, is one of the most important physicochemical methods applied to the structural determination of synthesized and natural products.[1–4] MS was first used to obtain molecular weights, but with the development of tandem MS it can also be used for the structural elucidation of compounds.[5–7] With the development of soft ionization techniques, it became possible to ionize polar molecules without thermal evaporation and mass spectrometry has become a powerful analytical tool of polar, non-volatile, and thermally labile classes of compounds.[8–11] Mass spectrometric methods such as electron ionization EI-MS,[12] fast atom bombardment FAB/MS,[13] atmospheric pressure chemical ionization APCI-MS[8,14,15] and electrospray ionization ESI-MS[16–19] have proved useful to characterize various types of synthesized products and natural compounds in herbs and other foods.[12,16,19–21] Modern soft ionization methods such as electrospray ionization (ESI) and matrix assisted laser desorption/ionization (MALDI) have been broadly used for exploring the structures of low and high molecular weight compounds containing polar functional groups.[22,23]

The coupling of liquid chromatography with APCI-MS and ESI-MS/MS techniques has been demonstrated to be a powerful tool to explore the chemical composition of complex natural and raw synthesis mixtures.[17,24–27] The combination of FAB ionization with collision-induced dissociation tandem mass spectrometric techniques has been shown to yield important structural information for the characterization of several natural and synthesized compounds.[13,28,29]

Beside the ionization source, the efficiency of a mass spectrometer depends also on the which analyzer, serving as an ion separation chamber, is used. In that field good results were obtained with quadrupole and time of flight (TOF) mass analyzers. The ability of a quadrupole–time-of-flight hybrid instrument (QTOF) to provide accurate mass determination of the product ions in MS/MS is yet another powerful tool in structure elucidation.[30] In a continuation of a program aimed at the isolation and identification of new molecules and biomolecules, we have used different MS techniques for the structural elucidation of various natural and synthesized compounds.[31–42] The aim of this review is to show the enormous potential of mass spectrometry in the structural elucidation of some natural and synthesized compounds. This will also shed light on our approach in identifying some natural and synthesized metabolites employing different MS techniques. For such purpose flavonoid glycosides were taken as example of natural products while 1,5-benzodiazepine and quinoxalinone derivatives were chosen as synthesized compounds.

2   Results and Discussion

2.1   Natural Flavonoids

Flavonoids are polyphenolic compounds which are important bioactive constituents in food and dietary supplements and play an important role in human nutrition. They are recognized as one of the largest and most widespread class of plant constituents occurring throughout the plant kingdom, and are also found in commonly consumed fruits, vegetables and beverages. These compounds have aroused considerable interest because of their potential beneficial biochemical and antioxidant effects on human health. Most of the experimental results demonstrate that flavonoids have several biological activities including radical scavenging, anti-inflammatory, anti-mutagenic, anti-cancer, anti-HIV, anti-allergic, anti-platelet and anti-oxidant activities.[43]

The basic structure of a flavonoid consists of a 15-carbon (C_6–C_3–C_6) skeleton containing one oxygenated (C) and two aromatic rings (A and B). Flavonoids are grouped together into subclasses based on their basic chemical structures; the most common ones being flavones, flavonols, isoflavones, flavanones, anthocyanins and chalcones. Flavonoids can exist as free aglycones but most of them commonly occur as C- or O-glycosides. Disaccharides are also often found in association with flavonoids and occasionally tri- and even tetrasaccharides.

Various approaches have been proposed for the use of mass spectrometric analyses for the structural characterization of flavonoids.[2,8,44,45] Furthermore, the low-energy collision CID-MS/MS analyses of various flavonoid aglycones and glycosides in mixtures were described in by using LC-ESI-MS/MS[19,20,46–48] and LC-APCI-MS/MS.[10,14] It has been demonstrated that product ion scans allow the establishment of the distribution of the substituents between the A- and B-rings. A careful study of the product ions patterns in CID-MS/MS can also be of a particular value in the structural elucidation of *O*- and *C*-glycosides.[13,49–54] Eventhough many advances have been achieved in the structural identification of flavonoids, it is still a challenge in food chemistry to identify these compounds in foodstuff or those derivatives arising during biotransformation.

The studied flavonoid glycosides (Fig. 1) were isolated from an aqueous methanolic extract of *Globularia alypum* growing in Morocco.[55]

2.2 Nomenclature

The major diagnostic CID-MS/MS product ions for flavonoid identification are those involving the cleavage of two C–C bonds of the C-ring. These product ions provide information on the number and type of substituent's in A- and B-rings. These product ions are usually designated according to the nomenclature previously proposed by Ma *et al.*[28] For free aglycone, the $^{i,j}A$ and $^{i,j}B$ labels refer to the product ions containing intact A- and B-rings, respectively, in which the superscripts i and j indicate the C-ring bonds that have been broken. For the flavonoid glycosides, the classical nomenclature proposed by Domon and Costello[56] was adopted. However, we have used, in this rationale, the designation: $^{k,l}X_j$, Y_j, Z_j for the product ions containing the aglycone, where j is the number of the interglycosidic bonds broken (starting from the aglycone) and k and l denote the cleavage within the carbohydrate rings (Scheme 1).

Full scan ion ESI mass spectra of the four studied flavonoids were recorded in the positive and negative modes and the obtained results are summarized in Table 1.

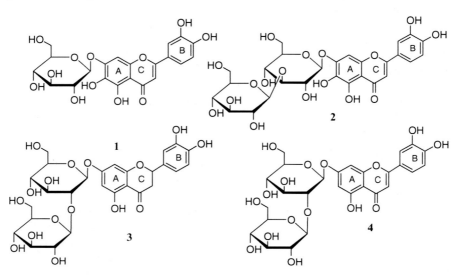

Fig. 1 Structure of the studied flavonoid glycosides

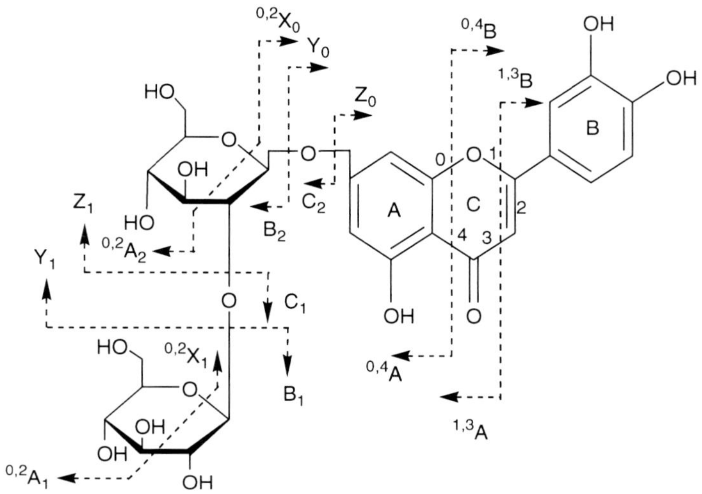

Scheme 1 Ion nomenclature adopted for the flavonoid glycosides product ions

Table 1 Mass spectrometric data for the flavonoids **1–4** obtained using positive and negative ESI-MS analyses

		1	2	3	4
+Mode	[M + H]$^+$	465	627	613	611
	Product ions	303	465, 447, 303	451, 289	449, 287
−Mode	[M − H]$^-$	463	625	611	609
	Product ions	301	463, 445, 301	475, 287	437, 285
Aglycone (Da)		302	302	288	286
Mol. wt. (Da)		464	626	612	610

Positive ESI-MS of each studied flavonoid showed the protonated and cationized molecules in addition to fragment ions corresponding to the aglycone ions (Y_0^+). This was also observed when the analyses were conducted in the negative ion mode where the deprotonated molecules and the Y_0^- ions were observed.

In order to characterize the product ions pathways of these compounds, CID-MS/MS analyses of the protonated, sodiated and deprotonated molecules were explored and the obtained results are discussed below.

2.3 Compound 1

The CID-MS/MS the protonated molecule [M + H]$^+$ m/z 465 of flavonoid **1** was recorded and the obtained results are shown in Fig. 2. The product ion Y_0^+ at m/z 303 corresponded to the aglycone moiety. The mass difference of m/z 162 Da between the [M + H]$^+$ peak and the aglycone was in agreement with the glucoside

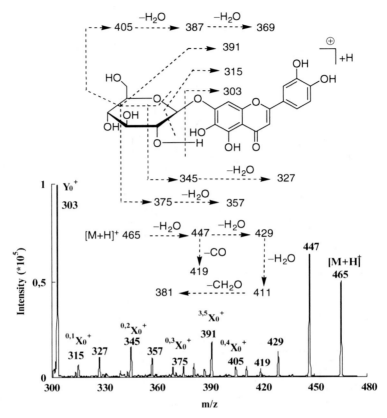

Fig. 2 CID-MS/MS of the protonated 6-hydroxyluteolin 7-*O*-glucoside at *m/z* 465 and its proposed gas-phase fragmentation routes

structure form of compound **1**. In addition other product ions were observed: [M + H − H$_2$O]$^+$ at *m/z* 447, [^{0,4}X$_0^+$] at *m/z* 405 [^{0,3}X$_0^+$], at *m/z* 375, [^{0,3}X$_0^+$ − H$_2$O] at *m/z* 357, [^{0,2}X$_0^+$] at *m/z* 345 [^{0,2}X$_0^+$] and[^{0,1}X$_0^+$] at *m/z* 315.

CID-MS/MS of the sodiated [M+Na]$^+$ ion (*m/z* 487) afforded the product ions: [M + Na − H$_2$O]$^+$ at *m/z* 469, [M + Na − CO]$^+$at *m/z* 459, [M + Na − 2H$_2$O]$^+$ at *m/z* 451 and the following respective product ions at *m/z* 427 (^{0,4}X$_0^+$), *m/z* 397 (^{0,3}X$_0^+$), *m/z* 371 (^{0,3}X$_0^+$-CO), *m/z* 367 (^{0,2}X$_0^+$), *m/z* 353 (^{1,5}X$_0^+$), *m/z* 325 (Y$_0^+$), *m/z* 307 (Z$_0^+$), *m/z* 297 (Y$_0^+$ − CO), *m/z* 269 (Y$_0^+$ − 2CO), and *m/z* 185 (B$_1^+$).

In order to characterize the aglycone part of the flavonoid **1**, the Y$_0^+$ product ion scan was more suited for providing data comparable to other flavonoid aglycones. Thus, the presence of the product ion led us to suppose that the aglycoine of compound **1** was a quercetin-based derivative. Thus, the product ion scan of the Y$_0^+$ at *m/z* 303 was recorded and Fig. 3 illustrates the obtained results. The ^{i,j}A$^+$ and ^{i,j}B$^+$ product ions ions allowed us to conclude that compound **1** aglycone was not quercetin. Although both aglycones product ions

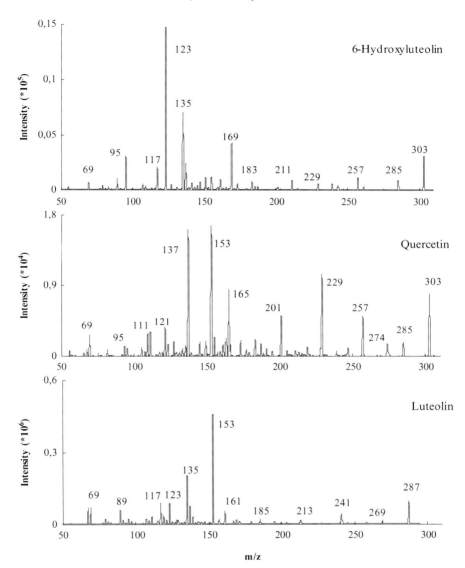

Fig. 3 CID-MS/MS of the protonated molecules [M+H]⁺ of 6-hydroxyluteolin, quercetin and luteolin

(6-hydroxyluteolin and quercetin have the same m/z 303 value, their CID-MS/MS spectra under similar low-energy conditions are significantly different (Fig. 3). In particular, the product ion (base peak) at m/z 153 for quercetin and corresponding to the $^{1,3}A^+$ product ion was not observed in the case of compound **1** where the $^{1,3}A^+$ product ions at m/z 169 and the complementary product ion was observed at m/z 135 ($^{1,3}B^+$). This indicated that the A ring was substituted with three OH groups. It has been reported that the $^{1,3}A^+$ observed for all flavonoid groups, is generally the most readily formed product ion and constitutes the most abundant product ion.[57] The $^{1,3}A^+$ product ion is most often found at m/z 153 for compounds with a 5,7-dihydroxyl group.[28,58] The absence of such product ion in the CID spectra of compound **1**, in addition to the presence of the product ion at m/z 169, clearly demonstrates the presence of a,trihydroxylated A ring in agreement with the proposed structure.

Figure 3 represents the CID-MS/MS of the protonated molecule m/z 303 and its proposed fragmentation routes are described in Scheme 2. The cleavage of the C-1 and C-3 bonds afforded the product ion $^{1,3}A^+$ at m/z 169 and the product ion $^{1,3}B^+$ at m/z 135 . The presence of this pair of product ions clearly provides the substitution pattern in the A (3 OH) and B (2 OH) rings. The obtained product ions can also undergo further fragmentation to create other product ions by successive losses of H_2O and CO. The product ion at m/z 123 (base peak) was formed from the product ion at m/z 169, whereas the product ion at m/z 117 was formed form the product ion m/z 135 by loss of water. An RDA-type (Retro Diels Alder) product ion, corresponding to the cleavage of the C-0 and C-4 bond, produced the product ions $^{0,4}A^+$ at m/z 125 and $^{0,4}B^+$ at m/z 179. These latter afforded the product ions: $[^4B^+ - H_2O]$ at m/z 161 and $[^{0,4}B^+ - CO]$ at m/z 151 and $[^{0,4}A^+ - H_2]$ at m/z 123. Cleavage of the C-0 and C-2 bonds afforded the product ion $^{0,2}B^+$ at m/z 137. Cleavage of the C-1 and C-2 bond produced the product ions $^{1,2}B^+$ and $^{1,2}A^+$ +2H at m/z 123 and 183 respectively. Finally cleavage of the C-0 and C-3 bonds afforded the product ions $^{0,3}B^+$ and $^{0,3}A^+$ at m/z 153 (low relative abundance).

In addition to the $^{i,j}A^+$ and $^{i,j}B^+$ product ions discussed above, the loss of small molecules from the protonated molecule of the aglycone were also observed (Scheme 2). Thus, the following product ions were observed at m/z 285 ($-H_2O$), 275 ($-CO$), 261 ($-C_2H_4O$) and 259 ($-CO_2$) mass units. In addition, we observed other product ions formed from the precursor ion at m/z 303 such as the product ions at m/z 257 ($-CO - H_2O$) and 243 ($-C_2H_4O - H_2O$).

The ESI-MS compound **1** ($-$ ion mode) afforded the deprotonated molecule [M $-$ H]$^-$ at m/z 463. The CID-MS/MS analysis of this latter ion afforded the diagnostic product ion at m/z 301, corresponding to the deprotonated aglycone moiety obtained by the loss of 162 Da in agreement with the monoglucoside structure form (Fig. 4). Other product ions were also observed [M $-$ H $-$ H$_2$O]$^-$ at m/z 445, $^{0,4}X_0^-$ at m/z 403, $^{0,3}X_0^-$ at m/z 373, $[^{0,3}X^- - H_2O]$ at m/z 355 and $^{0,2}X_0^-$ at m/z 343 which were formed by common fragmentation route. The presence of a product ion at m/z 300 corresponds to the radical aglycone product ion $[Y_0 - H]^-$ which was also observed as previously reported.[57,59–62] The formation of this radical product ion depended on the structure of the flavonoid glycoside and the nature and position of the sugar substitution.[60,63]

Figure 5 shows the second-generation CID-MS/MS analysis of aglycone **2** compared to that of the aglycone of the known quercetin and luteolin. The gas-phase fragmentation route is shown in Scheme 3. The obtained results indicated that 6-hydroxyluteolin and quercetin, have the same m/z 301 value albeit their negative CID-MS/MS analyses are different. In particular, the presence of the product ion

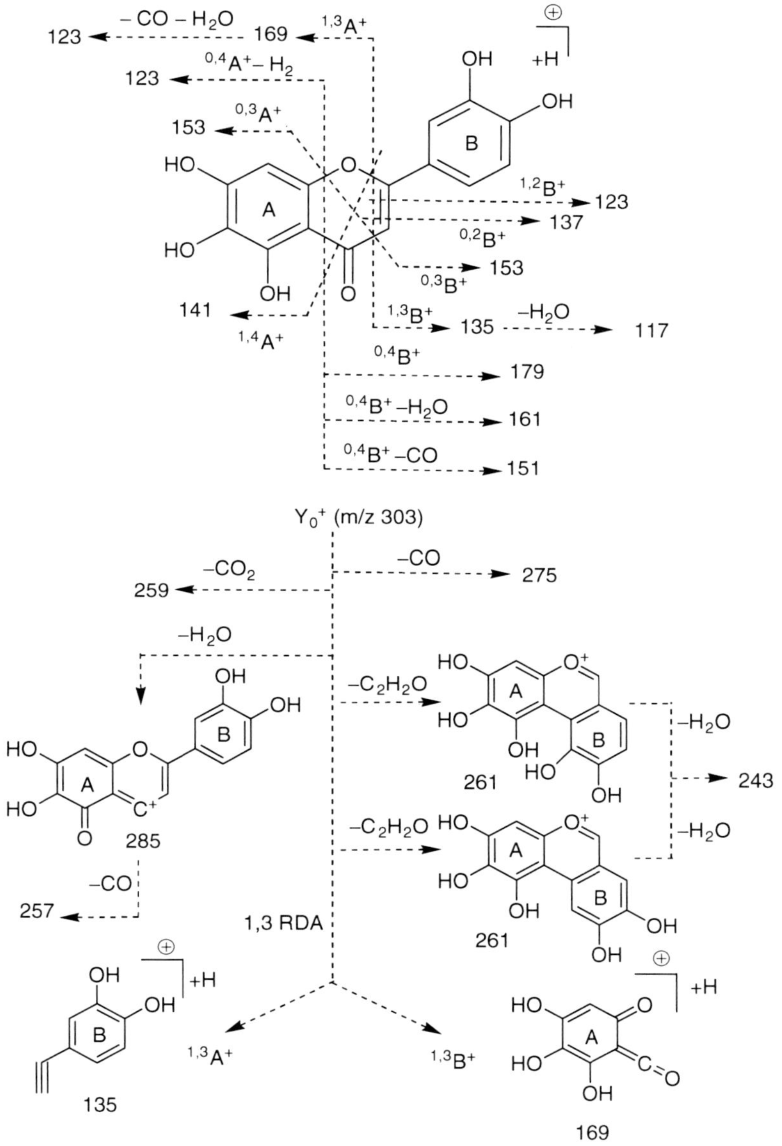

Scheme 2 CID-MS/MS data and proposed fragmentation pattern of the protonated aglycone **1**

(base peak) at m/z 151 for quercetin and the observed $^{1,3}A^-$ product ion at m/z 167, which was not observed in the case of the aglycone of compound **1.** In addition, the product ion at m/z 135 (base peak) for the 6-hydroxyluteolin was observed and the product ions $^{1,3}A^-$, $^{1,3}B^-$ at m/z 167, 133 are in agreement with a substitution of the A ring by three OH groups.

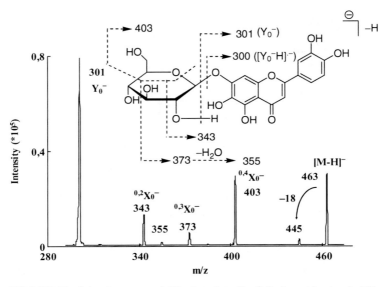

Fig. 4 CID-MS/MS of the deprotonated 6-hydroxyluteolin 7-O-glucoside at m/z 463 and the proposed gas-phase fragmentation routes

The product ion $^{1,3}A^-$ ion further eliminated of CO and CO_2 to form the product ions at m/z 139 and 123 in agreement with literature data concerning the luteolin product ions.[64] We also observed the product ion $[^{1,3}A^- - 2H]$ at m/z 165 (Scheme 3). This product ion can further lose CO to afford the product ions at m/z 137 and 121. The product ion observed at m/z 135 was formed either from the $^{0,2}B^-$ product ion or from the product ion at m/z 165 by loss of H_2CO. The latter being more probable, due to the fact that this was not observed in the case of luteolin. Further product ions at m/z 139 and 137 yield the ions at m/z 111 and 109 by elimination of H_2O. The latter gave, by loss of CO_2, the ion observed at m/z 65. This product ion could also arise from the ion at m/z 135 by successive losses of CO and a ketene moiety C_2H_2O.[64]

In addition, the second generation CID-MS/MS analysis of the precursor ion at m/z 301, produced from the deprotonated aglycone **2**, gave the following diagnostic product ions at m/z 283, 273, 255, 237, 229, 227, 211, 201, 183 and 173 (Scheme 3). The product ion at m/z 283 resulted from the loss of H_2O. The product ion at m/z 273 was formed by the loss of CO. The product ion at m/z 273 lost CO_2 to yield the product ion at m/z 229. Both the product ions at m/z 283 and 273 yielded the product ion at m/z 255, as shown in Scheme 3; the former eliminated a CO while the latter lost a H_2O. The product ion at m/z 255 created the product ions at m/z 237, 227 and 211 by elimination of H_2O, CO and CO_2 respectively. The product ion at m/z 237 afforded the product ion at m/z 201 by loss of two water molecules. Both the product ions at m/z 211 and 201 afforded the product ion at m/z 183, as shown in Scheme 3. The former eliminated a CO, while the

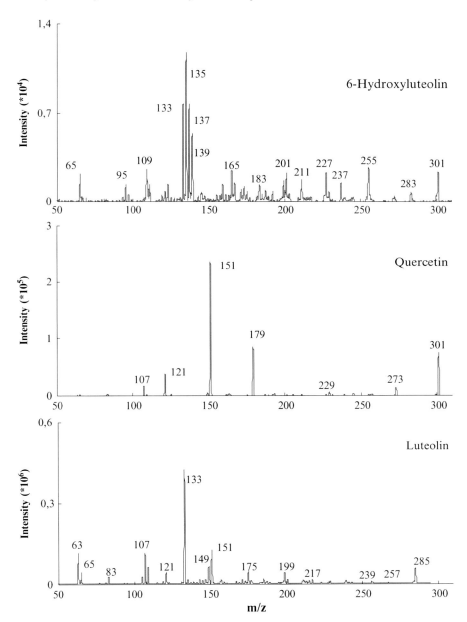

Fig. 5 CID-MS/MS of the deprotonated molecules [M – H]⁻ of 6-hydroxyluteolin, quercetin and luteolin

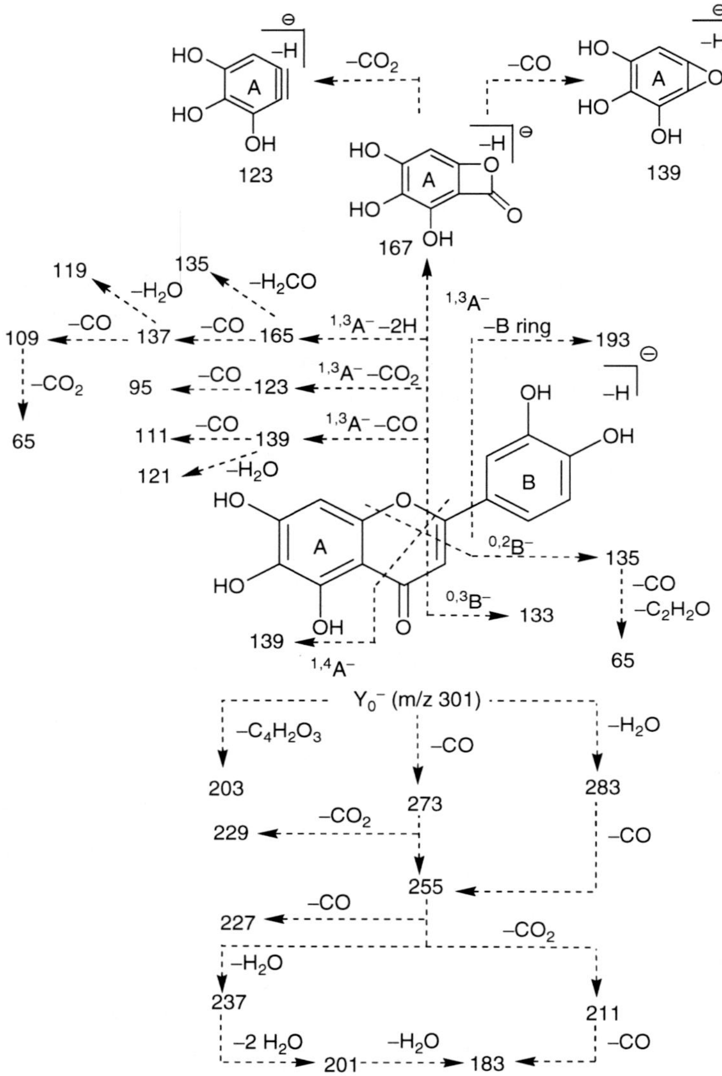

Scheme 3 CID-MS/MS and proposed fragmentation routes of the deprotonated molecule [M-H]⁻ of 6-hydroxyluteolin

latter lost a water molecule to produce the ion at m/z 183. Another product ion was observed at m/z 203 by the loss of a 98 Da corresponding to the $C_4H_2O_3$ unit. This loss of 98 Da was reported to be characteristic of compound containing tri-hydroxyl groups on the A ring in agreement of compound 1 structure.[65] The [M-B-ring]⁻ prodcut ion at m/z 193 resulted from the direct cleavage of the bond between the B- and C-rings.

2.4   *Compound* 2

The ESI-MS fragmentations observed for protonated molecule $[M + H]^+$ generated from compound **2** (Fig. 6), represent the typical characteristic fragmentation of flavonoid *O*-diglycosides.[13,66–68] The most intense fragment-ion was observed at *m/z* 303, corresponding to the protonated aglycone moiety formed by loss of 324 Da (two hexose residues). The CID-MS/MS analysis of the precursor ion extracted from the aglycone of compound **2** indicating that it had the same 6-hydroxyluteolin aglycone skeleton. Another less abundant product ion was observed at *m/z* 465 formed by loss of the external glucose unit (Y series of fragmentation in polysaccharide chains), while the complementary B type ion was absent. Further fragmentations of the glucose moiety gave rise to the product ions at *m/z* 567 ($^{0,4}X_1^+$), 507 ($^{0,2}X_1^+$), 479 ($^{0,2}X_1^{3,4}X_0^+$ or $^{0,2}X_1^{2,3}X_0^+$), 449 ($^{2,5}X_1^{4,5}X_0^+$), 447 (Z_1^+), 419 ($^{1,5}X_1^{4,5}X_0^+$), 345 ($^{0,2}X_0^+$) which are in agreement with the 1→3 diglucosidic linkage.

The CID-MS/MS analysis conducted on the ion Y_1^+ at *m/z* 465 gave the ion Y_0^+ at *m/z* 303 as the main product ion. Other product ion were also observed, these were observed at *m/z* 447 ($Y_0^+ - H_2O$), *m/z* 429 ($Y_0^+ - 2H_2O$), *m/z* 424 ($Y_1^+ - HC_2O$), *m/z* 411 ($Y_1^+ - 3H_2O$), *m/z* 405 ($^{0,4}X_0^+$), *m/z* 399 ($Y_1^+ - 2H_2O - H_2C_2O$), *m/z* 381 ($Y_1^+ - 3H_2O - H_2C_2O$), *m/z* 369 ($Y_1^+ - 2H_2O - 2H_2C_2O$), *m/z* 345 ($^{0,2}X_0^+$) and *m/z* 315 ($^{0,1}X_0^+$).

The CID-MS/MS was also conducted on the sodiated molecule $[M+Na]^+$ at *m/z* 649 is shown in Fig. 7. The most striking feature analysis is the high relative abundance of the ion at *m/z* 347 $[B_0 + Na]^+$ corresponding to the loss of the aglycone moiety, which forms the cationized adduct of the diglucoside moiety as in agreement

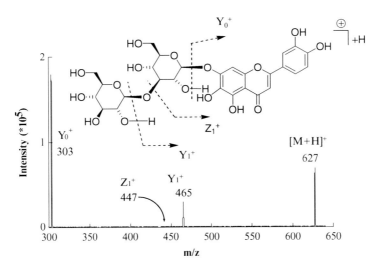

Fig. 6 CID-MS/MS of the protonated molecule of 6-hydroxyluteolin 7-*O*-laminaribioside at *m/z* 627 and the proposed gas-phase fragmentation routes

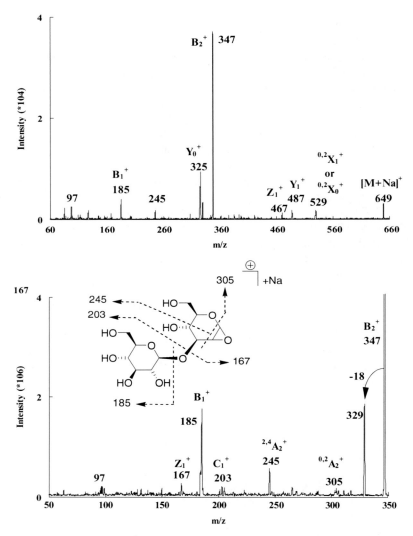

Fig. 7 CID mass spectra of the sodiated 6-hydroxyluteolin 7-*O*-laminaribioside (*m/z* 649) and of the sodiated diglycosyl moiety (*m/z* 347)

with the diglucoside structure of compound **2**. The aglycone moiety was observed at *m/z* 325 $[Y_0 + Na]^+$ in addition to the product ion radical aglycone observed at *m/z* 324. Second-generation CID-MS/MS analysis was conducted on the precursor ion at *m/z* 347 and the results obtained are shown in Fig. 7 and are in agreement with the 1→3 interglycosidic linkage. Thus product ion were observed at *m/z* 347 (B_2^+), *m/z* 329 ($B_2^+ - H_2O$), *m/z* 305 ($^{0,2}X_0^+$), *m/z* 289 ($^{2,5}X_0^+$), *m/z* 257 ($^{0,3}X_0^{0,3}X_1^+$), *m/z* 245 ($^{2,4}X_0^+$), *m/z* 215 ($^{2,3}X_0^+$), *m/z* 205 ($^{0,2}X_0^{0,3}X_1^+$), *m/z* 203 (C_1^+), *m/z* 185 (B_1^+), *m/z* 184 (Y_1^+) and *m/z* 167 (Z_1^+).

The CID-MS/ analysis of the deprotonated molecule at m/z 625 $[M - H]^-$ extracted form compound **2** is shown in Fig. 8 and represents typical MS/MS characteristic product ions of flavonoid O-diglycosides. This CID-MS/MS yielded anions, identical to the one obtained for compound **1** at m/z 301 and an identical subsequent fragmentation pattern indicates that they share the same basic structural characteristic of 6-hydroxyluteolin. In addition to the base peak located at m/z 301, other two product ions at m/z 463 and 445 were also observed (Fig. 8). These two product ions were formed by loss of the non-reducing glucose end unit according to the Y_1^- and Z_1^- series.[56] In the the following product ions at m/z 607 ($[M - H - H_2O]^-$), 505 ($^{0,2}X_1^-$), 389, 390 ($^{0,3}X_0 {}^{0,1}X_1^-$), 343 ($^{0,2}X_0^-$) and 325 ($^{0,2}X_0^- - H_2O$) were also observed in agreement with the oligosaccharide structure (Fig. 8).

The second-generation CID-MS/MS analysis conducted on the Y_1^- precursor ion at m/z 463 extracted from compound **2**, gave the following product ions at m/z 445 ($Y_1^- - H_2O$), m/z 403 ($^{0,4}X_0^-$), m/z 373 ($^{0,3}X_0^-$), m/z 355 ($^{0,3}X^- - H_2O$), m/z 343 ($^{0,2}X_0^-$) and m/z 301 (Y_0^-). As indicated in the positive ion mode, the difference observed in the product ions pathways, could arise from the position and the further delocalisation of the charge on each compound skeleton conducting to different product ions pathways.

Comparing the CID-MS/MS product ions formed from the individual precursor ions containing the 1→2 and 1→6 interglycosidic bonds, the 1→3 linkage presents a gradual oligosaccharide breakdown yielding the Y_1^- and the Z_1^- ions which is more likely similar to the 1→2 linkage. Such linkage was defined by the high relative abundance (13–79%) of the Y_1^- and/or the Z_1^- ions while the 1→6 isomer the ion Y_1^- could be found at very low abundance (1–3%) and the Z_1^- ion was not detected.[69] However and in the opposite of the 1→2 linkage which was reported to

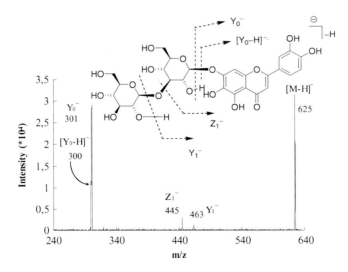

Fig. 8 CID-MS/MS of the deprotonated 6-hydroxyluteolin 7-O-laminaribioside (m/z 627) and its proposed gas-phase fragmentaion pattern routes

be characterized by a relatively high abundance of the $^{0,2}X_0^-$ ion considered as characteristic of such linkage,[70] this ion was observed with a relatively low abundance in the 1→3 linkage. It may be kept in mind that due to the high lability of the glycosidic linkage of the oligosaccharides at the 7 position, the presence of these characteristic product ions could not always be observed.[71]

2.5   Compound 3

The molecular mass of compound **3** was confirmed as 612 Da by recording the ESI-MS (+ion mode) which showed the protonated molecule [M + H]$^+$ at m/z 613. A loss of 324 Da was also observed indicating the presence of a two hexose residues. This results in the formation of the product-ion at m/z 289, suggesting that compound **3** was a flavanone-based compound.

The CID-MS/MS analysis of the protonated molecule [M + H]$^+$ at m/z 613 is shown in Fig. 9. The formation of the the characteristic product ion $^{0,2}X_0^{0,2}X_1^+$ at m/z 493 was noted. This result is in agreement with the loss of 120 Da corresponding to the cleavage of the O-glycoside moiety. Relatively abundant Y type product ions were observed at m/z 289 (Y_0^+, base peak) and 451 (Y_1^+). In addition other product ions: [M + H − H$_2$O]$^+$, at m/z 595, [M + H − 2H$_2$O]$^+$ at m/z 577, Z_1^+ at m/z 433, $^{0,2}X_0^{0,2}X_1^+$ at m/z 373, $^{0,2}X_0Y_1^+$ at m/z 313, B_2^+ at m/z 325 and B_1^+ at m/z 163; ther ions were formed by common fragmentation routes. Thus, the product ions observed for compound **3** are in agreement with previously reported data for flavanones[72] and they are indicative a 3′,4′,5,7-tetrahydroxyflavanone.[58]

The 1→2 interglucosidic linkage was supported by the presence of the $^{0,2}X_0^{0,2}X_1$ ion which can be seen in the CID-MS/MS of compound **3**. This product ion can be considered as characteristic of the 1→2 interglycosidic linkage in the 7-O-glucoglucoside adduct, since it can not be formed in the case of other interglycosidic types. Additional product ions at m/z 415 ($Y_1^+ − 2H_2O$), 397 ($Y_1^+ − 3H_2O$) and 355 ($Y_1^+ − 2H_2O − 60$), which are also characteristic of flavanone O-diglycosides,[67,72] were also observed. The formation of Z_1^+ ion at m/z 433 and absence of the corresponding radical Z_1^+ ions are useful for establishing that the eliminated terminal glycosyl unit is linked to another carbohydrate and not directly to the aglycone and could be of analytical value for the differentiation of O-diglycosyl and di-O-glycosyl flavonoids.[13] In the CID-spectrum shown in Fig. 9, the most striking feature is the high intensity of the signal for the Y_1^+ ion corresponding to a neutral loss of 162 Da. In previously reported data concerning flavonoid rutinoside and neohesperoside MS/MS spectra, an internal loss of the glucose moiety leading a labeled Y* product ion signal was observed in addition to the presence of the Y_0^+ and Y_1^+ ions.[66,67] The presence of this unusual glucose loss is easily distinguished when the two carbohydrate moieties are different, as in the case of the rhamnoglycoside derivatives. This was not the case for compound **3**, which is comprised of two glucosyl moieties where it is impossible to distinguish between the terminal and the inner glucose. This may explain

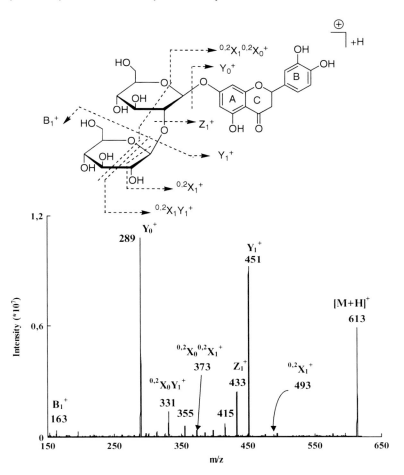

Fig. 9 CID-MS/MS analyses of the protonated eriodictyol 7-*O*-sophoroside **3** at *m/* 613 and its proposed gas-phase fragmentation route

the high abundance of the observed ion located at *m/z* 451, which may include either the ion resulting from the Y_1 product ion [M + H − 162]$^+$ corresponding to the terminal glucose loss and the Y* product ion corresponding to the inner glucose loss.

Second-generation CID-MS/MS of the precursor ion Y_0^+ at *m/z* 289 extracted from compound **3** afforded a series of five diagnostic product ions: at *m/z* 271 (−H_2O), 179 (−B ring), 163 ($^{1,4}B^+$ − 2H), 153 ($^{1,3}A^+$), 145 ($^{1,4}B^+$ − 2H − H_2O) and 135 ($^{1,4}B^+$ − 2H − CO) (Scheme 4).

The CID-MS/MS spectra of the the [M + Na]$^+$ sodiated molecule extracted from compound **3** at *m/z* is shown in Fig. 10 where the high relative abundance of the product ion at *m/z* 347 (B_2^+) corresponding to the loss of the aglycone moiety in agreement *O*-diglycosyl flavonoid should be noted. The Y_0^+ product ion at *m/z* 311

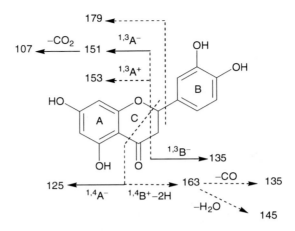

Scheme 4 CID-MS/MS analyses of the protonated (*dashed arrows*) and deprotonated (*full arrows*) molecules of the aglycone of eriodictyol

was observed with a relatively low abundance, as was the radical aglycone product ion observed at m/z 310. Another important product ion $^{1,3}A_0^+$ was observed at m/z 499, while the complementary $^{1,3}B_0^+$ was observed at m/z 159 with a low abundance. This showed that the product ions involving the C ring could occur before the one involving the diglucoside unit. This is of analytical value since it indicates that the two glycoside moieties are located on the A ring. Further successive losses from the $^{1,3}A_0^+$ product ion of the terminal and inner glucose units, give ions at m/z 337 $(Y_1^{1,3}A^+)$ and 175 $(Y_0^{1,3}A^+)$, respectively. Additional signals were observed at m/z 379 $(^{0,2}X_1^{1,3}A^+)$ and 259 $(^{0,2}X_1^{0,2}X_0^{1,3}A^+)$ and were in full agreement with the 1→2 interglycosidic linkage.

When the CID-MS/MS experiments were conducted on the the deprotonated molecular ion at m/z m/z 611 generated from compound **3**, the presence of the Y_0^- product ion (base peak) at m/z 287 was observed resulting from loss of 324 mass units, in agreement with the presence of two hexose residues (Fig. 11). In this CID-MS/MS the Y and Z type product ions were also observed at m/z 449 (Y_1^-), 431 (Z_1^-) and 287 (Y_0^-), along with other product ions at m/z 593 $[M - H - H_2O]^-$, 491 $(^{0,2}X_0^{0,2}X_1^-)$, 329 $(^{0,2}X_0Y_1^-)$ and 311 $(^{0,2}X_0Z_1^-)$. In contrast to rhamnoglucoside disaccharides where the presence $^{0,2}X0^-$ was characteristic of the 1→2 isomer, this product ion could be formed in all kinds of interglycosidic linkages between two glucose moieties. Therefore, its presence could not be considered as indicative of a peculiar interglycosidic linkage.

The presence of a product ion ion at m/z 475 with a high relative abundance that exceeds that of Y_0^- obtained at lower CID collision energy should be noted. The complement of this ion was also detected at m/z 135 and corresponds to the $^{1,3}B^-$ product ion ion. This means that the product ion m/z 475 corresponds to the $^{1,3}A^-$ product ion ion, which is an indication that the breakdown of the aglycone skeleton occurs before that of the glycosidic bonds. This is of great

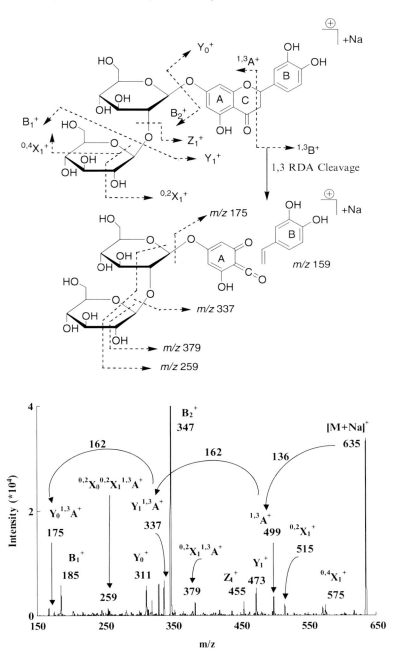

Fig. 10 CID-MS/MS of the sodiated eriodictyol 7-*O*-sophoroside **3** molecule at *m/z* 635 and its proposed gas-phase fragmentation routes

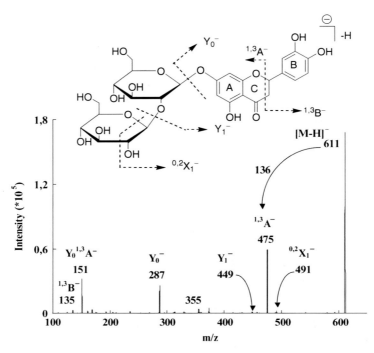

Fig. 11 CID-MS/MS of deprotonated eriodictyol 7-O-sophoroside **3** at m/z 611 and its proposed gas-phase fragmentation routes

importance since it confirms that the two glycoside moieties are not located on the B ring. Further CID-MS/MS analyses were conducted on the product ion ion at m/z 475 at different collision energy values (Fig. 12). The product ions corresponding to the loss of the terminal glucose unit and corresponding to the Y_1^- and Z_1^- ions were observed at m/z 313 and 295 respectively. Further loss of the inner glucose residue yields the Y_0^- ion at m/z 151 corresponding to the $^{1,3}A^-$ aglycone part of the molecule, in agreement with the structure of compound **3**. Additional product ions at m/z 457 (- H_2O), 415 ($^{0,4}X_0^-$ or $^{0,4}X_1^-$), 373 ($^{2,5}X_0$ $^{2,5}X_1^-$), 355 ($^{0,2}X_0$ $^{0,2}X_1^-$), 253 ($^{0,4}X_0$ Y_1^-), 235 ($^{0,2}X_1$ $^{0,2}X_0^-$), 217 ($^{0,2}X_1$ $^{0,2}X_0^-$ - H_2O) and 193 ($^{0,4}X_0Y_1^-$). All the product ion pathways, corresponding to theses ions, are schematized in Fig. 12. The 1→2 interglycosidic linkage was supported by the presence of the ions m/z 355 and 235 and corresponding to successive losses of 120 units from the 475 ion.

In order to explore the product ions of the aglycone compound **3**, the CID-MS/ MS of the Y_0^- precursor ion at m/z 287 was recorded and showed main diagnostic product ions (Y_0^- – H_2CO) at m/z 257, (Y_0^- – H_2CO – H_2O) at m/z 239, (Y_0^- – H_2CO – $2H_2O$) at m/z , 211, ($^{1,3}A^-$) at m/z 151, ($^{1,3}B^-$) at m/z 135, ($^{1,4}A^-$) at m/z 125 and ($^{1,3}A^-$ – CO_2) at m/z 107 (Scheme 4), in agreement with a 3′,4′,5,7-tetrahydroxyflavonone.[10,58,59,64]

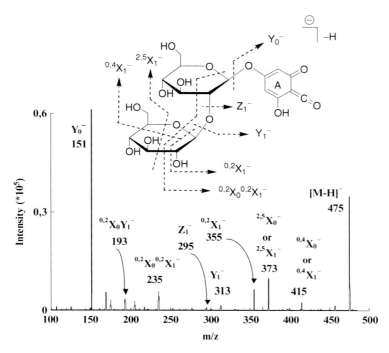

Fig. 12 Second-generation CID-MS/MS of the $^{1,2}A^-$ at m/z 475 extracted from compound **3** and its proposed gas-phase frgamentation routes

2.6 *Compound* **4**

In order to characterize the aglycone part of compound **4**, the CID-MS/MS analyses were recorded for the $[M + H]^+$ and Y_0^+ ions. The CID-MS/MS of protonated molecule $[M + H]^+$ of compound **4** afforded characteristic product ions of the flavonoid *O*-diglycosides[13,66–68] giving two major product ions Y_0^+ and Y_1 at m/z 287 and 449 respectively (Fig. 13). The Y_0^+ at m/z 287 was the base peak, while the Y_1^+ product ion at m/z 449 was observed (low abundance) suggesting the existence of a diglycoside moiety, as it is known that compounds which exhibeit a high abundance of m/z 449 have two glucose units attached to the different positions of the aglycone.[66,71,73] In addition, the presence of product ions at m/z 593 $[M + H - 18]^+$, 551 ($^{0,4}X_0^+$), 521 ($^{0,3}X_0^+$), 491 ($^{0,2}X_0^+$), 431 (Z_1^+) and 371 ($^{0,2}X_0{}^{0,2}X_1^+$) were also observed. The $^{0,2}X_0{}^{0,2}X_1^+$ product ion is of a great importance since it results from two successive losses of 120 mass units, indicating a 1→2 interglycosidic linkage.

The CID-MS/MS of the aglycone of compound **4** Y_0^+ at m/z 287 is presented in Scheme 5 and afforded the following product ionss at m/z 269 ($Y_0^+ - H_2O$), 241 ($Y_0^+ - H_2O - CO$), 179 ($^{0,4}B^+$), 161 ($^{0,4}B^+ - H_2O$), 153 ($^{1,3}A^+$), 137 ($^{0,2}B^+$), 135 ($^{1,3}B^+$), 123 ($^{1,3}A^+ - H_2O$), and 117 ($^{1,3}B^+ - H_2O$), which are in agreement with previously reported data for luteolin.[28,74]

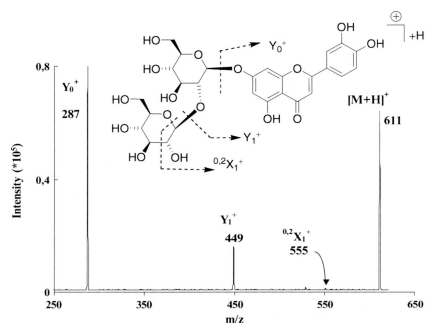

Fig. 13 CID-MS/MS of the protonated luteolin 7-*O*-sophoroside **4** (*m/z* 611) and its proposed gas-phase fragmentation routes

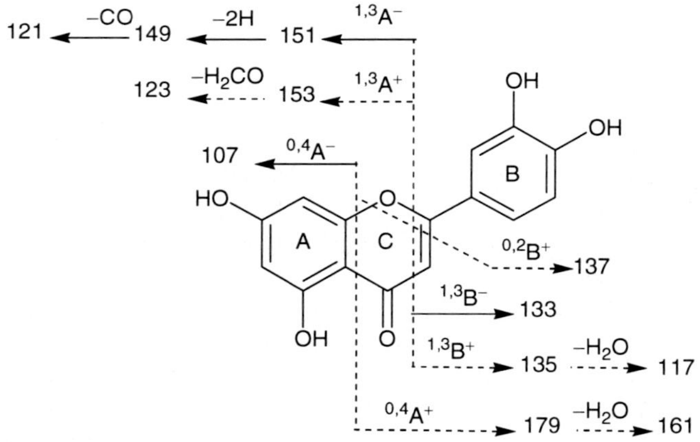

Scheme 5 Main product ionss observed in CID-MS/MS of the [M + H]$^+$ (*dashed arrows*) and [M − H]$^-$(*full arrows*) ions of luteolin

The CID-MS/MS of the [M − H]$^-$ ion at *m/z* 609 exhibited the product ions typical of glycosyl derivatives (Fig. 14). The loss of the terminal and the inner sugars were successively observed giving a low abundance of the product ion Y$_1^-$ at *m/z* 447 and an intense product ion Y$_0^-$ at *m/z* 285 (base peak) corresponding to the

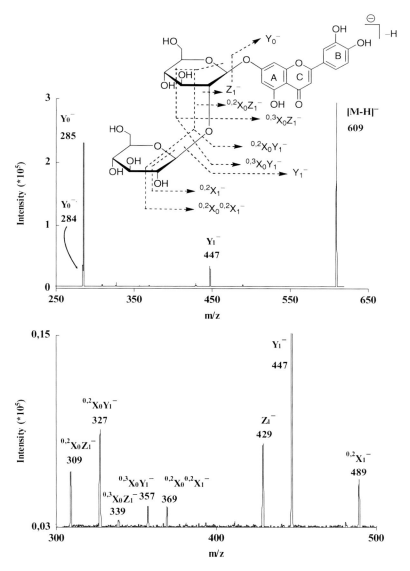

Fig. 14 CID-MS/MS of the deprotonated luteolin 7-O-sophoroside **4** at m/z 609 and its proposed gas-phase fragmentation routes

aglycone moiety. In addition, the presence of other product ions were observed at $^{0,2}X_1^-$ at m/z 489, Z_1^- at m/z, $^{0,2}X_0^{0,2}X_1^-$ at m/z 429369, $^{0,3}X_0Y_1^-$ at m/z 357, $^{0,3}X_0Z_1^-$ at m/z 339, $^{0,2}X_0Y_1^-$ at m/z 327, $^{0,2}X_0Z_1^-$ at m/z 309 and $^{0,1}X_0$ at m/z 298. The product ion $^{0,2}X_1^{0,2}X_0$ at m/z 369 was formed by successive loss of two 120 mass units indicating a 1→2 dihexoside isomer, since it can not be formed in the case of other interglycosidic linkages. The fact that the 1→2 isomer exhibits more product ions could be related to the presence of the hydroxyl group nearest to the acetal linkage (between the aglycone and the sugar part, *i.e.,* the 2 OH group of glucose). Formation of Y_0^- appears to be more difficult compared to the rutinose analogues. This explains why the formation of other product ions is favored in flavonoids having interglycosidic linkages like the neohesperidoside derivatives.[67]

The second-generation CID-MS/MS of the aglycone Y_0^- ion at m/z 285 exhibited the following product ions: m/z 257 (Y_0^- – CO), 243 (Y_0^- – C_2H_2O), 241 (Y_0^- – CO$_2$), 175 (Y_0^- – C_3O_2 – C_2H_2O), 151 ($^{1,3}A^-$), 149 ($^{1,3}A^-$ – 2H), 133 ($^{1,3}B^-$), 121 ($^{1,3}A^-$ – H_2CO), and 107 ($^{0,4}A^-$) (Scheme 5), in agreement with a luteolin unit.[10,64] It is worth noting that in contrast to the behavior of eriodictyol aglycone, where no radical product ion was observed in the negative CID-spectra, the glycoside 4 showed both a collision-induced homolytic and heterolytic cleavage of the O-glycosidic bond producing a deprotonated radical aglycone ion [Y_0 – H]$^-$ at m/z 284 and an aglycone ion Y_0^- at m/z 285 (Fig. 14).

3   Synthesized 1,5-Benzodiazepine Derivatives

Benzodiazepines represent a large group of organic compounds which have attracted much attention as an important class of heterocyclic derivatives in the field of drugs and pharmaceuticals. The wide spectrum of the pharmacological effects exhibited by these compounds makes them one of the most versatile classes of drugs used in psychopharmacology.[75,76] They are also widely used as anticonvulsant, antianxiety, analgesic, sedative, anti-depressive, hypnotic and anti-inflammatory agents.[77–81]

Among the analytical techniques used for the structural exploration of benzodiazepine derivatives, ESI-MS has been shown to be a powerful tool for their characterization. The low-energy collision-induced dissociation CID-MS/MS analyses of the selected protonated molecules were measured with a QhQ-MS/MS tandem instrument. The three novel 1,5-benzodiazepine derivatives 5–7 are shown in Fig. 15.[82]

The high-resolution ESI-QqToF mass spectra (+ ion mode) of the three 1,5-benzodiazepines 5–7 were recorded and the obtained results are summarized in Table 2. The protonated and the sodiated product ions were observed for the three studied compounds. It is worth noting that compound 7 afforded two protonated molecules located at m/z 448.1386 ($C_{25}H_{23}N_3O_3Cl$ (^{35}Cl)) and m/z 450.1384 ($C_{25}H_{23}N_3O_3Cl$ (^{37}Cl)), and two sodiated molecules at m/z 470.1219 ($C_{25}H_{23}N_3O_3NaCl$ (^{35}Cl) and at m/z 472.1198 ($C_{25}H_{23}N_3O_3NaCl$ (^{37}Cl)). This is in agreement of the presence of the chlorine atom in this compound which was easily deduced from the isotopic profile (3:1) of the protonated and sodiated ions.

In order to characterize the product ions pathways of these benzodiazepines 5–7, low-energy CID-MS/MS analyses of the corresponding protonated molecules 5a–7a were carried out on the conventional QhQ-MS/MS instrument, which resulted in the formation of a series of the diagnostic product ions discussed below.

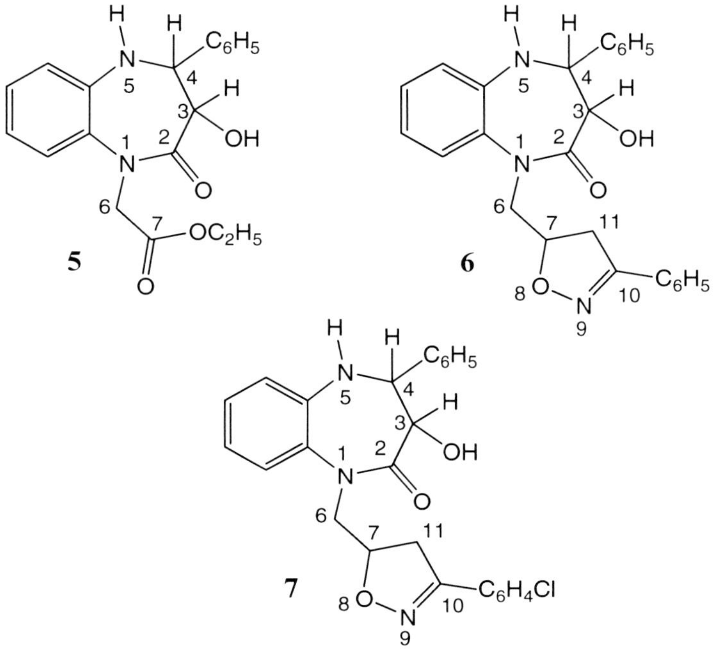

Fig. 15 Structure of the studied benzodiazepines

Table 2 Mass spectrometric data for the benzodiazepines **5–7** obtained using positive full scan ESI-MS analyses

Cpd.	[M + H]⁺	[M + Na]⁺	Other observed product ions	Mol. wt.
5	341	363	119, 195, 209, 221, 267, 283, 295, 323	340
6	414	436	119, 173, 278, 323, 356, 396	413
7	448	470	119, 173, 194, 357, 390	447

3.1 Compound 5

The product ion scan of the protonated molecule $[M + H]^+$ ion **5a** at m/z 341 was recorded and is shown in Fig. 16a. It showed major product ions at m/z 107, 119, 147, 195, 207, 209, 221, 249, 267, 283, 323 and 295 (Table 3).

Plausible product ion pathways for the major observed product ions formed from the ion **5a** are proposed in Scheme 6. The protonated molecule **5a** at m/z 341 eliminates glyoxal to afford the benzimidazolinium ion **5b** at m/z 283. The latter can lose ethanol, ethyl acetate or ethyl formate generating respectively the ions **5c** at m/z 237), **5d** at m/z 195 or **5e** at m/z 209. Finally, the successive elimination of benzene and ethyl acetate from **5b** yields the ion **5f** at m/z 119, which could also be obtained from the protonated molecule by successive elimination of glyoxal, benzene and ethyl acetate.

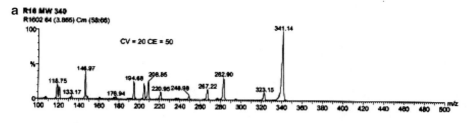

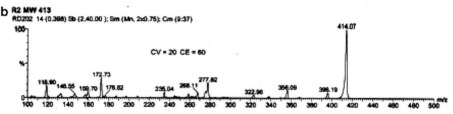

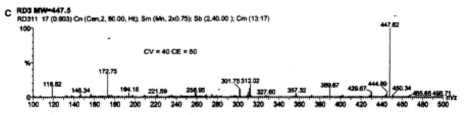

Fig. 16 CID-MS/MSa of the protonated molecules **5a** (**a**), **6a** (**b**), **7a** (**c**) generated from the benzodiazepines **5**, **6** and **7** respectively

Table 3 CID-MS/MS analyses of the protonated molecules generated from the benzodiazepines 5–7

Selected [M + H]+ ion	Observed product ion ions
341 (**5a**)	341 (**5a**), 323 (**5i**), 295 (**5g**), 283 (**5b**), 267 (**5h**), 237 (**5c**), 221 (**5j**), 209 (**5e**), 207 (**5m**), 195 (**5d**), 175 (**5k**), 147 (**5l**), 119 (**5f**)
414 (**6a**)	414 (**6a**), 396 (**6h**), 356 (**6b**), 338 (**6o**), 323 (**6f**), 278 (**6d**), 268 (**6k**), 262 (**6p**), 259 (**6n**), 235 (**6l**), 173 (**6e**), 160 (**6g**), 147 (**6m**), 119 (**6c**)
448 (**7a**)	448 (**7a**), 430 (**7h**), 390 (**7b**), 357 (**7f**), 312 (**7d**), 302 (**7k**), 259 (**7i**), 194 (**7g**), 173 (**7e**), 119 (**7c**)

The protonated molecule **5a** at m/z 341 can also eliminate EtOH giving the product ion **5g** at m/z 295, which loses CO to give **5h** at m/z 267. This latter ion loses glyoxal, forming the product ion **5e** at m/z 209, which further affords **5m** at m/z 207. The ion **5a** can also lose water giving **5i** at m/z 323, which eliminates phenylacetylene to give the benzimidazolonium **5j** at m/z 221. The protonated molecule **5a** can also successively eliminate water and phenylacetylene to give the ion **5j** (m/z 221). The latter gives the ion **5k** at m/z 175 through the loss of ethanol. Finally, the ion **5k** eliminates CO to yield **5l** at m/z 147 which could also be produced from **5j** through successive losses of ethanol and CO.

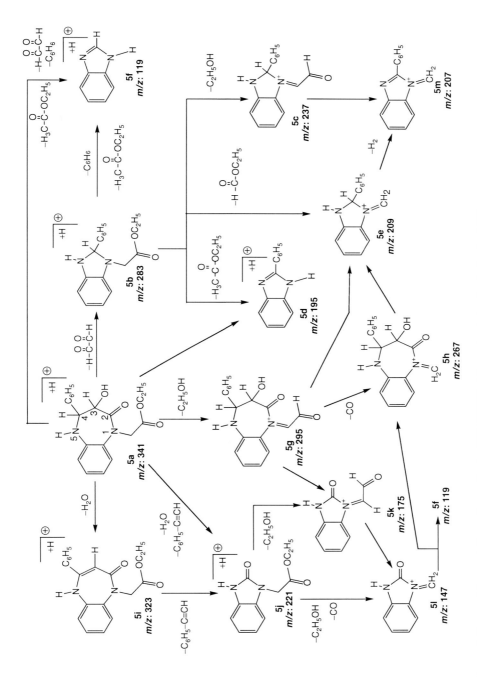

Scheme 6 Proposed fragmentation pathways for the CID-MS/MS of the protonated molecule **5a** at *m/z* 341

It should be noted that the genesis of product ions generated from compound **5** were confirmed through precursor scan analysis. In addition, these product ions were identified as product ions formed during the single-stage ESI-QqTOF-MS which produced the expected accurate masses corresponding to the proposed chemical structures of the CID-MS/MS product ions. Thus, the diagnostic product ions **5a**, **5i**, **5g**, **5b**, **5h**, **5j**, **5e**, **5d** and **5f**, respectively, at m/z 341.1464, 323.1384, 295.1061, 283.1340, 267.1094, 221.0874, 209.1062, 195.0909 and 119.0589, were measured as product ions in single-stage ESI-QqToF-MS and afforded the expected accurate masses corresponding for the proposed chemical structures.

It may be noted that all the observed major product ions for the protonated benzodiazepine **5a** occurred in the saturated seven membered ring containing the nitrogen atoms. Various product ions were thus formed through different breakdown routes with concomitant eliminations of neutral molecules like glyoxal, benzene and ethyl formiate, forming the product ion at m/z 119. In addition, a unique simultaneous CID-MS/MS product ions occurred by a pathway dictated by the substituent on the N-1-position to form the product ions **5g** at (m/z 295 and **5h** at m/z 267. The aromatic ring portion of the 1,5-benzodiazepines was obviously resistant to CID-MS/MS fragmentation.

3.2   Compound 6

The CID-MS/MS of the protonated molecule **6a** (m/z 414) is shown in Fig. 16b and in Table 3. The tentative fragmentation pathways of the protonated molecule is proposed in Scheme 7.

The precursor ion **6a** at m/z 414 affords the benzimidazolinium product ion **6b** at m/z 356 by breaking of the N-1–C-2 and C-3–C-4 linkages with elimination of glyoxal. This latter ion affords the product ion **6c** at m/z 119by consecutive elimination of benzene and methylidene isoxazoline. The product **6b** at m/z 356 can also yield the product ion **6d** at m/z 278 by elimination of benzene. The latter loses benzonitrile and hydrogen molecules to yield the ion **6e** at m/z 173. The ion **6b** at m/z 356 can also eliminate hydroxylamine to give the product ion **6f** at m/z 323 or afford the product ions **6o** at m/z 338, **6p** at m/z 262, **6k** at m/z 268 and **6l** at m/z 235 which structures are shown in Scheme 7.

In addition, the protonated molecule **6a** can eliminate a neutral benzodiazepine or water molecules to give the ions **6g** at m/z 160 or **6h** at m/z 396 respectively. The latter affords the product ion **6n** at m/z 659 or eliminates phenylacetylene to give the product ion **6i** at m/z 294, which further affords the product ions **6j** at m/z 276 and **6m** at m/z 147.

All the discussed diagnostic product ions **6a**, **6h**, **6b**, **6f**, **6d**, **6e** and **6c**, respectively, at m/z 414.1804, 396.1700, 356.1699, 323.1499, 278.1199, 173.0698 and 119.0588, were measured as fragment ions in the single-stage ESI-QqToF mass spectra and gave the expected accurate masses. In addition precursor scan analyses were performed to confirm the formation of the observed product ions.

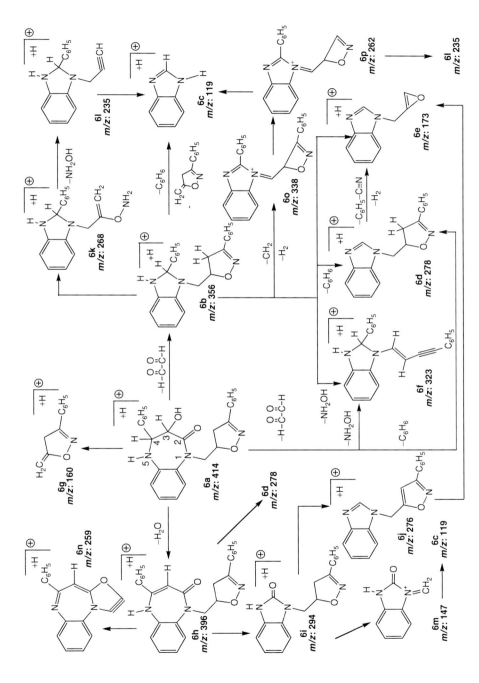

Scheme 7 Proposed fragmentation pathways for the CID-MS/MS of the protonated molecule **6a** at *m/z* 414

3.3 Compound 7

The CID-MS/MS of the protonated molecule **7a** at *m/z* 448 generated afforded a series of the product ions shown in Fig. 16c and summarized in Table 3. The formation routes of the obtained product ions are tentatively rationalized in Scheme 8.

The product ion **7b** at *m/z* 390 could be formed from **7a** by elimination of a glyoxal molecule through the breaking of the *N*-1-*C*-2 and *C*-3-*C*-4 bonds. The generated product ion **7b** can lose benzene and methylidene isoxazoline or benzene to yield the product ions **7c** at *m/z* 119 or **7d** at *m/z* 312 respectively. The latter loses benzonitrile and hydrogen molecules to yield the ion **7e** at *m/z* 173. The ion **7b** at *m/z* 390 can also eliminate hydroxylamine to give the ion **7f** at *m/z* 357. The latter could also be formed from the protonated molecule **7a** by consecutive elimination of glyoxal and hydroxylamine.

The protonated molecule **7a** can also yield **7g** (*m/z* 194) and **7h** (*m/z* 430) through elimination of benzodiazepine and water molecules respectively. The latter generates the product ion **7i** (*m/z* 259). Finally, the product ion **7a** could suffer a cleavage of the *C*-2–*C*-3 and C-4–N-5 bonds giving the product ion **7j** at *m/z* 330 which further yields the ion **7k** at *m/z* 302 by elimination of carbon monoxide molecule.

It may be noted that the genesis of the product ions discussed above was confirmed through precursor scan analyses. In addition, the diagnostic product ions **7a**, **7b**, **7f**, **7g**, **7e** and **7c**, respectively, at *m/z* 450.1384, 448.1386, 392.1289, 390.1291, 359.1028, 357.1084, 196.0298, 194.0297, 173.0689 and 119.0596, were also measured as product ion ions in the single-stage ESI-QqToF mass spectra and afforded the expected accurate masses corresponding to the proposed chemical structures.

The structural similarity between the 1,5-benzodiazepines **6** and **7** may be noted and explain the similarity observed in their CID-MS/MS spectra and their fragmentation pathways. Thus, the product ion pairs **6c**, **7c** and **6e**, **7e**, lacking the chlorine atom in their structures, were identified in the CID-MS/MS spectra of both compounds, while the diagnostic product ions containing the chlorine atom were unambiguously observed. This is a very interesting finding, which permits us to suggest that benzodiazepine product ions pathways could be generalized and thus serve as a diagnostic tool for the systematic structural characterization of unknown 1,5-benzodiazepines.

The obtained CID-MS/MS results indicated that the product ions involved the cleavage of the N-1–C-2 and C-3–C-4 linkages of the seven-membered ring and/or the substituents attached to this saturated ring which also contains the nitrogen atom in agreement with reported data on 1,4-benzodiazepine derivatives.[83–89]

4 Synthesized Quinoxalinone Derivatives

Glycoquinoxalinone derivatives have received much research attention owing to their excellent biological and medical properties. They have been used as antibiotics, anti-inflammatory, antiviral, anticancer, antagonists, and for hypertension treatment.[90–96] The two studied glycoquinoxalinone derivatives, 8 and 9 are shown in Fig. 17.[97,98]

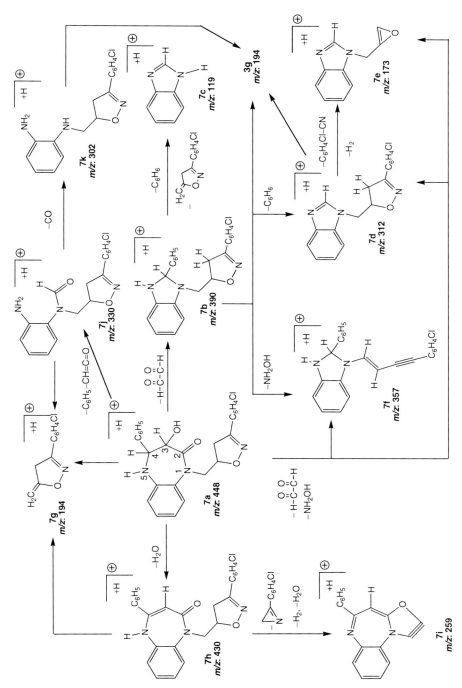

Scheme 8 Proposed fragmentation pathways for the CID-MS/MS of the protonated molecule **7a** at *m/z* 448

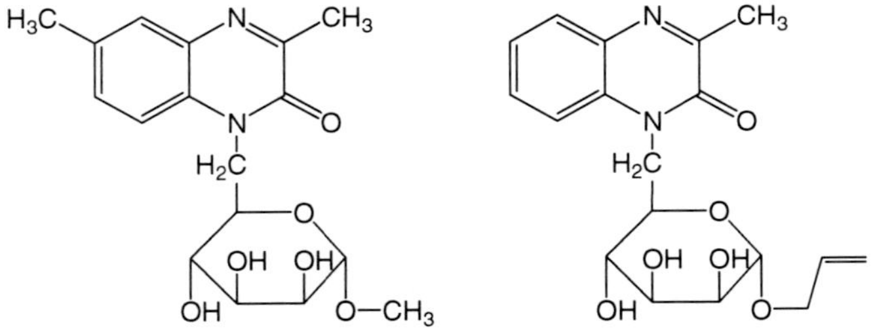

Fig. 17 Structure of the quinoxalinone derivatives

The ESI-QqTOF-MS (+ion mode) of compounds **8** and **9** produced the protonated and the sodiated ions at m/z 351.1120 and 373.0951. CID-ESI-MS/MS analyses conducted on the protonated molecules generated from both quinoxalinone derivatives and the obtained results are discussed below.

4.1 Compound 8

CID-MS/MS analysis of the protonated molecule **8a** at m/z 351.1120 showed a series of product ions shown in Scheme 9.

The product ion **8a** can either undergo an $^{2,5}X_0$ type cleavage or can eliminate the sugar moiety. This yields respectively the product ions **8b** at m/z 259.0805 and **8c** at m/z 175.0630. The latter ion can further form the product ions **8d** at m/z 147.0791 (loss of CO), **8e** at m/z 159.0748 (loss of CH_4) and **8f** at m/z 145.0615 (consecutive losses of CO and H_2). The product ion **8a** can also lose water and methanol. This generates respectively the three product ion ions **8h** at m/z 333.0316, **8k** at m/z 301.0911 and **8i** at m/z 319.0904. The latter can also be formed from **8h** at m/z 333.0316 through a loss of a methanol. It can also release the product ion **8j** at m/z 283.0821 by loss of a water. Other product ions were formed from the product ion **8k** at m/z 301.0911. The loss of one or two water molecules affords thus respectively the product ions **8i** at m/z 301.0834 and **8j** at m/z 283.0821. The prodcut ion **8k** can undertake 'A', $^{2,5}X_0$ or 'B' cleavage type yielding respectively the ions **8c** at m/z 175.0630, **8b** at m/z 259.0805 and **8l** at m/z 187.0753. The latter could also be formed from the ion **8h** at m/z 333.1115 through a 'B' type breaking.

4.2 Compound 9

The CID-QqTOF-MS/MS product ions of the protonated molecule **9a** at m/z 363.1349 afforded product ions generated by four main breakdown pathways as shown in Scheme 10.

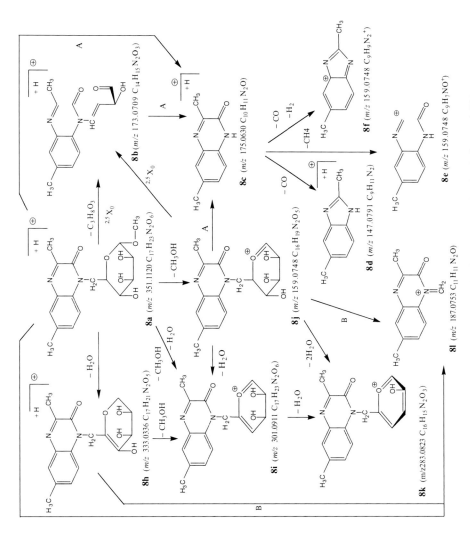

Scheme 9 Proposed fragmentation pathways observed for the CID-MS/MS of the protonated molecule **8a** at *m/z* 351

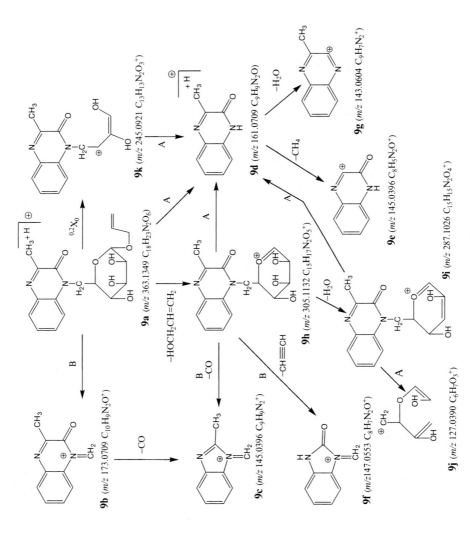

Scheme 10 Proposed fragmentation pathways for the CID-MS/MS of the protonated molecule **9a** at *m/z* 363

The product ion **9a** froms through a 'B' type product ion which loses the mannose moiety to afford the product ion **9b** at m/z 173.0541.

This ion loses CO to afford the benzimidazolium ion **9c** at m/z 145.0615. The 'A'-type breakdown of the ion **9a** with loss of the glycosyl residue yields the ion **9d** at m/z 161.0833). The latter eliminates either methane or water to afford either the ion **9e** at m/z 145.0615 or **9g** at m/z 143.0604 respectively. The protonated molecule **9a** can also yield the ion **9k** at m/z 245.0543 through an $^{0,2}X_0$ fission. The latter generates the ion **9d** at m/z 161.0633 through an 'A' type product ions. The precursor ion **9a** can also lose an allyl alcohol to form the ion **9h** at m/z 305.0829., which forms the ion **9d** at m/z 161.0633 through a 'A'-type fission . It can also lose water to yield the ion **9i** at m/z 287.0814 which afford the ion **9j** at m/z 127.0205 by further 'A' type product ions. The product ion **9h** can finally suffers a 'B' type fission followed by a loss of acetylene or carbon monoxide yielding respectively the ions **9f** at m/z 147.0553 or **9c** and m/z 145.0396.

The study of the ESI-MS and CID-MS/MS product ions routes of these two examples suggested that the quinoxaline derivatives fragment according to a general principle. Breaking between N-1 of the quinoxaline ring and C-6 of the sugar moiety to afford ions *via* the 'A' product ions route. (Scheme 10).

5 Conclusion

The present study showed the ability of mass spectrometry to analyze natural and synthesized molecular structures due to the advances gained by the development of soft ionization techniques. Analyses of flavonoid glycosides showed the potential of mass spectrometry in the structural elucidation of natural products, particularly glycosides. It was thus possible to know the number, sequence and characteristics as ring size, and linkages of the studied mono and diglycosides. The structures of four falvonoid were thus elucidated through the combined use of positive and negative electrospray ionization, collision-induced dissociation and tandem mass spectrometry. The low-collision energy CID-spectra of $[M+H]^+$, $[M+Na]^+$ and $[M-H]^-$ ions showed extensive product ions of the carbohydrate moiety, loss of the glycan residue, and product ions of the aglycones units that permit characterization of the substituents in the aglycones A-, B and C-rings and the interglycosidic linkages. Extensive product ions of the different aglycones were obtained allowing their identification. The different flavonoids families (flavone, flavanone, flavonol) were thus easily diffrentiated and their ring hydroxylation degrees were determined. The differentiation between quercetin and 6-hydroxyluteolin aglycones was thus achieved by product ion analysis of the protonated and deprotonated aglycone, that showed the characteristic product ions $^{1,3}A$ at m/z 151 and m/z 153 for quercetin, and m/z 167 and 169 for 6-hydroxyluteolin, consistent with the trihydroxylated A-ring skeleton.

Product ions of the different studied diglucosides showed that both the inner and the terminal glucose residues were involved in the product ionss, giving useful information on the interglycosidic linkage. The $1 \rightarrow 2$ interglycosidic linkage was thus easily assigned through the presence of the $^{0,2}X_0{}^{0,2}X_1$ ion. The latter is characteristic of the $1 \rightarrow 2$ interglycosidic linkage in the glucoglucoside adducts, since it can not be formed in the case of other interglycosidic types. In the case of the eriodictyol diglucoside the 1,3 product ions of the C-ring was observed before those involving the carbohydrates thus allowing the position determination of the diglucoside moiety on the A-ring. CID-MS/MS analysis of the sodiated diglucoside gave complementary information for the structural characterization of the studied compound. The $B_2{}^+$ product ion which is useful for establishing that the terminal carbohydrate unit is linked to another carbohydrate and not directly to the aglycone was obtained as base peak. This result is of analytical value for the differentiation of O-diglycosyl and di-O-glycosyl flavonoids. Further CID-MS/MS of the $B_2{}^+$ product ion gave results consistents with the involved interglycosidic linkage. Product ions mechanisms and ion structures consistent with the obtained data have been proposed.

The product ions of three synthetic benzodiazepine derivatives were explored through CID-MS/MS giving a multitude of product ions. Breakdown pathways of the different observed product ions were proposed and confirmation of the various geneses of the product ions was achieved by conducting a series of precursor ion scans. The obtained results showed that major product ions occurred in the saturated seven-membered ring containing the nitrogen atoms. All the major product ions involved cleavages of the N-1–C-2 and C-3–C-4 bonds. These occurred with concomitant eliminations of glyoxal, benzene and ethyl formate, forming the product ion at m/z 119, which was observed in all the studied compounds. In addition, an unique simultaneous CID-MS/MS product ion was noticed for the 1,5-benzodiazepines **5** and **7**, which occurred by a pathway dictated by the substituent located on the N-1-position. It was evident that the aromatic ring portion of the 1,5-benzodiazepines was resistant to CID-MS/MS product ions. CID-MS/MS analyses have thus proven to be a specific and very sensitive method for the structural identification of these novel 1,5-benzodiazepine derivatives.

Finally the mass spectral analysis of synthesized quinoxalinone derivatives was conducted by electrospray ionization tandem mass spectrometry with a QqTOF-MS/MS hybrid instrument. Further CID-MS/MS spectra using low-energy CID permitted the rationalization of the exact gas-phase product ions pathways of the studied glycoquinoxalinone derivatives. Among the obtained results, fission between N-1 of the quinoxaline ring and C-6 of the sugar moiety was observed. This afforded ions formed through the 'A' product ions route. Another fission between the C-5 and C-6 sugar carbones, was also observed giving ions produced by the 'B' product ions route.

This study clearly demonstrates the potential of mass spectrometry and it enormous capabilities for the structural elucidation of both natural and synthesized organic compounds. However MS it still not fully employed as a normal tool for structural determination even a large number of papers have been published. We hope that this study prompts, initiates and opens perspectives for similar exploration on other derivatives.

References

1. Wolfender JL, Hostettmann K (1996) Importance of LC/MS in plant analysis. Spectrosc Eur 8:7–12
2. Stobiecki M (2000) Application of mass spectrometry for identification and structural studies of flavonoid glycosides. Phytochemistry 54:237–256
3. Waridel P, Wolfender JL, Ndjoko K, Hobby KR, Major HJ, Hostettmann K (2001) Evaluation of quadrupole time-of-flight tandem mass spectrometry and ion-trap multiple-stage mass spectrometry for the differentiation of *C*-glycosidic flavonoid isomers. J Chromatogr A 926:29–41
4. El-Aneed A, Cohen A, Banoub J (2009) Mass spectrometry, review of the basics: Electrospray, MALDI, and commonly used mass analyzers. Appl Spec Rev 44:210–230
5. Takayama M, Fukai T, Noruma T, Yamauchi T (1991) Formation and product ions of the $[M + Na]^+$ ion of glycosides in fast atom bombardment mass spectrometry. Org Mass Spectrom 26:655–659
6. Wolfender JL, Maillard M, Marston A, Hostettmann K (1992) Mass spectrometry of underivatized naturally occurring glycosides. Phytochem Anal 3:193–214
7. Grossert JS (2001) A retrospective view of mass spectrometry and natural products – sixty years of progress, with a focus on contributions by R. Graham Cooks. Int J Mass Spectrom 212:65–79
8. Justesen U, Knuthsen P, Leth T (1998) Quantitative analysis of flavonols, flavones, and flavanones in fruits, vegetables and beverages by high-performance liquid chromatography with photo-diode array and mass spectrometric detection. J Chromatogr A 799:101–110
9. Swatsitang P, Tucker G, Robards K, Jardine D (2000) Isolation and identification of phenolic compounds in *Citrus sinensis*. Anal Chim Acta 417:231–240
10. Justesen U (2000) Negative atmospheric pressure chemical ionisation low-energy collision activation mass spectrometry for the characterisation of flavonoids in extracts of fresh herbs. J Chromatogr A 902:369–379
11. Cabrera GM (2006) Mass spectrometry in the structural elucidation of natural products: glycosides. In: Imperato F (ed) Phytochemistry: advance in research. Research Signpost, Kerala, pp 1–22
12. Franski R, Bednarek P, Wojtaszek P, Stobiecki M (1999) Identification of flavonoid diglycosides in yellow lupin (*Lupinus luteus* l.) with mass spectrometric techniques. J Mass Spectrom 34:486–495
13. Li QM, Clayes M (1994) Characterization and differentiation of diglycosyl flavonoids by positive fast atom bombardment and tandem mass spectrometry. Biol Mass Spectrom 23:406–416
14. Stevens JE, Taylor AW, Deinzer ML (1999) Quantitative analysis of xanthohumol and related prenylflavonoids in hops and beer by liquid chromatography–tandem mass spectrometry. J Chromatogr A 832:97–107
15. Nielsen SE, Freese R, Cornett C, Dragsted LO (2000) Identification and quantification of flavonoids in human urine samples by column-switching liquid chromatography coupled to atmospheric pressure chemical ionization mass spectrometry. Anal Chem 72:1503–1509
16. Hakkinen S, Auriola S (1998) High-performance liquid chromatography with electrospray ionization mass spectrometry and diode array ultraviolet detection in the identification of flavonol aglycones and glycosides in berries. J Chromatogr A 829:91–100
17. Ryan D, Robards K, Lavee S (1999) Determination of phenolic compounds in olives by reversed-phase chromatography and mass spectrometry. J Chromatogr A 832:87–96
18. Ryan D, Robards K, Prenzler P, Antolovich M (1999) Applications of mass spectrometry to plant phenols. Trends Anal Chem 18:362–372
19. Huck CW, Huber CG, Ongania KH, Bonn G (2000) Isolation and characterization of methoxylated flavones in the flowers of *Primula veris* by liquid chromatography and mass spectrometry. J Chromatogr A 870:453–462

20. Raffaelli A, Monetti G, Mercati V, Toja E (1997) Mass spectrometric characterization of flavonoids in extracts from *Passiflora incarnata*. J Chromatogr A 777:223–231

21. Peter-Katalini J (1994) Analysis of glycoconjugates by fast atom bombardment mass spectrometry and related ms techniques. Mass Spectrom Rev 13:77–98

22. March RE, Li H, Belgacem O, Papanastasiou D (2007) High-energy and low-energy collision-induced dissociation of protonated flavonoids generated by MALDI and by electrospray ionization. Int J Mass Spectrom 262:51–66

23. Madeira PJA, Florêncio MH (2009) Flavonoid–matrix cluster ions in MALDI mass spectrometry. J Mass Spectrom 44:1105–1113

24. Mauri P, Pietta P (2000) Electrospray characterization of selected medicinal plant extracts. J Pharm Biomed Anal 23:61–68

25. Cuyckens F, Claeys M (2002) Optimization of a liquid chromatography method based on simultaneous electrospray ionization mass spectrometric and ultraviolet photodiode array detection for analysis of flavonoid glycosides. Rapid Commun Mass Spectrom 16:2341–2348

26. Carbone V, Montoro P, de Tommasi N, Pizza C (2004) Analysis of flavonoids from Cyclanthera pedata fruits by liquid chromatography/electrospray mass spectrometry. J Pharm Biomed Anal 34:295–304

27. Ferreres F, Valentão P, Pereira JA, Bento A, Noites A, Seabra RM, Andrade PB (2008) HPLC-DAD-MS/MS-ESI Screening of Phenolic Compounds in Pieris brassicae L. Reared on Brassica rapa var. *rapa* L. J Agric Food Chem 56:844–856

28. Ma YL, Li QM, Van den Heuvel H, Claeys M (1997) Characterization of flavone and flavonol aglycones by collision-induced dissociation tandem mass spectrometry. Rapid Commun Mass Spectrom 11:1357–1364

29. Niessen WMA (1999) Liquid chromatography – mass spectrometry, 2nd edn. Marcel Dekker Inc., New York

30. Hopfgartner G, Chernushevich IV, Covey T, Plomley JB, Bonner R (1999) Exact mass measurement of product ions for the structural elucidation of drug metabolites with a tandem quadrupole orthogonal-acceleration time-of-flight mass spectrometer. J Am Soc Mass Spectrom 10:1305–1314

31. Banoub J, Cohen A, El Aneed A, Martin P, Lequart V (2004) Characterization and *de novo* sequencing of Atlantic salmon vitellogenin protein by electrospray tandem and matrix-assisted laser desorption/ionization mass spectrometry. Eur J Mass Spectrom 10:121–134

32. Banoub J, Boullanger P, Lafont D, Cohen A, El Aneed A, Rowlands E (2005) In situ formation of C-glycosides during electrospray ionization tandem mass spectrometry of a series of synthetic amphiphilic cholesteryl polyethoxy neoglycolipids containing *N*-acetyl-D-glucosamine. J Am Soc Mass Spectrom 16:565–570

33. Es-Safi N, Kerhoas L, Ducrot PH (2005a) Application of positive and negative ESI-MS, CID/MS and tandem MS/MS to a study of product ions of 6-hydroxyluteolin 7-*O*-glucoside and 7-*O*-glucosyl-(1→3)-glucoside. Rapid Commun Mass Spectrom 19:2734–2742

34. Es-Safi N, Kerhoas L, Einhorn J, Ducrot PH (2005b) Application of ESI-MS, CID/MS and tandem MS/MS to the product ions study of eriodictyol 7-*O*-glucosyl-(1→2)-glucoside and luteolin 7-*O*-glucosyl-(1→2)-glucoside. Int J Mass Spectrom 247:93–100

35. Es-Safi N, Guyot S, Ducrot PH (2006) NMR, ESI-MS and MALDI-TOF/MS analysis of pear juice polymeric proanthocyanidins with potent free radical scavenging activity. J Agric Food Chem 54:6969–6977

36. Es-Safi N, Kerhoas L, Ducrot PH (2007) Product ions study of iridoid glucosides through positive and negative ESI-MS, CID/MS and tandem MS/MS. Rapid Commun Mass Spectrom 21:1165–1175

37. Es-Safi N, Kerhoas L, Ducrot PH (2007) NMR Analysis and Product ions Study of Globularin through Positive and Negative ESI-MS, CID/MS and Tandem MS/MS. Specrosc Lett 40:695–714

38. Es-Safi N, Le Guernevé C, Ducrot PH (2008) Application of NMR and MS Spectroscopy to the Structural Elucidation of Modified Flavan-3-ols and their Coupling Reaction Products. Specrosc Lett 41:41–56

39. Ghomsi NT, Ahabchane NH, Es-Safi N, Guarrigue B, Essassi EM (2007) Synthesis and spectroscopic stuctural elucidation of new quinoxaline derivatives. Spectrosc Lett 40:741–751

40. Joly N, El Aneed A, Martin P, Cecchelli R, Banoub J (2005) Structural determination of the novel product ions routes of morphine opiate receptor antagonists using electrospray ionization quadrupole time-of-flight tandem mass spectrometry. Rapid Commun Mass Spectrom 19:3119–3130

41. Joly N, Vaillant C, Cohen AM, Martin P, Essassi EM, Massoui M, Banoub J (2007) Structural determination of the novel product ions routes of zwitteronic morphine opiate antagonists naloxonazine and naloxone hydrochlorides using electrospray ionization tandem mass spectrometry. Rapid Commun Mass Spectrom 21:1062–1074

42. Joly N, Jarmouni C, Massoui M, Essassi EM, Martin P, Banoub J (2008) Electrospray tandem mass spectrometric analysis of novel synthetic quinoxalinone derivatives. Rapid Commun Mass Spectrom 22:819–833

43. Harborne JB, Williams CA (2000) Advances in flavonoid research since 1992. Phytochemistry 55:481–504

44. Dugo P, Lo Presti M, Ohman M, Fazio A, Dugo G, Mondello L (2005) Determination of flavonoids in citrus juices by micro-HPLC-ESI-MS. J Sep Sci 28:1149–1156

45. Zhang J, Brodbelt JS (2005) Silver complexation and tandem mass spectrometry for differentiation of isomeric flavonoid diglycosides. Anal Chem 77:1761–1770

46. Kerhoas L, Aouak D, Cingoz A, Routaboul JM, Lepiniec L, Einhorn J, Birlirakis N (2006) Structural characterization of the major flavonoid glycosides from Arabidopsis thaliana seeds. J Agric Food Chem 54:6603–6612

47. Kachlicki P, Einhorn J, Muth D, Kerhoas L, Stobiecki S (2008) Evaluation of glycosylation and malonylation patterns in flavonoid glycosides during LC/MS/MS metabolite profiling. J Mass Spectrom 43:572–586

48. Justino GC, Borges CM, Florêncio MH (2009) Electrospray ionization tandem mass spectrometry product ions of protonated flavone and flavonolaglycones: a re-examination. Rapid Commun Mass Spectrom 23:237–248

49. Ma YL, Cuyckens F, Van den Heuvel H, Claeys M (2001) Mass spectrometric methods for the characterisation and differentiation of isomeric O-diglycosyl flavonoids. Phytochem Anal 12:159–165

50. Liu R, Ye M, Guo H, Bi K, Guo D (2005) Liquid chromatography/electrospray ionization mass spectrometry for the characterization of twenty-three flavonoids in the extract of *Dalbergia odorifer*. Rapid Commun Mass Spectrom 19:1557–1565

51. Ablajan K, Abliz A, Shang XY, He JM, Zhang RP, Shi JG (2006) Structural characterization of flavonol 3, 7-di-O-glycosides and determination of the glycosylation position by using negative on electrospray ionization tandem mass spectrometry. J Mass Spectrom 41:352–360

52. Zhou DY, Xu Q, Xue XY, Zhang FF, Liang XM (2006) Identification of O-diglycosyl flavanones in Fructus aurantii by liquid chromatography with electrospray ionization and collision-induced dissociation mass spectrometry. J Pharm Biomed Anal 42:441–448

53. Shi PY, He Q, Song Y, Qu HB, Cheng YY (2007) Characterization and identification of isomeric flavonoid O-diglycosides from genus Citrus in negative electrospray ionization by ion trap mass spectrometry and time-of-flight mass, spectrometry. Anal Chim Acta 598:110–118

54. Abad-Garcia B, Garm' On-Lobato S, Berrueta LA, Gallo B, Vicente F (2009) ractical guidelines for characterization of O-diglycosyl flavonoid isomers by triple quadrupole MS and their applications for identification of some fruit juices flavonoids. J Mass Spectrom 44:1017–1025

55. Es-Safi N, Khlifi S, Kerhoas L, Kollmann A, El Abbouyi A, Ducrot PH (2005) Antioxidant constituents of the aerial parts of *Globularia alypum* growing in Morocco. J Nat Prod 68:1293–1296

56. Domon B, Castello CE (1988) A systematic nomenclature for carbohydrate product ionss in FAB–MS/MS spectra of glycoconjugates. Glycoconj J 5:397–409

57. Cuyckens F, Claeys M (2004) Mass spectrometry in the structural analysis of flavonoids. J Mass Spectrom 39:1–15
58. Wolfender JL, Waridel P, Ndjoko K, Hobby KR, Major HJ, Hostettmann KJ (2000) Evaluation of QTOF-MS/MS and multiple stage IT-MS^n for the dereplication of flavonoids and related compounds in crude plant extracts. Analusis 28:895–906
59. Hvattum E (2002) Determination of phenolic compounds in rose hip (Rosa canina) using liquid chromatography coupled to electrospray ionisation tandem mass spectrometry and diode-array detection. Rapid Commun Mass Spectrom 16:655–662
60. Hvattum E, Ekeberg D (2003) Study of the collision-induced radical cleavage of flavonoid glycosides using negative electrospray ionization tandem quadrupole mass spectrometry. J Mass Spectrom 38:43–49
61. March RE, Miao XS, Metcalfe CD (2004) A product ions study of a flavone triglycoside, kaempferol-3-O-robinoside-7-O-rhamnoside. Rapid Commun Mass Spectrom 18:931–934
62. March RE, Miao XS, Metcalfe CD, Stobiecki M, Marczak L (2004) A product ions study of an isoflavone glycoside, genistein-7-O-glucoside, using electrospray quadrupole time-of-flight mass spectrometry at high mass resolution. Int J Mass Spectrom 232:171–183
63. Cuyckens F, Claeys M (2005) Determination of the glycosylation site in flavonoid mono-O-glycosides by collision-induced dissociation of electrospray-generated deprotonated and sodiated molecules. J Mass Spectrom 40:364–372
64. Fabre N, Rustan I, de Hoffmann E, Quetin-Leclercq J (2001) Determination of flavone, flavonol, and flavanone aglycones by negative ion liquid chromatography electrospray ion trap mass spectrometry. J Am Soc Mass Spectrom 12:707–715
65. Wu W, Liu Z, Song F, Lui S (2004) Structural Analysis of Selected Characteristic Flavones by Electrospray Tandem Mass Spectrometry. Anal Sci 20:1103–1105
66. Cuyckens F, Ma YL, Pocsfalvi G, Claeys M (2000) Tandem mass spectral strategies for the structural characterization of flavonoid glycosides. Analusis 28:888–895
67. Cuyckens F, Rozenberg R, de Hoffmann E, Claeys M (2001) Structure characterization of flavonoid O-diglycosides by positive and negative nano-electrospray ionization ion trap mass spectrometry. J Mass Spectrom 36:1203–1210
68. Ferreres F, Llorach R, Gil-Izquierdo A (2004) Characterization of the interglycosidic linkage in di-, tri-, tetra- and pentaglycosylated flavonoids and differentiation of positional isomers by liquid chromatography/electrospray ionization tandem mass spectrometry. J Mass Spectrom 39:312–321
69. Zhang J, Brodbelt JS (2004) Screening flavonoid metabolites of naringin and narirutin in urine after human consumption of grapefruit juice by LC-MS and LC-MS/MS. Analyst 129:1227–1233
70. Careri M, Elviri L, Mangia A (1999) Validation of a liquid chromatography ionspray mass spectrometry method for the analysis of flavanones, flavones and flavonols. Rapid Commun Mass Spectrom 13:2399–2405
71. Vallejo F, Tomas-Barberan FA, Ferreres F (2004) Characterisation of flavonols in broccoli (Brassica oleracea L. var. italica) by liquid chromatography–UV diode-array detection–electrospray ionisation mass spectrometry. J Chromatogr A 1054:181–193
72. Ma YL, Vedernikova I, Van den Heuvel H, Claeys M (2000) Internal glucose residue loss in protonated O-diglycosyl flavonoids upon low-energy collision-induced dissociation. J Am Soc Mass Spectrom 11:136–144
73. Sanchez-Rabaneda F, Jauregui O, Lamuela-Raventos RM, Viladomat F, Bastida J, Codina C (2004) Qualitative analysis of phenolic compounds in apple pomace using liquid chromatography coupled to mass spectrometry in tandem mode. Rapid Commun Mass Spectrom 18:553–563
74. Waridel P, Wolfender JL, Lachavanne JB, Hostettmann K (2004) Identification of the polar constituents of Potamogeton species by HPLC-UV with post-column derivatization, HPLC-MSn and HPLC-NMR, and isolation of a new ent-labdane diglycoside. Phytochemistry 65:2401–2410
75. Lal H, Fielding S (1979) Anxiolytics. Futura Publishing Co., New York

76. Möhler H, Okada T (1977) Benzodiazepine receptor: demonstration in the central nervous system. Science 198:849–851
77. Schutz H (1982) Benzodiazepines. Springer, Heidelberg
78. Landquist JK (1984) In: Katritzkyand AR, Rees CW (eds) Comprehensive heterocyclic chemistry, vol 1. Pergamon, Oxford, pp 166–170
79. Fryer RI (1991) Bicyclic diazepines. In: Taylor EC (ed) Comprehensive heterocyclic chemistry. Wiley, New York, p 50, Chapter II
80. Randall LO, Kappel B (1973) Benzodiazepines. Raven Press, New York
81. De Baun JR, Pallos FM, Baker DR (1976) US Patent; 3: 978, 227
82. Rida M, El Meslouhi H, Es-Safi N, Essassi EM, Banoub J (2008) Gas-phase product ions study of novel synthetic 1, 5-benzodiazepine derivatives using electrospray ionization tandem mass spectrometry. Rapid Commun Mass Spectrom 22:2253–2268
83. Ghezzo E, Traldi P, Minghetti G, Cinellu MA, Bandini AL, Banditelli G, Zecca L (1990) Fast-Atom bombardment mass spectrometry and collisional spectroscopy in the structural characterization of underivatized 1, 4-benzodiazepines. Rapid Commun Mass Spectrom 4:314–317
84. Smyth WF, McClean S, Ramachandran VN (2000) A study of the electrospray ionisation of pharmacologically significant 1, 4-benzodiazepines and their subsequent product ions using an ion-trap mass spectrometer. Rapid Commun Mass Spectrom 14:2061–2069
85. Smyth WF, Joyce C, Ramachandran VN, Kane EO, Coulter D (2004) Characterisation of selected hypnotic drugs and their metabolites using electrospray ionisation with ion trap mass spectrometry and with quadrupole time-of-flight mass spectrometry and their determination by liquid chromatography-electrospray ionisation–ion trap mass spectrometry. Anal Chim Acta 506:203–214
86. Smyth WF, McClean S, Hack CJ, Ramachandran VN, Doherty B, Joyce C, O'Donnell F, Smyth TJ, O'Kane O, Brooks P (2006) The characterisation of synthetic and natural-product pharmaceuticals by electrospray ionisation-mass spectrometry (ESI-MS) and liquid chromatography (LC)-ESI-MS. Trends Anal Chem 25:572–582
87. Jourdil N, Bessard J, Vincent F, Eysseric H, Bessard G (2003) Automated solid-phase extraction and liquid chromatography–electrospray ionization-mass spectrometry for the determination of flunitrazepam and its metabolites in human urine and plasma samples. J Chromatogr B 788:207–219
88. Kratzsch C, Tenberken O, Peters FT, Weber AA, Kraemere T, Maurer HH (2004) Screening, library-assisted identification and validated quantification of 23 benzodiazepines, flumazenil, zaleplone, zolpidem and zopiclone in plasma by liquid chromatography/mass spectrometry with atmospheric pressure chemical ionization. J Mass Spectrom 39:856–872
89. Risoli A, Cheng JBL, Verkerk UH, Zhao J, Ragno G, Hopkinson AC, Siu KWM (2007) Gas-phase product ions of protonated benzodiazepines. Rapid Commun Mass Spectrom 21:2273–2281
90. Went GR, Ledig K (1969) U.S. Patent, 3: 431
91. Tanaka A, Usui T (1981) Studies on furan derivatives. X. Preparation of 2-substituted 3-(5-nitro-2-furyl) quinoxaline 1, 4-dioxides and determination of their antibacterial activity. Chem Pharm Bull 29:110–115
92. Monge A, Palop JA, Fernandez E (1989) New quinoxaline and pyrimido[4, 5-b]quinoxaline derivatives. Potential antihypertensive and blood platelet antiaggregating agents. J Heterocycl Chem 26:1623–1626
93. Balzarini J, Declercq E, Carboez A, Burt V, Kleim JP (2000) Long-term exposure of HIV type 1-infected cell cultures to combinations of the novel quinoxaline GW420867X with lamivudine, abacavir, and a variety of nonnucleoside reverse transcriptase inhibitors. AIDS Res Hum Retroviruses 16:517–528
94. Catarzi D, Colotta V, Varano F, Cecchi L, Filac C (2000) 7-Chloro-4, 5-dihydro-8-(1, 2, 4-triazol-4-yl)-4-oxo-1, 2, 4-triazolo[1, 5-a]quinoxaline-2- carboxylates as novel highly selective AMPA receptor antagonists. J Med Chem 43:3824–3826
95. Carta A, Sanna P, Gherardini L, Usai D, Zanetti S (2001) Novel functionalized pyrido[2, 3-g] quinoxalinones as antibacterial, antifungal and anticancer agents. Farmaco 56:933–938

96. Antonio C, Mario L, Stefania Z, Leonardo AS (2003) Quinoxalin-2-ones: Part 5. Synthesis and antimicrobial evaluation of 3-alkyl-, 3-halomethyl- and 3-carboxyethylquinoxaline-2-ones variously substituted on the benzo-moiety. Farmaco 58:1251–1255

97. Bouhlal D, Gode P, Goethals G, Massoui M, Villa P, Martin P (2001) Glycosyl-pyrrolo[2, 1-c] [1, 4]benzodiazepine-5, 11-diones. Synthesis, Tensioactivity and Anitbacterial Activity. Heterocycles 55:303–312

98. Lakhrissi B, Essassi EM, Massoui M, Goethals G, Lequart V, Monflier E, Cecchelli R, Martin P (2004) Synthesis and amphiphilic behavior of N, N-Bis-glucosyl-1, 5-benzodiazepin-2, 4-dione. J Carbohydr Chem 23:389–40